Project-Based
AutoCAD®

Darren J. Manning

McGraw Hill **Glencoe**

New York, New York   Columbus, Ohio   Chicago, Illinois   Peoria, Illinois   Woodland Hills, California

Cover and Interior Design: Squarecrow Creative Group

The McGraw·Hill Companies

Send all inquiries to:
Glencoe/McGraw-Hill
3008 W. Willow Knolls Drive
Peoria, IL 61614-1083

ISBN 0-07-828732-4
Printed in the United States of America
1 2 3 4 5 6 7 8 9 10   079   07 06 05 04 03 02

Contents in Brief

Reviewers

B. Russell Alexander, P.E.
Kathryn E. Alexander, P.E.
The Design Factory, Inc.
Columbus, OH

Kathleen Dapprich
CAD Instructor
Alpena Community College
Alpena, MI

Hollis E. Driskell
CAD Instructor
Trinity Valley Community College
Athens, TX

George Gibson
CAD Instructor
Athens Technical College
Athens, GA

Brian Jeker
Industrial Technology
Greene, NY

Dean R. Kerste
CAD Instructor
Monroe County Community College
Monroe, MI

Darrell Ledbetter
CAD Instructor
Lincoln County School of Technology
Lincolnton, NC

Eddie Ray Lee
CAD Instructor
Vincennes University
Vincennes, IN

James B. Locken
CADD Instructor
Mid Florida Tech
Orlando, FL

Olen K. Parker
President
American Design Drafting
    Association
Newbern, TN

Alex Prayson
Assistant Professor, Design
    Engineering Technology
Tulsa Community College
Tulsa, OK

Phillips D. Rockwell, P.E.
Owner, Bay Machine Design
Berkeley, CA

Daniel Stover
Technical Supervisor/
CAD Professional
Seattle, WA

Contributors

Real-World Projects

Hydroelectric Turbine Nozzle Assembly
Canyon Industries, Inc.
5500 Blue Heron Lane
Deming, WA 98244

Structural Field Assembly of Tire Conveyor
Emery Energy Company, LLC
444 E. 200 S.
Salt Lake City, UT 8411

Compensator Swivel Bracket: 3D CAD/CAM
Browning Component, Inc.
7900 NE Day Road West
Bainbridge, WA 98110

Mechanical Shaft Assembly
Final Draft CAD/CAM
5984 North Fork Road
Deming, WA 98244

Technical Writing

Timothy M. Looney
Adjunct Professor of Mechanical
    Engineering Technology
University of Massachusetts Lowell
Lowell, MA

Contents

Project 1   Hydroelectric Turbine Nozzle Assembly   11

Courtesy of Canyon Industries, Inc.

Courtesy of Browning Component, Inc./Patented by Browning Automatic Transmission, LP

CAD Reference

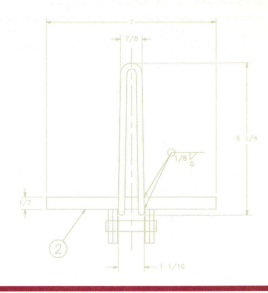

Appendixes

Appendix A

Appendix B

Appendix C

Appendix D

Before you begin . . .

Project-Based AutoCAD approaches CAD from the drafter's perspective. Instead of trying to include all of the many commands and options available in the AutoCAD software, this textbook focuses on how the drafter might use AutoCAD on the job.

Real-World Projects

Each of the four projects in this book demonstrates a different aspect of drafting in the real world. Project 1 concentrates on the development of working drawings for a mechanical assembly. Project 2 recreates the process of documenting a prototype field assembly. Project 3 introduces 3D modeling and the concepts of CAD/CAM and CNC. The concept drawing in Project 4 demonstrates the process of designing and developing a new concept to fill a market need.

The Importance of Drafting Standards

Drafting skill and knowledge of drafting standards are extremely important in any drafting job. Although *Project-Based AutoCAD* includes some drafting instruction, its primary focus is on using AutoCAD to accomplish real-world drafting projects. As you work on these projects, it is vital that you follow established drafting guidelines. If you are not familiar with the specifications in ASME Y14 and the ADDA guidelines, you should endeavor to learn them well.

Drafting Today

In today's world, it is not enough simply to learn how to use AutoCAD, nor is it enough to know the drafting standards. You must be able to apply drafting standards correctly and use AutoCAD or other CAD software efficiently to accomplish the drawings and models that drafters are called upon to create.

Hydroelectric Turbine Nozzle Assembly

When you have completed this project, you should be able to:

- Identify the parameters of a two-dimensional design assignment.
- Use a predefined ANSI template to prepare an AutoCAD drawing file for use with an entire set of technical drawings.
- Create a reference assembly drawing.
- Use an assembly drawing as the basis for a cross-section detail.
- Apply weld symbols appropriately to a working drawing.
- Identify parts of the assembly that need dimension details and create them accordingly.
- Dimension a detail drawing using geometric dimensioning and tolerancing (GD&T).

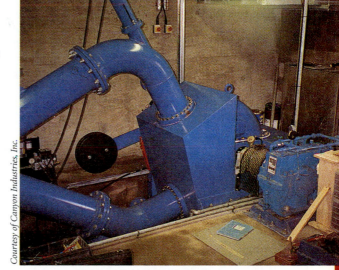

Courtesy of Canyon Industries, Inc.

Fig. 1.1 *A finished hydroelectric turbine assembly installed in a powerhouse.*

Introduction

Two-dimensional (2D) CAD drawings are often used by manufacturers to define new designs and ideas in detail so that they can be manufactured to very specific requirements. In this project, you will create a set of working drawings for a nozzle assembly that controls the flow of water into the housing of a hydroelectric turbine.

Before you begin the actual drawings, you will need to become familiar with the concepts behind power generation and hydroelectric turbines. You must understand the function of the nozzle assembly and how it fits into the overall turbine design. Knowing these details will help you design the nozzle assembly to fit the customer's needs. Then you will create a complete set of working drawings to describe the nozzle assembly for the manufacturer. Your drawings will guide the manufacture and assembly of the nozzle in the hydroelectric turbine and will provide the basis for quality control checks. Your drawings must therefore be both precise and accurate.

Design Parameters

The guidelines by which a new design is created are known as **design parameters**. To define the parameters of a design problem, you must first know something about how the finished product will be used. In this project, courtesy of Canyon Industries, the assignment is to create a set of working drawings for the nozzle assembly of a hydroelectric turbine. What must you find out before you can begin planning the design?

A good starting point is to explore the environment in which the nozzle assembly will be used. Also, you should gather information from the customer and from engineers to determine exactly what function the nozzle assembly is expected to perform, and under what physical conditions. The purpose of this section is to familiarize you with the basic concepts so that you can begin planning the design and forming strategies to solve the design problem.

Section Objectives

- Explore the uses of hydroelectric turbines in power generation.
- Identify the function of the nozzle assembly.
- Brainstorm design elements that will meet the functional needs.
- Plan the necessary drawings and specifications.

Key Terms

- design parameters
- flow
- head
- hydroelectric turbine
- nozzle
- runner

Initial Research

How do you go about finding out the background information you will need to complete a design assignment? You have several resources at your disposal. First, you should meet with the customer, by appointment if necessary. Ask very specific questions about the assignment. For the nozzle assembly, for example, you might ask the following questions:

- Exactly what is the nozzle assembly expected to do?
- To what specific physical conditions will the nozzle assembly be subjected? What are the extremes of cold and hot temperatures, humidity, etc.?
- What equipment will interface with the nozzle assembly?
- Are there any specific material requirements due to the equipment interfaces?

If you are completely unfamiliar with the type of product to be designed, you should also look for background information about the product in general. The Internet is one convenient common source of background information. For the nozzle assembly project, use a search engine with keywords such as *hydroelectricity*, *power generation*, and *turbine*. When you find information that might help you in the project, save it and read it carefully. **Caution:** When using materials from the Internet and other external sources, be sure to read critically. Compare the information with that from your customer. If any discrepancies arise, ask the customer for clarification. Information from the customer should always take precedence over information from third-party sources. This is important because information from external sources may refer to a slightly different system or setup, or it may reflect the laws or regulations of another state or country.

For this project, you will not be able to sit down with the customer to discuss the project at length. Therefore, much of the background information you would ordinarily obtain from the customer is summarized in this section. However, it is important for the designer/drafter to understand how to

The purpose of this exercise is to practice finding and using various resources for background material that may be useful as you design a new product. The steps outlined below are general in nature. Take the time to work through each step. The information resources you discover may be useful not only throughout your work in this textbook, but also as you accept and work on design projects for employers.

1. List the most likely sources of information. For a given project, these may include some or all of the following:
   - The customer's public relations department
   - The customer's engineering department and/or the person who hired you
   - The Internet, using megasearch engines
   - Textbooks and reference books
   - For local customers and projects, local libraries, university libraries, newspapers, and sometimes even museums (for pertinent historical, political, and other information)

   The last two items in step 1 are the most likely sources of information for this project. Keep in mind that available sources may vary from project to project.

2. Follow up each likely source. In this case, your followup should consist of a minimum of calling local libraries/museums and performing a thorough Internet search.

3. Gather and organize the information that you have obtained. Compare the information provided in this textbook with the information you found using other sources. Clear up any questions that arise.

4. Document your findings and summarize them on paper to help orient yourself to the design task before you begin planning the drawings.

find and use background information. Practice Exercise 1.1A will guide you through a typical search for background information for a new project.

Background: Hydroelectric Turbines

The nozzle assembly to be designed in this project will control the flow of water to the turbines in a hydroelectric dam. A **hydroelectric turbine** is a type of rotary engine that consists of a series of curved vanes connected to a central shaft. Running water, usually from a river, is directed toward the vanes by one or more **nozzles**. As the water runs over the vanes, it moves them in the direction of the curvature of the vanes. As the vanes move, they cause the central shaft to rotate. In this way, the turbine transforms the energy of the moving water into mechanical energy. The shaft can be connected to a power generator to produce electrical energy.

Canyon Industries is a major manufacturer of utility-grade hydroelectric systems for power generation, energy recovery projects, and other uses. The company designs and manufactures each system to specific site requirements, which increases the general efficiency of the systems. This customization process requires a large amount of custom drafting for each job undertaken.

Turbine efficiency is extremely important to the overall operation of a hydroelectric system. The turbine must be designed to use the available water source to produce the greatest possible amount of energy.

The turbines used in the Canyon Industries systems are designed to be used with a water source that has a high head and/or high flow. **Head** is defined as the total of vertical feet over which the water falls, and **flow** is the quantity of water by volume. These two factors determine the water power that can be converted into radial, or rotary, power by the turbine. The volume of water is critical to the nozzle assembly design.

The nozzle assembly in this project will be used with a turbine that is based on the Pelton water wheel, which was developed in 1878 by Lester Allen Pelton for use in California gold mines. The original Pelton wheel used the gravitational force of falling water to turn curved "buckets." In today's version, water is funneled into a pressurized pipeline with a narrow nozzle at one end. The water sprays out of the nozzle and into the turbine housing under force. When the water strikes the double-cupped buckets, the impact causes the wheel to rotate with a high rate of efficiency. Fig. 1.1.1.

Nozzle Assembly

In creating the working drawings for the nozzle assembly, it will help you to understand exactly how the assembly fits in the system as a whole. The assembly elevation in Fig. 1.1.2 shows the complete system. In this drawing, the nozzle assembly is shown on the right, behind the turbine housing. The couplings in the center of the drawing show how the shaft is connected to the generator on the left.

The turbine to be used in this project is a double-nozzle Pelton hydroelectric turbine. In the double-nozzle design, two nozzles feed water to the turbine housing. Fig. 1.1.3. The use of two nozzles allows a smaller Pelton wheel (also called a **runner**), to be used with a given water flow. It also increases the rotational speed of the runner, which increases efficiency. The force of the water causes the runner to turn a shaft that

Courtesy of Canyon Industries, Inc.

Fig. 1.1.1 *A modern Pelton wheel of the type used by Canyon Industries.*

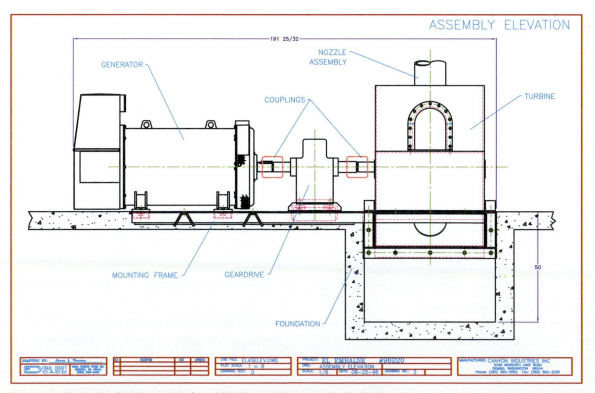

Fig. 1.1.2 *The hydroelectric system for which the nozzle assembly is designed.*

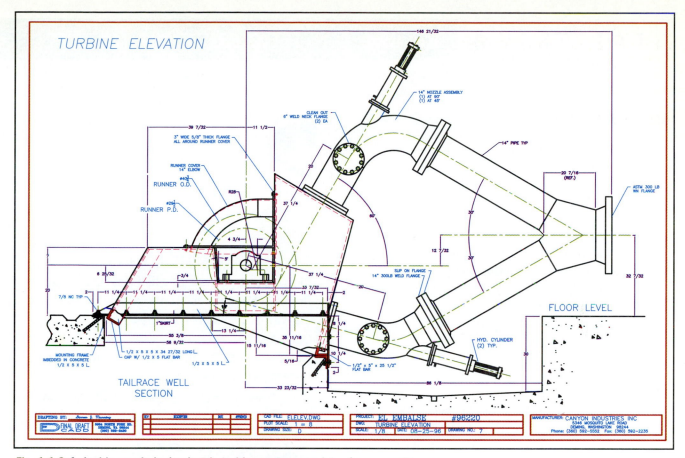

*Fig. 1.1.3 **A double-nozzle hydroelectric turbine system consists of upper and lower nozzle assemblies that feed the water to the turbine.***

is connected to a generator, which converts the mechanical energy into electricity. The nozzle is designed so that the outlet can be opened and closed by operating a hydraulic cylinder. This also allows precise opening of the nozzle to regulate the flow of water into the turbine housing.

Project Parameters

In this project, we will concentrate on creating a set of working drawings for the lower nozzle assembly. The purposes of the drawings include:

- showing the overall construction of the nozzle assembly
- showing how the nozzle assembly interfaces with the main turbine housing and with other assemblies in the turbine
- assisting with cost estimation
- providing a basis for ordering materials
- providing information needed for engineering validation and approval

Some manufacturers do not break down their drawings into detailed parts. In these situations, the design department works closely with the manufacturing departments to provide necessary information as needed to create each part. However, it is common for the detail parts of an assembly to be contracted to independent manufacturers. If this method is used, both the designers and the manufacturers must comply with the same set of guidelines and standards.

This project is intended to guide you through the process of creating a partial nozzle assembly for use by an independent manufacturer. The nozzle assembly will need to be broken down into enough detail so that all the information is available for the manufacturer to produce the assembly. By the time you have finished this project, you will have created the following drawings:

- assembly drawing of the lower nozzle assembly, including a detail section and a parts list
- bell housing detail drawing

Review Questions

1. Explain why it is important for the designer/drafter to obtain background information before beginning a new project.
2. List at least three ways to obtain background information about an assigned project.
3. Explain the basic operation of a hydroelectric turbine.
4. What is the purpose of the nozzle assembly for which you will create working drawings in this project?
5. Explain the advantages of having two nozzles to supply water to the turbine housing.
6. What is another name for the Pelton hydroelectric turbine used in this project?

Practice Problems

1. XYZ Corporation has hired you to design and create a complete set of working drawings for a cell phone for use in various rugged terrains, including mountains and deserts. Make a list of the information you will need before you can begin this project. Perform a search for background information using all available resources. Record the resources and the information you obtained from each. Summarize your findings.
2. Research various kinds of turbines that could be used in a hydroelectric power generation system. Sketch each type of turbine. Next to each sketch, list the advantages and disadvantages of using the turbine for the application discussed in this section.

Portfolio Project

Needle Valve

Design and develop the drawings for a needle valve for an industrial chemical manufacturer. Needle valves provide an alternative for controlling the flow of a liquid or gas through a hydraulic system. The valve you design will control the flow of acetic acid into a mixing chamber. It must meet the following specifications:

- The valve must provide both precise flow (metering) control and complete shutoff capability.
- Maximum flow through the valve will be 8 gallons per minute (GPM).
- Maximum operating pressure will be 4,000 pounds per square inch (PSI).

Research various types of needle valves. Decide what characteristics the valve must have to meet the customer's requirements. Operate on the assumption that none of the needle valves currently on the market meet the customer's requirements exactly, so this cannot be a purchased part, and you cannot copy a design from another source. Characteristics must include the size of the valve and its orifice, a method of controlling flow, and the material from which the valve should be manufactured.

Document your research by recording the sources you use for reference and the information you found at each source. List the types of drawings that will be required to complete this portfolio assignment.

Nozzle Assembly Drawing

The nozzle assembly drawing provides general identification of all of the parts of the lower nozzle assembly. It shows overall dimensions and specifications for the finished nozzle, and it includes a parts list. It also shows how the nozzle interfaces with adjoining pieces. In addition, the information on this drawing provides the basis for estimating costs and ordering materials.

With this section, you will begin the actual AutoCAD work. Pay careful attention to drawing file setup as discussed in the first part of this section. A little preplanning can save you much-needed time after the project gets underway, especially in a concurrent engineering environment. After you have set up the file properly, the drawing process should proceed smoothly, even if you are using commands and techniques that are new to you.

Section Objectives
- Set up a drawing file for a set of technical drawings.
- Apply basic AutoCAD techniques to create a reference assembly drawing.
- Create a parts list.
- Set up plot styles to plot a drawing in compliance with ANSI standards.

Key Terms
- hatch
- model space
- paper space
- parts list
- plot style table
- section lines
- title block

Drawing Setup

Before you can begin working on a drawing in AutoCAD, you must set up the drawing file. This includes defining the drawing limits, units, text styles, and layers. Drawing setup for files that contain drawings for use in industry requires a certain amount of planning ahead.

Limits and Units

You should be familiar with the concept of setting the limits for a drawing using the LIMITS command. If you need a quick review, refer to the *CAD Reference* at the end of this textbook.

Before you can set the limits for a specific drawing, you need to know the following things:

- system of units to be used (U.S. customary or metric)
- number and arrangement of the views to be placed
- scale of the views on the page
- number of drawing sheets needed to show all of the views when the working drawings are plotted

For the nozzle assembly project, U.S. customary units will be used. As determined in Section 1.1, the set will consist of two drawings: an assembly drawing and a detail of the bell housing. Each of the drawings will be placed on a separate drawing sheet.

There are two ways to approach the limit calculations for a set of working drawings. Sometimes the drafter creates an individual drawing file for each drawing sheet. If you use this approach, you need to know exactly what views you will place on each plotted sheet. Then you can calculate the space you will need to place the views.

The other approach, which is used in this project, keeps all the working drawings together in a single drawing file. This approach is possible because the latest versions of AutoCAD allow you to create as many layouts as you need to display all of the pages in a set of working drawings. When you use this approach, you should calculate the amount of space needed to place all of the views in the drawing file in the configuration you will use for plotting. Figure 1.2.1 shows an example of how to calculate the limits for the nozzle assembly drawings if you plan to place the two drawings side by side in the drawing file. The limits will be different if you choose to place one drawing above the other. *Note:* The completed drawings are shown in place on the drawing to give you an idea of how to calculate the space required. You will create the actual drawings in this and the following sections.

Text Styles

The basic text style for a set of working drawings must be clear and easy to read. ASME Y14.2M-1992, R 1998, specifies the Gothic font. This font is similar to the Roman Simplex (romans) font that is loaded automatically by AutoCAD, so many drafters use the romans font. Gothic is also available for AutoCAD, however.

Although you could set up a variety of text styles based on this font with different sizes for titles, notes, and special uses, it is easier to set up a single style with a height of 0. This causes AutoCAD to prompt for the height at the time of text insertion, which allows you to use the same style for all of the text in the drawing. You simply need to make a note of the sizes you use so that you can keep the sizes consistent within the drawing.

Layers

There is nothing wrong with setting up layers as you need them while you are working on a drawing. However, if you know in advance what layers you will need, it's generally more efficient to set them up at this stage. You can always add more layers later if you discover a need for them.

For the nozzle assembly drawings, you will need layers for object lines, dimensions, hidden lines, centerlines, and so on. The layers that will be used in this project are shown in Fig. 1.2.2. (The layers for the border and title block will be inserted when you use the appropriate drawing template.)

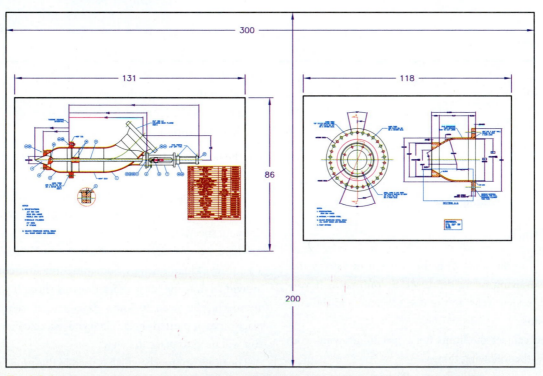

Fig. 1.2.1 *If the two drawings for the nozzle assembly are placed side by side in a drawing file, the drawing must be at least 259" wide and at least 86" high. To allow for additional maneuvering space, this drawing was set up with limits of 300,200.*

By default, lines in AutoCAD have a width, or line weight, of 0. Although AutoCAD allows you to set up different line widths for each layer, you should not do so. Many companies prefer line widths of 0 because their CAD/CAM machinery and equipment will not accept any other width. It is therefore a good general practice to leave the line widths at 0 in your drawings. For plotted drawing sheets, on which different line widths are standard drafting practice, you can set up a plot style table that adds line width to the various layers. The plot style will be set up later in this project.

File Creation

Now that you have all the pieces, you can create a new drawing in AutoCAD and set it up correctly. Practice Exercise 1.2A steps you through the process of setting up the nozzle assembly drawing. Work through the exercise before you continue.

Name	Color	Linetype
0	☐ White	Continuous
Center	■ 74	Center
Dimensions	■ Blue	Continuous
Hatch	■ Red	Continuous
Hidden	■ Magenta	Hidden
Object	☐ White	Continuous
Phantom	■ 11	Phantom
Text	■ 140	Continuous
Title Block	■ 14	Continuous
Viewport	■ 14	Continuous

Fig. 1.2.2 Suggested layers, layer colors, and linetypes for use in the Nozzle Assembly Project. Note that layer and color specifications may differ from company to company.

Practice Exercise 1.2A

AutoCAD provides several ways to create a new drawing file. See the *CAD Reference* at the end of this book for more information about the various methods. For this project, use one of AutoCAD's standard templates to set up the file for an ANSI D drawing file. *Note:* The ANSI D template applies to the layouts from which you will later print the drawing. It does not set up the drawing limits, units, layers, etc.

1. Choose to create a new drawing using a template. Select the **ANSI D - Named Plot Styles** template. When the drawing file appears, switch to the **Model** tab.
2. Use the **UNITS** command to specify decimal units with a precision of **0.00**.
3. Calculate the drawing limits using the procedure discussed in "Limits and Units." Use the **LIMITS** command to set the limits according to your calculations. *Note:* For purposes of illustration in this textbook, drawing limits for this project were set as shown in Figure 1.2.1. If you decide to lay out your drawings differently, your limits will not necessarily be the same.
4. (Optional) If you use AutoCAD's grid, you may want to set it at 5 units for visual reference.
5. Enter **ZOOM All** to see the entire drawing area.
6. Enter the **STYLE** command. Create a new style named **ROMANS** and choose the **romans.shx** font for it. Be sure to specify a height of **0**. Pick the **Apply** button to make ROMANS the current style. Then pick the **Close** button to close the Text Style window.
7. Pick the **Layers** button and set up the layers as shown in Figure 1.2.2. Notice that AutoCAD has already set up layers named Title Block and Viewport. Change the color of these layers to 14 if you want to match the drawings shown in this textbook. Be sure to load and specify the appropriate linetypes for the Center, Hidden, and Phantom layers.
8. Save the drawing as **PROJECT 1**.

Courtesy of Canyon Industries, Inc.

Fig. 1.2.3 *This project will concentrate on the lower nozzle assembly.*

Draw the Bell Housing

For this project, we will ignore the upper nozzle assembly and concentrate on a partial assembly drawing that shows only the lower nozzle. A photograph of the portion relevant to this project is shown in Fig. 1.2.3. The complete working drawing for both nozzles is shown for reference in Fig. 1.2.4.

The first part to be drawn is the bell: the area where the lower pipe narrows to feed the water into the valve that forces water onto the turbine. This portion of the drawing is shown in Fig. 1.2.5.

To create this part of the drawing, you will need to use basic AutoCAD commands such as LINE, PLINE, TRIM, EXTEND, and OFFSET, as well as object snaps. Refer to the *CAD Reference* if you need to review the use of object snaps. Then work through Practice Exercise 1.2B to create the bell. Refer to Fig. 1.2.5 as necessary.

Fig. 1.2.4 *The complete assembly drawing used by Canyon Industries showing variations for both the upper nozzle and the lower nozzle.*

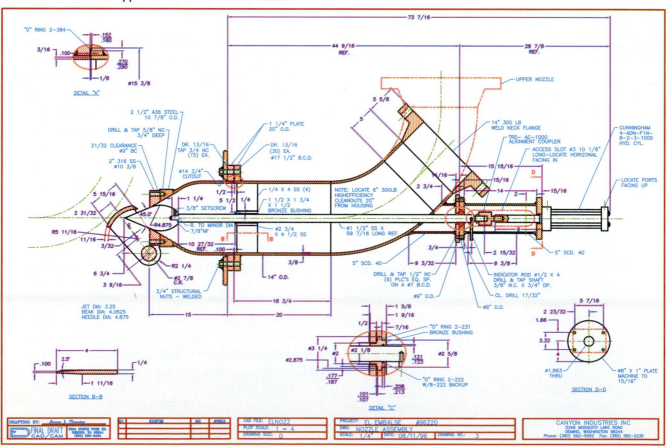

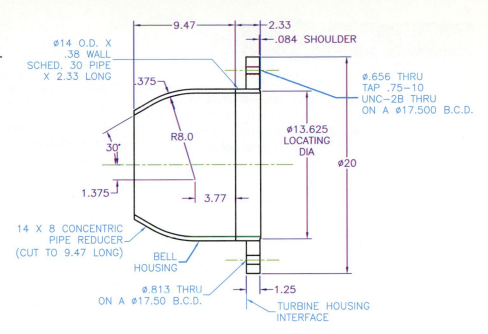

Fig. 1.2.5 Construction of the bell for the lower nozzle assembly. The dimensions and notes are for reference only. Do not include them in your drawing.

Ø14 O.D. X
.38 WALL
SCHED. 30 PIPE
X 2.33 LONG

.375

9.47

2.33

.084 SHOULDER

Ø.656 THRU
TAP .75−10
UNC−2B THRU
ON A Ø17.500 B.C.D.

R8.0

Ø13.625
LOCATING
DIA

Ø20

30°

1.375

3.77

14 X 8 CONCENTRIC
PIPE REDUCER
(CUT TO 9.47 LONG)

BELL
HOUSING

Ø.813 THRU
ON A Ø17.50 B.C.D.

1.25

TURBINE HOUSING
INTERFACE

Practice Exercise 1.2B

The engineers have determined that a 14″ O.D. Schedule 30 pipe is required for the nozzle assembly.

1. As shown in Fig. 1.2.5, the thickness should be .375″. However, you should find and confirm this value before using it. Accuracy is of extreme importance in technical drawings, so make a habit of confirming scheduled dimensions before using them. Tables of pipe schedules that specify the thickness of scheduled pipe are available in many places. You can find them by searching the Internet or by looking them up in your local or school library.

2. Make **Object** the current layer. Lay out the **2.33″** section of pipe, making it long. Use the **LINE** command with polar coordinates to create the top line. Consult the *CAD Reference* if you need to review the use of polar coordinates. Zoom in to enlarge the line on the screen. Then use the **OFFSET** command with an offset distance of **14**

to create the bottom line that shows the 14″ diameter. Use an offset distance of .375 to show the pipe thickness. Add vertical lines from the top line to the bottom line on both sides of the pipe to finish this portion of the pipe. Use the **Endpoint** object snap to set the lines accurately.

3. Create the centerline. For construction purposes, it is a good idea to run the centerline all the way across the screen. To do this efficiently, enter the **XLINE** command and select the **Offset** option. Set the offset distance to **7** (half of the outer diameter of the pipe). Select the new line and then use the Layer Control on the Object Properties toolbar to change the centerline to the **Center** layer. *Hint:* Set the linetype scale to 2.0 to show the centerline at an appropriate scale for the drawing.

A 14″ × 8.0 concentric pipe reducer will be placed on the left end of the 14″ pipe. Pipe reducers can be manufactured in various shapes and sizes. The reducer to be used in this project will be drawn with a 30° taper.

4. Begin construction of the pipe reducer by constructing the inside diameter line at the top of the pipe. The construction dimensions are shown in blue in Figure 1.2.5. Use the **PLINE** command to create the first line segment. Snap to the end of the *inside* top line of the 14″ pipe and use polar coordinates to extend the line **3.77** to the left. End the PLINE command.

5. To locate the center of the arc for the next segment, offset the centerline down by **1.375** and offset the left vertical line of the 14″ pipe to the left by **3.77**. The intersection of these two temporary lines marks the center point. The easiest method of creating the arc is to use the **CIRCLE** command to create a circle with a center at the location you have just identified and a radius of **8.0**. Erase the two temporary lines.

6. To locate the left end of the pipe reducer, offset the left vertical line of the 14″ pipe to the left by **9.47**. Trim the circle to leave only the arc between the location line you just created and the left end of the polyline you created in step 4, as shown in Fig. 1.2.6.

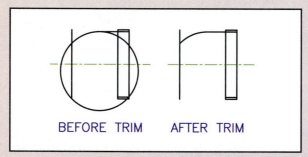

BEFORE TRIM   AFTER TRIM

Fig. 1.2.6 **Trim the circle as shown here to create the curved portion of the inner wall of the concentric pipe reducer.**

7. Enter the **PEDIT** command and pick the polyline segment you created in step 4 to select it for editing. Enter **J** to activate the Join option, and add the arc you created in steps 5 and 6. End the PEDIT command. By converting the entire length of the pipe reducer into a single polyline, you make it easier to develop the other pipe walls.

8. Enter the **OFFSET** command and offset the polyline by **.375** to create the top wall.

9. Enter the **MIRROR** command and select the two polylines that represent the upper part of the concentric pipe reducer. Use the centerline as the mirror line. Enter the **TRIM** command, press **ENTER** to select all of the geometry as cutting edges, and trim the left edges of the pipe reducer to finish it.

Next, a flange needs to be added to the right side of the 14″ pipe. The flange will be made of 1.25″ plate and will have a diameter of 20″. A hole will be cut through the center, concentric with the outside of the 14″ pipe. The hole will be the same diameter as the outside of the 14″ pipe so that the flange can be slipped onto the end of the pipe and then welded into place.

10. To begin the flange, offset the centerline by **10** both above and below the current centerline to locate the 20″ concentric flange. Select the two new lines and change them to layer **Object**.

11. Offset the right vertical line of the 14″ pipe by **.084** to the left to locate the right edge of the flange. Use the **EXTEND** command to extend this line to the two lines you created in step 10.

12. Offset the right edge of the flange by **1.25** to the left to create the left edge of the flange.

13. Enter the **TRIM** command and trim the construction lines to finish the upper and lower portions of the flange. Use Fig. 1.2.5 as a guide. Don't forget to trim the section of line that lies inside the nozzle.

The flange must contain holes for mounting to the turbine housing. The mounting holes will be placed in a pattern of Ø.813″ through holes on a 17.500″ bolt circle diameter (B.C.D.) alternating with a pattern of .75–10 UNC-2B (Ø.656) through holes. To indicate this on the drawing, show a Ø.656 hole in the top part of the flange and a Ø.813 hole in the bottom part. (The number and placement of each type of hole will be placed in a note in a later procedure.)

14. To find the centerlines of the top and bottom holes, offset the main centerline up and down by half of the B.C.D. Then offset the hole centerlines by half of the hole diameters to locate the holes. Change the lines that define the holes onto the **Object** layer and trim these and the centerlines appropriately. Show the tap drill diameter on the upper hole by offsetting the .75″ tap drill diameter from the centerline of the upper hole. (Again, use half the tap drill diameter and offset the centerline both up and down.) Change the lines representing the tap drill diameter to the **Hidden** layer, and trim them to the sides of the flange.

The final step in creating the bell housing is to create a locating diameter on the end of the 14″ pipe. The purpose of the locating diameter is to locate the bell housing precisely to the lower nozzle housing.

15. Refer once more to Fig. 1.2.5 to draw the locating diameter. Note that it is Ø13.625″, with a shoulder of .084″. Use the **OFFSET** command to offset the main centerline up and down by half of the locating diameter to find the location of the shoulder. Change both of the new lines to the

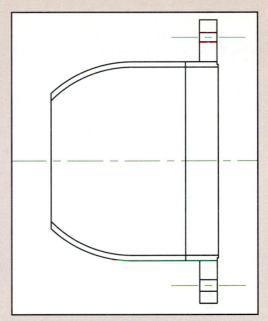

Fig. 1.2.7 **The finished bell housing. Do not add dimensions or notes at this time.**

Object layer. Extend the left edge of the upper and lower flange representations to the shoulder and trim the flange lines as necessary to finish the bell housing. The result of this Practice Exercise is shown in Fig. 1.2.7. *Note:* Do not trim the main centerline. You will need it later.

Add the Beak

At the left end of the concentric pipe reducer, a "beak" will be installed. This is where the water exits the nozzle assembly. The beak funnels the water from the pipe through the needle, which forces the water onto the turbine. You will draw the needle later in this project.

To install the beak, a plate must first be welded onto the bell housing. The plate has two purposes:

- provide the locating diameter for the beak
- house a series of tapped holes that provide a means of fastening the beak to the bell housing

The beak will have tapped holes that match those on the plate, so the engineers have decided that the beak and plate will be machined together on a lathe.

Continue building the nozzle assembly drawing by adding the beak to the PROJECT 1.dwg file you started in Practice Exercises 1.2A and 1.2B. Before you begin, however, it is a good idea to make a backup copy of the PROJECT 1.dwg file. Store the backup in a safe place. Then open your working copy and proceed with the steps described in Practice Exercise 1.2C.

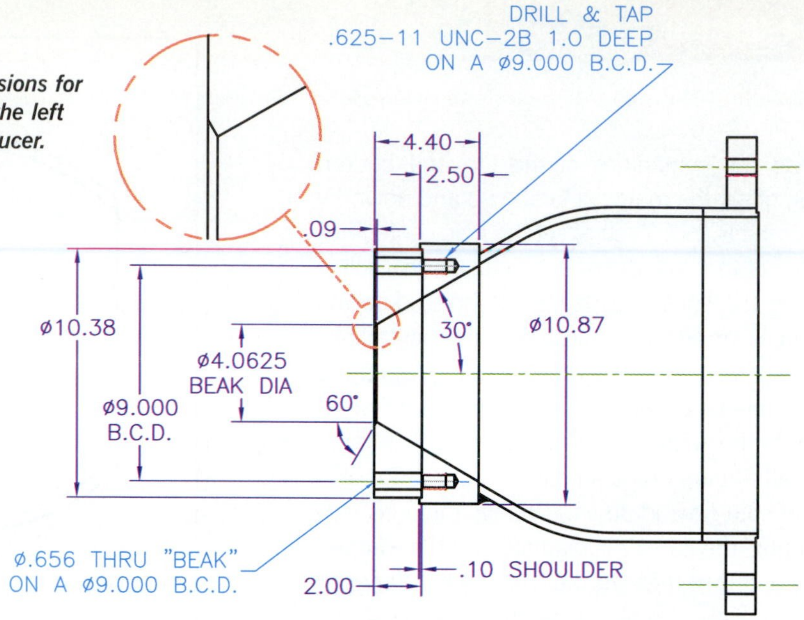

Fig. 1.2.8 *Construction dimensions for adding the plate and beak to the left end of the concentric pipe reducer.*

DRILL & TAP
.625–11 UNC–2B 1.0 DEEP
ON A Ø9.000 B.C.D.

4.40

2.50

.09

Ø10.38

Ø4.0625
BEAK DIA

Ø9.000
B.C.D.

30°

Ø10.87

60°

Ø.656 THRU "BEAK"
ON A Ø9.000 B.C.D.

2.00

.10 SHOULDER

Practice Exercise

This exercise steps you through the addition of the plate and beak to the left side of the concentric pipe reducer. Refer to Fig. 1.2.8 as you follow the steps in this exercise.

The plate that attaches the beak to the bell is 2.50" thick and has a diameter of 10.87". The basic outline of the plate can be created entirely using the OFFSET, TRIM, and EXTEND commands.

1. Using the **OFFSET** command, offset the left side of the concentric pipe reducer by **2.50** to locate the right side of the plate.
2. Offset the main centerline up and down by half the diameter of the plate. Change these lines to the **Object** layer.
3. Enter the **EXTEND** command to extend the right and left sides of the plate to meet the lines that locate the top and bottom of the plate. Then trim the top and bottom lines to the sides of the plate.

4. Offset the left side of the plate to the left by **.10** to find the location of the locating diameter. Then offset the main centerline up and down by half of the locating diameter. Change the two offset lines to the **Object** layer and trim the lines forming the left side of the plate and the locating diameter to finish them.

The plate also contains eight .625–11 UNC-2B tapped holes on a 9.000" bolt circle diameter. You can create them using the same procedure described in Practice Exercise 1.2B for the bolt holes in the flange, or you can use the mirroring procedure described below. The mirroring procedure works because these holes are exactly the same size.

5. Offset the main centerline upward by half the bolt circle diameter to find the centerline of the upper bolt hole.
6. Use the detail shown in Fig. 1.2.9 for guidance in drawing the upper bolt hole. Use the **LINE**, **OFFSET**, **TRIM**, and **EXTEND** commands in the same manner as in previous sequences.

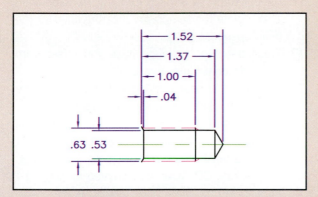

Fig. 1.2.9 *Detail showing the bolt hole for the plate that fastens the beak to the bell housing.*

7. Trim the bolt-hole centerline appropriately.
8. Because you will need this bolt hole representation more than once in the drawing, use the **BLOCK** command to create a block of it. Choose the intersection of the bolt-hole centerline and the leftmost vertical line for the insertion point.
9. Use the **MIRROR** command to place the lower bolt hole. Select the bolt hole block as the object to mirror, and use the centerline of the hole as the mirror line. Do not delete source objects.

The plate is now complete. Follow the steps below to attach the beak on the left side of the plate.

10. Any one of several existing lines can be offset to locate the left side of the beak. The most convenient is probably the right side of the plate. Offset this line by **4.40** to find the leftmost edge of the beak.
11. The top and bottom of the beak align with the locating diameter of the plate. Therefore, to create the top and bottom of the plate, use the **EXTEND** command to extend the short horizontal lines that mark the locating diameter, as shown in Figure 1.2.10. Trim the right edge to these lines.

The beak has bolt holes spaced to match those in the plate. The diameter of the holes is Ø.656″, as shown in Fig. 1.2.8, and are located on a bolt circle diameter of Ø9.000″.

12. Because the centerlines for the bolt holes are the same for the plate and the beak, use AutoCAD's grip feature to extend the length of the centerlines of the plate holes to the left. Do this by picking the centerline, picking the blue grip box on the left side to activate it (it turns red when activated), and moving it to the left so that it extends through the beak. Turn **Ortho** on before you do this so that the centerline remains perfectly straight. Do this for both the upper and the lower bolt hole.
13. Offset the bolt-hole centerlines up and down by half the diameter of the bolt hole. Change the object lines to the **Object** layer and trim them to the right and left edges of the beak.
14. The left side of the beak has a Ø4.0625 opening set back from the face (left side) of the beak by .09″. To create the opening, first offset the left side of the beak to the right by **.09** to locate the depth. Then offset the main centerline by half

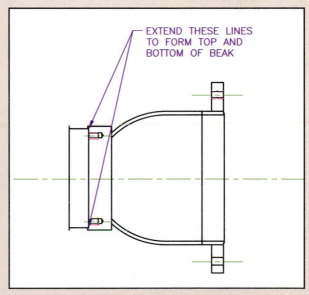

EXTEND THESE LINES TO FORM TOP AND BOTTOM OF BEAK

Fig. 1.2.10 *Extend these lines to form the top and bottom of the beak.*

the diameter of the opening to locate its upper and lower edges. Enter the **LINE** command and use the **Intersection** object snap to start the line at the intersection shown in Fig. 1.2.11. Use the **Tangent** object snap to place the other end of the line tangent to the 30° curve on the line that defines the inside of the upper wall of the bell housing. Trim the curve back to its intersection with the line to create a smooth line.

15. Offset the line you placed in step 14 up by the thickness of the pipe wall (.375"). Trim the curve

to this line to match the lower line, and then trim the upper line to the right edge of the plate.

16. Repeat steps 14 and 15 for the lower pipe wall. Then delete the temporary locating lines for the beak opening.

17. Erase the two reference lines you offset from the main centerline in step 14.

18. To finish the beak, add a 60° taper from the inside of the opening to the left edge of the beak. Refer to the inset in Fig. 1.2.8 (p. 24). Do this for both the upper and lower edge of the beak. Trim the vertical line that represents the .09" inset to the 30° lines you created in steps 14 through 16.

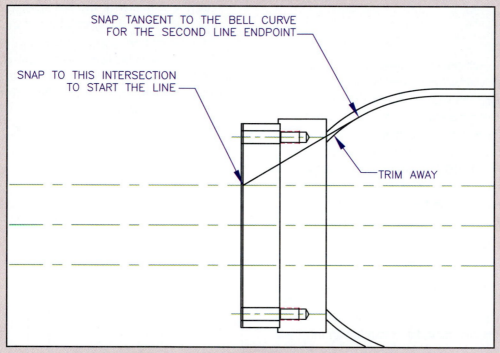

Fig. 1.2.11 Snap the endpoints of the line as shown here. Then trim away the unneeded part of the curve on the bell.

Attach the Lower Nozzle Housing

The next step in creating the nozzle assembly drawing is to draw the lower nozzle housing. Both the upper and the lower nozzle housings are shown in Fig. 1.2.12. Refer again to Fig. 1.1.4 in the previous section for a general assembly drawing that shows how these pieces fit into the overall assembly. In this section, you will draw the lower nozzle housing.

The nozzle consists of the same 14″ pipe that was used in the bell housing. We will add a flange on the left side to mate with the similar flange on the bell housing. On the right end, a weld-neck flange allows connection to the next section of pipe. Refer to Fig. 1.2.13 for the dimensions and specifications needed to draw the nozzle housing. Notice that the drawing is a broken section, so that the weld-neck flange and part of the pipe appear from the outside only. Follow the steps in Practice Exercise 1.2D to create the lower nozzle housing and flanges. Refer to Figure 1.2.13 as you work through the exercise.

Courtesy of Canyon Industries, Inc.

Fig. 1.2.12 The upper and lower nozzle housing assemblies, showing the welded flanges that connect to further sections of pipe.

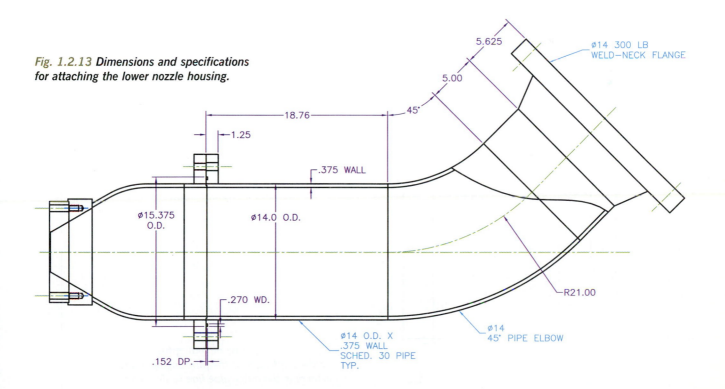

Fig. 1.2.13 Dimensions and specifications for attaching the lower nozzle housing.

1. Begin by adding the next section of Ø14″ pipe. As you can see in Fig. 1.2.13, this is a straight segment that is 18.76″ long. Create this segment by offsetting the left side of the upper 1.25″ flange representation **18.76** to the right. Then extend the outer pipe wall to the offset line, as shown in Fig. 1.2.14.

2. Mirror the outer diameter line you created in the previous step to create the lower boundary for the pipe segment. Use the main centerline as the mirror line.

3. Offset both of the outer pipe lines to the inside by **.375** to create the inner walls of the pipe.

4. Extend the line you offset in step 1 (labeled "NEW OFFSET LINE" in Fig. 1.2.14) downward to the line that represents the lower outer diameter of the pipe. Trim the top of this line to the outer wall at the top of the pipe.

On the left side of this section of pipe, a flange is needed to mate with the flange on the bell housing. In Fig. 1.2.13, you can see that this flange is exactly the same as the one on the bell housing, except that this one has a groove in it, called a *gland*, for an O-ring. The O-ring keeps a seal to prevent the water from leaking at the joint. Because the two flanges are almost identical, you can save time by copying the existing flange and then adding the O-ring groove.

5. Enter the **COPY** command and use crossing windows to select both the upper and lower flange representations. Enter the **R** option and pick both centerlines to remove them from the selection set. Use the **Endpoint** or **Intersection** object snap to select the lower left corner of the upper flange on the bell housing as the base point. For the second point of displacement, use the **Endpoint** or **Intersection** object snap to select the lower right corner of the same flange. This operation places the new upper and lower flanges in their correct position using a single operation.

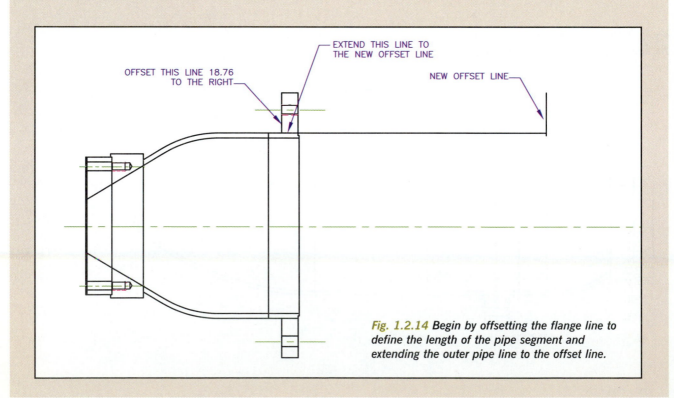

OFFSET THIS LINE 18.76 TO THE RIGHT

EXTEND THIS LINE TO THE NEW OFFSET LINE

NEW OFFSET LINE

Fig. 1.2.14 Begin by offsetting the flange line to define the length of the pipe segment and extending the outer pipe line to the offset line.

6. Trim the right side of the upper and lower flanges to the outer wall of the pipe.

7. Use AutoCAD's grips feature to center the flange centerlines across the two adjoining flanges.

8. Add the O-ring groove. Begin by offsetting the main centerline up and down by half of the outer diameter of the groove. (Refer to Fig. 1.2.13 for the exact diameter.) Change both of the offset lines to the **Object** layer. Then reenter the **OFFSET** command and change the offset distance to the width of the groove, which is **.270**. Offset the outer diameter of both grooves to the inside. Reenter the **OFFSET** command again with an offset distance of **.152**. Offset the line to the right by **.152** as shown in Fig. 1.2.15 to create the depth of the groove. Trim the construction lines to finish the groove.

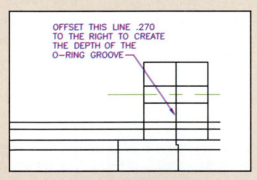

Fig. 1.2.15 *Offset the line between the two flanges to create the depth of the O-ring groove.*

Now you can add the next section of pipe, which is a 45° elbow. The inside radius of the elbow is 14″, and the curve must meet the straight section of pipe exactly. There are many ways to accomplish this accurately. Perhaps the simplest is to create a circle and trim it to create the necessary curve. Follow these steps:

9. At any location on the screen, create a circle with a radius of **14**. Select the circle and enter the **MOVE** command. For the base point, use the **Quadrant** object snap to select the bottom quadrant of the circle. For the second point of displacement, use the **Endpoint** object snap to select the top line on the right side of the straight section of pipe to position the circle as shown in Fig. 1.2.16.

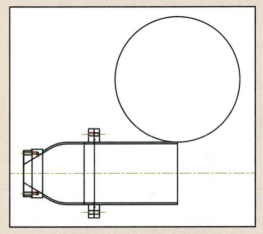

Fig. 1.2.16 *Snap the lower quadrant of the circle to the upper right corner of the straight section of pipe.*

10. Offset the circle by **.375** (the thickness of the pipe wall) to the outside to create the inner wall diameter.

11. The radius at the bottom of the elbow is larger than the top radius. Rather than calculate it exactly, you can use the OFFSET command to place the inner and outer pipe walls accurately. Enter **OFFSET**, but instead of entering an offset value, use the default **Through** option. Using the **Intersection** object snap, select the intersection of the inner circle (the one you created in step 9) and the right end of the adjoining straight pipe section. Then use the **Endpoint** object snap to snap to the bottom right corner of the straight pipe section. Select the inner circle as the object to offset, and offset it to the outside. This creates the curve for the lower outside wall of the pipe. Create the lower inside wall by offsetting the circle that represents the lower outside wall by **.375** to the inside. Your drawing should now look like the one in Fig. 1.2.17.

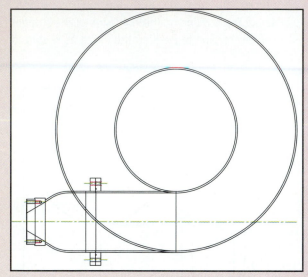

Fig. 1.2.17 *Circles in place to form the walls of the 45° pipe section.*

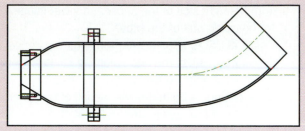

Fig. 1.2.18 *Nozzle assembly with 45° elbow and 5″ straight pipe section in place.*

12. Before trimming the circles, use them as a basis for creating the curved centerline for the elbow. Enter the **CIRCLE** command. For the center point of the circle, use the **Center** object snap to select the center of any of the four existing circles. Enter **21.00** for the radius. (Refer again to Fig. 1.2.13.) Change this circle to the **Center** layer.

13. To define the other end of the 45° elbow, create a temporary line. Enter the **LINE** command. Then use the **Center** object snap to select the center of any of the circles as the first point. Use polar coordinates of **@30<–45** to place the other end of the line across the circles at 45° from the left edge of the pipe. Enter the **TRIM** command and select the temporary line and the right edge of the straight pipe section as the cutting edges. Trim all five of the circles to create the elbow. Then reenter the **TRIM** command and trim the temporary line to the outside walls of the elbow to finish it.

14. At the right end of the elbow, we will place a straight piece of pipe that is 5″ long. To do this, enter the **LINE** command and snap to the right end of the upper wall of the elbow. Specify the length and direction using polar coordinates: **@5<45**. Create another 5″ line at the end of the bottom wall of the elbow using the same technique. Then create a third line to connect the right ends of these two lines to complete the 5″ section of straight pipe. Use the **Endpoint** object snap to join the lines exactly.

15. Offset either of the pipe wall lines to the inside of the pipe by half the pipe diameter to establish the centerline. Change this line to the **Center** layer. Your drawing should now look like the one in Fig. 1.2.18.

Referring again to Fig. 1.2.13 on page 27, notice that this view is actually a broken-out section. The rightmost part of the drawing shows only the outside of the nozzle assembly. In Fig. 1.2.13, the break line extends through part of the 45° elbow and part of the 5″ pipe section.

16. Use the **PLINE** command to create a break line similar to the one shown in Fig. 1.2.13. Trim the upper inside wall back to the section line. Offset the lower, outer wall line by the wall thickness (**.375**) and trim it back to the break line.

For the last piece of the lower nozzle assembly, the engineers have specified a 14″, 300-lb. weld-neck flange. Standard sizes for forged flanges are specified by ANSI and other standards-setting organizations. The tables in Appendix A are adapted from ANSI B16.5. Use these tables to determine the correct dimensions for a weld neck with a diameter of 14″.

From Fig. 1.2.13 you can see that the total length of the weld neck is 5.625. Use this number in the following steps to look up the specifications for the actual flange.

17. Check the table for the length of the weld neck. Offset the top line of the 5″ straight pipe by the length of the weld neck.

18. Extend the centerline from the 5″ straight pipe segment through the weld neck. To do this accurately, offset the rightmost line of the weld neck to the right by **3″**. Enter the **EXTEND** command and select the offset line as the boundary line. Pick the centerline to extend it to the offset line. Then erase the temporary offset line.

19. Offset the centerline to the left and right of the centerline by half the diameter of the flange, as determined from the ANSI table. Change the offset lines to the **Object** layer. Extend the top line of the weld neck (which you created in step 17) to meet these offset lines.

20. From the ANSI table, determine the width of the flange. Offset the upper line of the flange down by the width of the flange. Trim the two offset lines from step 19 to the edge of the lower line to finish the lip of the flange.

21. From the ANSI table, determine the diameter of the bolt hole circle. Offset the centerline to both sides by half the B.C.D. to create the centerlines for the bolt holes. Trim and extend them as necessary to finish the centerlines.

22. To locate the upper end of the neck, offset both ends of the flange **3.00** to the inside. To finish the weld neck, create lines from the 5″ straight pipe segment to the lower end of the offset lines. Erase the offset lines. When you finish this practice exercise, your drawing should look like the one in Fig. 1.2.19.

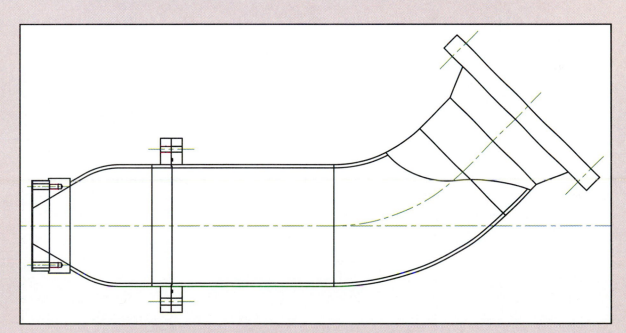

*Fig. 1.2.19 **The nozzle assembly drawing at the end of Practice Exercise 1.2D.***

Create the Rear Housing Interface

The next step is to add interface with the 5.0″ schedule 40 pipe at the rear of the lower nozzle assembly. See Fig. 1.2.20. This pipe will be in line with the main centerline of the lower nozzle. On one end, it will be cut to fit the outside profile of the 14″ elbow. On the other end, a round piece of plate will be welded, with a hole cut into it to allow the shaft to protrude through. The plate will have a 9.0″ O.D. and will be .69″ thick. It will be 44.57″ from the face of the flange it interfaces with on the turbine housing.

Another through hole must be cut through the outside curve of the lower nozzle housing elbow. This hole will also be in line with the main centerline of the 14″ pipe and will allow the shaft to protrude through. The hole will have a diameter of 2.25″.

Follow the steps in Practice Exercise 1.2E to create this part of the nozzle assembly drawing. Refer to Fig. 1.2.21 for necessary dimensions as you work.

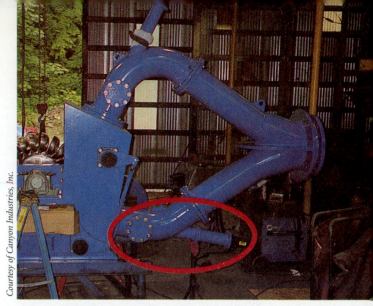

Courtesy of Canyon Industries, Inc.

*Fig. 1.2.20 **The rear housing assembly.***

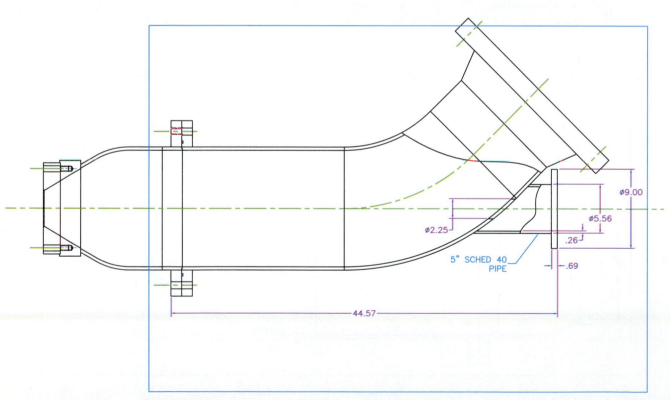

*Fig. 1.2.21 **Dimensions for creating the rear housing interface.***

Practice Exercise

1.2E

This practice exercise steps you through the procedure to add the rear housing interface to the nozzle assembly drawing. Begin by establishing the inner and outer walls of the 5″ schedule 40 pipe.

1. As you can see from Fig. 1.2.21, the outer diameter of the 5″ schedule 40 pipe is 5.56″. Check a pipe schedule to verify that this is so. Then offset the main centerline by half of this diameter to locate the outer walls of the pipe. Offset the main centerline by half of the inner diameter to establish the inner pipe walls. Change the four offset lines to the **Object** layer.

2. Trim left side of the inner and outer walls of the 5″ pipe to the outside of the 14″ 45° elbow.

3. Locate the upper and lower lines of the 9″ plate by offsetting the main centerline by half the diameter of the plate. Change the lines to the **Object** layer.

4. Locate the right side of the terminal plate by offsetting the left side of the flange on the bell housing by **44.57**. Extend the offset line to the line that defines the upper end of the 9″ plate. Trim the lower end of the offset line to the line that defines the lower end of the plate.

5. Offset the right side of the terminal plate by the thickness of the plate to establish its width. Refer to Fig. 1.2.21 for the width.

6. Trim the remaining construction lines to finish the outline of the 5″ pipe and 9″ plate.

7. Create the break line as shown in Fig. 1.2.20 and trim the inside walls of the pipe accordingly.

Create the hole in the outside curve of the 14″ pipe to allow the straight shaft to protrude through. The diameter of this hole is 2.25″. Notice on Fig. 1.2.21 that in this view, the hole shows only as horizontal lines within the wall thickness of the 14″ pipe.

8. Offset the main centerline by half the diameter of the hole. Change the offset lines to the **Object** layer.

9. Trim the offset lines using the inner and outer walls of the 14″ pipe as cutting edges. When you finish, your drawing should look like the one in Fig. 1.2.22.

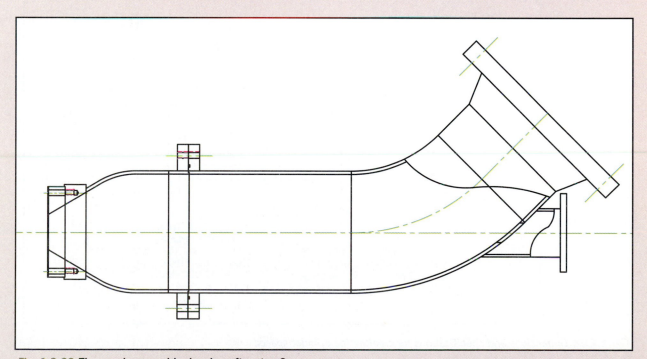

Fig. 1.2.22 The nozzle assembly drawing after step 9.

Fig. 1.2.23 *The cylinder spacer.*

Add the Cylinder Spacer

The cylinder spacer that links the 5″ pipe you created in Practice Exercise 1.2E and the hydraulic cylinder is shown in Fig. 1.2.23. The spacer consists of another piece of 5″ Schedule 40 pipe. A round flange made from Ø8.0″ plate will be welded on each end. The left flange, which is .94″ thick, will attach to the lower housing weldment with .50–13.0 × 1.75 socket-head screws and flat washers.

As you can see in Fig. 1.2.23, a horizontal slot runs most of the length on both sides of the cylinder spacer. This slot will be 3.0″ × 13.1″.

Create the cylinder spacer by following the steps in Practice Exercise 1.2F. Refer to Fig. 1.2.24 as necessary for dimensions and specifications.

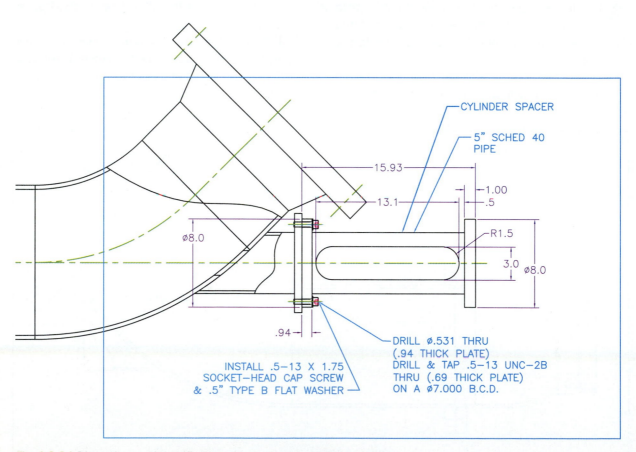

Fig. 1.2.24 *Dimensions and specifications for creating the cylinder spacer.*

In this exercise, you will create the cylinder spacer that connects the rear housing with the hydraulic cylinder.

1. First create the left flange, which mates with the 5″ pipe on the left side of the cylinder spacer. The flange is made from a plate that is 8″ in diameter. Offset the main centerline by half of this diameter to establish the ends of the plate. Change both offset lines to the **Object** layer.

2. Offset the right side of the existing flange on the 5″ pipe by **.94** to the left to create the thickness of the flange. Trim and extend the lines as necessary to define the shape of the flange.

3. Now add the body of the spacer. Offset the main centerline by half the diameter to create the outer walls of the spacer. Recall that the outer diameter of 5″ Schedule 30 pipe is Ø5.56″. Change the lines to the **Object** layer.

4. The overall length of the cylinder spacer, including the two flanges, is 15.93″. To establish this length, offset the left wall of the left spacer flange **15.93** to the right. To establish the width of the right flange, offset the line you just created back to the left by **1.00**.

5. Offset the main centerline up and down by half the diameter of the Ø8″ flange to establish the ends of the right flange. Change the lines to the **Object** layer.

6. Trim the right flange and the horizontal lines that represent the body of the cylinder spacer. After this operation, the cylinder spacer should look similar to the one shown in Fig. 1.2.25.

7. Create the slot through both sides of the cylinder. The slot is a total of 13.1″ long, and the ends have a radius of 1.5″. First use offset construction lines to box in the basic dimensions. Offset the main centerline up and down by the radius (which is half the diameter), and change the lines to the **Object** layer. Trim both lines to the inside edges of the flanges.

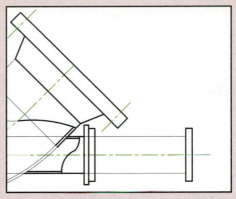

Fig. 1.2.25 *Interim drawing of the cylinder spacer after step 6.*

8. The slot is centered on the length of the cylinder spacer, so you can determine its location using the Midpoint object snap and a temporary line. Enter the **LINE** command and use the **Midpoint** object snap to select the midpoint of the top line of the slot as the first point. Then select the midpoint of the lower line of the slot as the second point. End the LINE command. This temporary line marks the center of the slot. Offset this line by half the width of the slot to locate the ends of the slot. Erase the temporary line.

9. At any point on the screen, create a circle that has a radius of **1.5**. Then enter the **MOVE** command, select the circle, and select the left quadrant of the circle as the base point. Use the intersection of the left slot boundary line (which you created in step 8) and the main centerline as the second point of displacement. Enter the **COPY** command and copy the circle. This time, use the right quadrant of the circle as the base point. Use the intersection of the right slot boundary line and the main centerline as the second point of displacement. The spacer should now look like the one in Fig. 1.2.26.

10. Delete the right and left slot boundary lines. Trim the two horizontal lines to the circles. Trim away the inside of both circles to complete the slot, as shown in Fig. 1.2.27.

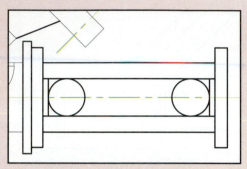

Fig. 1.2.26 *Circles in place to define the ends of the slot. Trim away the unneeded lines to finish the slot.*

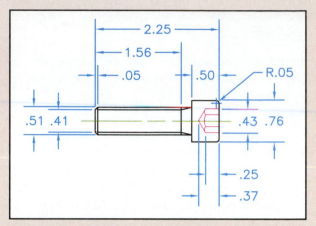

Fig. 1.2.28 *Draw the cap screw using the dimensions shown here.*

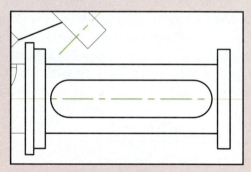

Fig. 1.2.27 *The finished slot.*

Recall that the left flange will be attached to the lower nozzle housing weldment using .50–13 × 1.75 socket-head cap screws with flat washers. The screws will be on a 7.000" B.C.D.

11. Offset the main centerline up and down by half of the B.C.D. to find the centerlines of the holes. Create a Ø.531 through hole at the top and bottom of the flange by offsetting the hole centerlines by half the diameter of the hole. Change the lines that represent the hole to the **Object** layer and trim them to the edges of the flange. Trim the hole centerlines so that they extend through the mating flanges. (You will need to select the two hole centerlines and use the Properties window to change their linetype scale so that the linetype appears correctly.)

12. At an empty place on the screen, draw the cap screw as shown in Fig. 1.2.28. Do not include the dimensions shown in the illustration; they are there only for your reference. After you have created the screw, create a block of it so that you can reuse it. For the insertion point of the block, use the intersection of the centerline and the leftmost line of the screw head. Name the block **.50-13 x 1.75 CS**. Refer to the *CAD Reference* if you need to refresh your memory regarding blocking techniques.

13. Insert the block into the drawing twice—once for each hole that appears in this view of the flange. (As an alternative, you can insert the block once, place it correctly, and then use the **MIRROR** command to place the other one.) When you have completed this step, the nozzle assembly drawing should look like the one in Fig. 1.2.29.

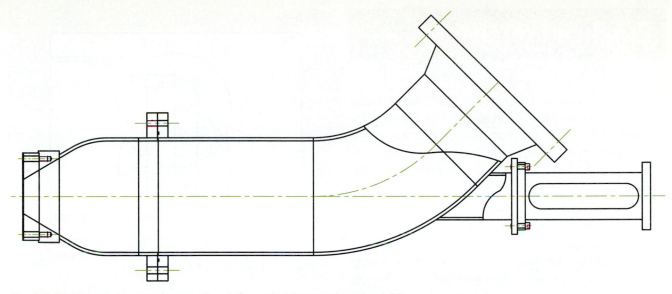

Fig. 1.2.29 The nozzle assembly drawing at the end of Practice Exercise 1.2F.

Create the Needle

Return your attention to the other end of the nozzle assembly. The next task is to draw the needle that regulates the stream of water into the turbine assembly. The needle will be 4.875″ in diameter at its widest point. It will have a 22.5° taper on the left side and a flat back (right side).

The needle will be drawn as a partial section to show the tapped .875–14 × 1.5 hole on the centerline of the needle. Notice also the Ø1.5 shaft that runs from the upper surface of the needle to the tapped hole on the centerline.

Follow the steps in Practice Exercise 1.2G to create the needle within the beak of the nozzle assembly. Follow the dimensions and specifications in Fig. 1.2.30.

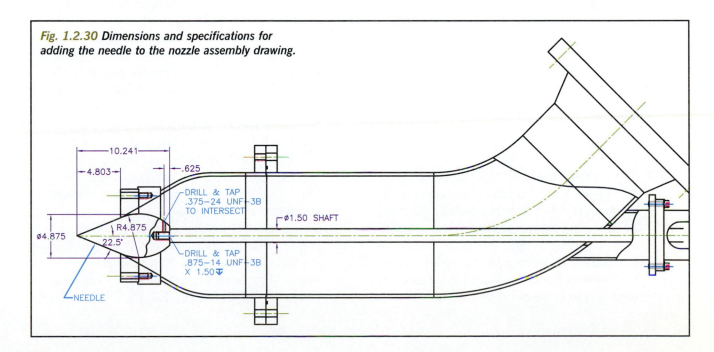

Fig. 1.2.30 Dimensions and specifications for adding the needle to the nozzle assembly drawing.

Because the widest part of the needle forms a true 4.875″ radius, it is best to begin by defining the upper and lower arcs that define the radius. These arcs can then be used as a basis for the various tangents that will be needed to finish the shape of the needle.

1. Zoom in on the beak end of the nozzle assembly drawing.
2. The center points of the arcs are in line with the interface between the beak and its adjoining plate. To establish the center points exactly, offset the main centerline by half of the total height of the needle. See Fig. 1.2.31. Use these center points for two circles. Make the radius of both circles **4.875**.

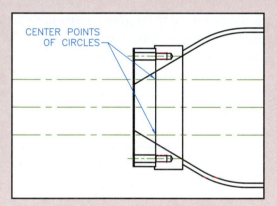

Fig. 1.2.31 *Locate the center points of the circles from which the upper and lower arcs will be created.*

3. Begin forming the 22.5° taper on the left end of the needle. Enter the **LINE** command and use the **Nearest** object snap to snap to a point anywhere on the main centerline to the left of the beak and plate. Use the polar coordinates **@10<22.5** to create a temporary line defining the shape of the upper half of the nose of the needle. With Ortho on, move the line so that it is

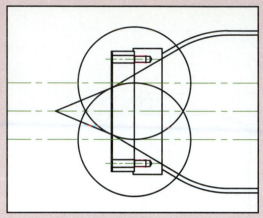

Fig. 1.2.32 *The result of the MIRROR command to create the nose of the needle.*

tangent to the upper circle from step 2. The lower end of the line must remain on the main centerline. Trim the upper end of the line to the point of tangency with the circle.

4. Use the **MIRROR** command to mirror the line about the centerline to create the lower nose of the needle. Figure 1.2.32 shows the result of this operation.
5. Now create the flat back for the needle. To do this, use the **From** object snap. Enter the **LINE** command and enter **From** at the keyboard. Then use the **Endpoint** object snap to specify the tip of the needle for the base point. At the Offset prompt, enter the polar coordinate **@10.241<0** to start the new line on the centerline exactly 10.241″ from the needle tip. With Ortho on, select any point above the lower circle, as shown in Fig. 1.2.33. The exact length of the line does not matter. Use the circle as a cutting edge to trim the upper part of the line away. Then mirror the line about the main centerline to complete the back of the needle.
6. Trim away all of the unneeded parts of both circles to complete the outline of the needle. Trim the plate and beak away from the inside of the needle. Also delete the two temporary lines you created in step 2. See Fig. 1.2.34.

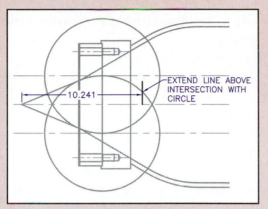

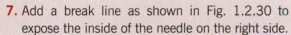

Fig. 1.2.33 Begin a new line at the main centerline, 10.241 from the needle tip.

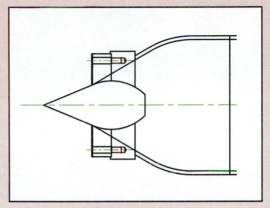

Fig. 1.2.34 Trim both circles away to leave only the radii needed for the needle.

7. Add a break line as shown in Fig. 1.2.30 to expose the inside of the needle on the right side.

8. Add a tapped hole and shaft of .875–14 on the centerline of the flat of the needle. Both will be 1.5″ deep, and the shaft will have a .875–14 × 1.25″ tapped end to thread into the needle. See Fig. 1.2.35 for construction details.

9. Now add a .375–24 tapped hole for a setscrew. Set the vertical centerline for the setscrew .625 to the left of the flat back of the needle. Do this by offsetting the back of the needle **.625** to the left. Change the offset line to the **Center** layer. With Ortho on, use AutoCAD's grips to extend the line upward to create the centerline. Offset the centerline **.1875** to the right and left; change these lines to the **Hidden** layer. Then offset the centerline by **.1563** to the right and left and change these lines to the **Object** layer. Trim the lines to the needle outline at the top and to the upper edge of the .875–14 hole at the bottom. When you finish, the needle should look like the illustration in Fig. 1.2.36.

10. Add a Ø1.50″ shaft along the main centerline of the nozzle by offsetting the centerline up and down by **.75**. Change the lines to the **Object** layer and trim them to the back of the needle on the left and the break line in the rear housing interface on the right, as shown in Fig. 1.2.30.

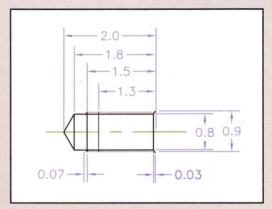

Fig. 1.2.35 Construction details for the .875 tapped hole and shaft.

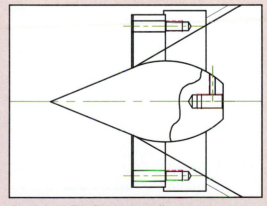

Fig. 1.2.36 The finished needle.

Add the Hydraulic Cylinder

The engineers have specified a Cunningham hydraulic cylinder for the nozzle assembly. Although this will be a purchased part, the cylinder must be shown as part of the assembly drawing.

The hydraulic cylinder assembly attaches to the right end of the cylinder spacer, as shown in Fig. 1.2.37. The cylinder is mounted on a 1.0″ rectangular flange that has four .538 clearance holes. Four .50–13 × 1.75 socket-head cap screws will be used with .5″ washers to attach the hydraulic cylinder to the cylinder spacer.

Practice Exercise 1.2H lists the steps to add the hydraulic cylinder assembly to the nozzle assembly drawing. Refer to Fig. 1.2.38 for dimensions and specifications as you complete this exercise.

Fig. 1.2.37 The hydraulic cylinder (black) attaches to the right end of the cylinder spacer.

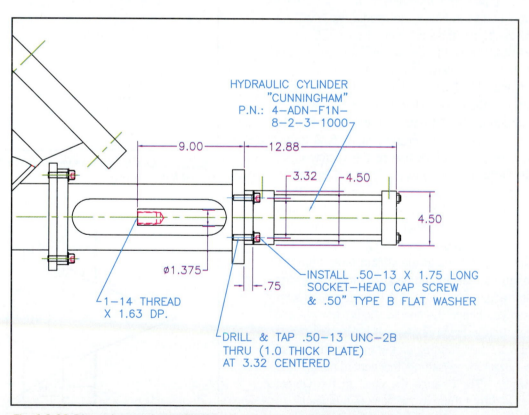

Fig. 1.2.38 Dimensions and specifications for drawing the hydraulic cylinder.

1. Begin by creating the rectangular flange. As you can see from Fig. 1.2.38, the flange is 4.50″. Offset the centerline by half of this to create the end lines. Offset the right side of the cylinder spacer by **.75** to create the width of the flange, and trim the lines as necessary.

2. Add **.5–13 UNC-2B** through holes in the flanges on both the cylinder spacer and the hydraulic cylinder for the cap screw. The centerlines of the holes should be on a diameter of **3.32**.

3. Add the **.5–13 × 1.75** socket-head cap screws and **.5** Type B flat washers, as shown in the detail in Fig. 1.2.39.

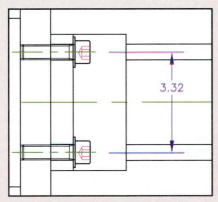

Fig. 1.2.39 **Detail of the holes and socket-head cap screws.**

4. Including the flanges, the total length of the hydraulic cylinder is **12.88.** Offset the right side of the flange on the cylinder spacer by this length to locate the right end of the hydraulic cylinder. Because the ends of the cylinder flanges are the same height as the rectangular flange, you can extend the upper and lower lines of the flange to the right side of the cylinder to form the basis for these ends.

5. The left end of the cylinder is 1.75″ wide. Create this width by offsetting the right side of the rectangular flange **1.75** to the right. The right end of the cylinder is 1.25″ wide. Create this width by offsetting the right side of the hydraulic cylinder **1.25** to the left. Trim the lines to finish the ends of the hydraulic cylinder.

6. The middle portion of the hydraulic cylinder has an inside diameter of **Ø2.8** and an outside diameter of **Ø3.8**. Create these lines by offsetting the centerline by half the diameter. Be sure to change the lines to the **Object** layer, and trim them to the ends of the cylinder.

7. The internal cylinder shaft has a diameter of Ø1.375″ and protrudes 9.00″ into the cylinder spacer. Use offset lines to add the part of the shaft that is visible through the slot in the cylinder spacer.

8. On the end of the shaft that extends into the cylinder spacer, there is a 1–14 × 1.63 threaded tapped hole. Add this hole as shown in the detail in Fig. 1.2.40.

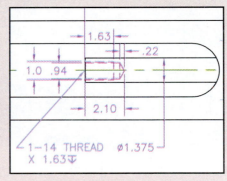

Fig. 1.2.40 **Detail of the hole on the left end of the cylinder shaft.**

9. Although it is not necessary to detail the ports in the cylinder that bring the hydraulic fluid in and out of the cylinder, you should mark their approximate locations on the drawing. There are two ports, one in each end piece of the hydraulic cylinder. Show them as unmarked vertical lines at the top of the cylinder. Refer to Fig. 1.2.38 to see the approximate placement of the lines. (They are green in the illustration.)

10. Add the nuts.

Add the Star Support

The 1.50″ O.D. shaft that runs from the needle to the cylinder spacer is supported within the nozzle by a star support. The support consists of four .25″ thick flat bars that are 4.0″ wide and welded 90° apart on the inside circumference of the Ø14″ pipe section of the lower nozzle housing. The bars are welded to a 2.75″ O.D. × 4.40″ center yoke that runs around the shaft in a radial, or "star," pattern.

The yoke of the star is bored precisely to fit two bronze bushings. These bushings keep the shaft and nozzle aligned with the opening in the beak. Their outside diameters are 1.75″ to fit the inside of the yoke, and their inside diameters are 1.500″ to fit the outside of the center shaft. Both bushings are 1.5″ long and are located at each end of the bored round stock. They will be filled with lubricant to reduce wear on the shaft.

Practice Exercise 1.2I lists the steps to add the star support and its associated bushings to the nozzle assembly. The 3D model in Fig. 1.2.41A should help you visualize the star support and the bushings. Refer to Fig. 1.2.41B for dimensions and specifications as you work through the steps.

A

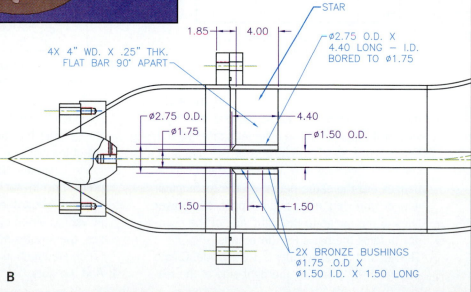

Fig. 1.2.41 *(A) Visualize the star support and bushings with the aid of this 3D model. (B) Dimensions and specifications for the star support for the central shaft.*

B

1. Begin by creating the 4.0″ × .25″ bars. Two of the bars are visible in the drawing, one above and one below the shaft. Place the left end of the bars by offsetting the left side of the leftmost flange **1.85**. Refer again to Fig. 1.2.41. Extend this line down to the inner wall at the bottom of the 14″ pipe. Trim away the excess parts to leave the left side of both support bars. Offset the left side of both support bars by **4.00** to create the right side of the bars.

As you can see from Fig. 1.2.41B, the round stock at the center of the star support is 4.40″ long, but the end facing the needle is tapered back to 4″ to meet the support bars.

2. Offset the main centerline both up and down by half of the outside diameter of the round stock. Change both lines to the **Object** layer and trim them to the sides of the bars.

3. Offset the main centerline by half the diameter of the hole to be bored in the round stock to locate the inside diameter of the stock. Change the lines to the **Object** layer, but do not trim them yet.

4. To establish the length of the inner (left) edge of the round stock, offset the left side of both bars **.4** to the left. Then enter the **LINE** command and use the **Endpoint** or **Intersection** object snap to draw a vertical line from the inner edge to the outer edge both above and below the shaft.

5. Now trim the lines that form the inner diameter of the stock. Also trim the left edge of each bar back so that it does not extend through the round stock. Also, delete the temporary lines you created in step 4 to establish the lower edge of the round stock.

The bronze bushings fit between the shaft and the round stock to keep the shaft and nozzle aligned with the opening in the beak. One bushing is located at each end of the round stock. The bushings are represented in the nozzle assembly drawing by short vertical lines between the shaft and the round stock.

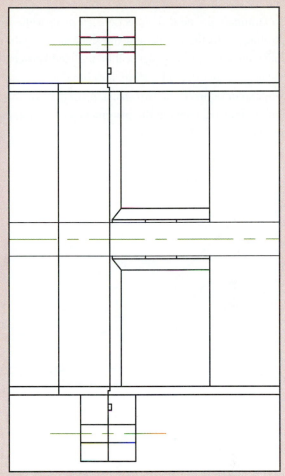

Fig. 1.2.42 *The star support at the end of Practice Exercise 1.2I.*

6. The right edge of the right bushing is already in place because you did not trim the right edge of the support bar in earlier steps. Offset this line to the left by the width of the bushing (**1.50**) and trim it to create the left side of the right bushing. Do this both above and below the cylinder.

7. Create a line from the right edge of the round stock to the central shaft. Use the **Perpendicular** object snap to locate the bottom point exactly. This is the left edge of the left bushing. Offset the left edge by **1.50** to create the right edge of the left bushing. Do this both above and below the cylinder. When you finish this practice exercise, the star support should look like the one in Fig. 1.2.42.

Install the Alignment Coupler

To eliminate the need for precise alignment of the main shaft attached to the needle and the shaft of the hydraulic cylinder, an alignment coupler will be installed between the two. The alignment coupler eliminates binding and accepts misalignment between the two shafts. A detail of the nozzle assembly drawing showing the placement of the alignment coupler is shown in Fig. 1.2.43.

The engineers have specified a standard coupler from TRD Manufacturing, Inc., with a 1.0–14 external thread on one end and a 1–14 internal thread on the other end. An excerpt from TRD's catalog showing the dimensions of their alignment couplers is shown in Fig. 1.2.44. Use the information in Figs. 1.2.43 and 1.2.44 as you follow the steps in Practice Exercise 1.2J to install the alignment coupler.

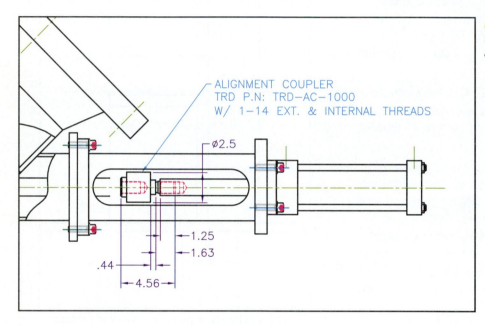

Fig. 1.2.43 *Detail showing the placement of the alignment coupler.*

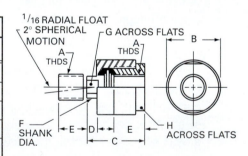

ALIGNMENT COUPLERS									
Part No.	A	B	C	D	E	F	G	H	Max. Pull at Yield
AC437	$7/16$-20	$1^1/4$	2	$1/2$	$1/4$	$5/8$	$1/2$	1	10,000
AC500	$1/2$-20	$1^1/4$	2	$1/2$	$1/4$	$5/8$	$1/2$	1	14,000
AC625	$5/8$-18	$1^1/4$	2	$1/2$	$1/4$	$5/8$	$1/2$	1	19,000
AC750	$3/4$-16	$1^3/4$	$2^5/16$	$1/2$	$1^1/8$	$11/32$	$11/16$	$1^1/2$	34,000
AC875	$7/8$-14	$1^3/4$	$2^5/16$	$1/2$	$1^1/8$	$11/32$	$11/16$	$1^1/2$	39,000
AC1000	1-14	$2^1/2$	$2^{15}/16$	$1/2$	$1^5/8$	$1^3/8$	$1^5/32$	$2^1/4$	64,000
AC1250	$1^1/4$-12	$2^1/2$	$2^{15}/16$	$1/2$	$1^5/8$	$1^3/8$	$1^5/32$	$2^1/4$	78,000
AC1375	$1^3/8$-12	$2^1/2$	$2^{15}/16$	$1/2$	$1^5/8$	$1^3/8$	$1^5/32$	$2^1/4$	78,000
AC1500	$1^1/2$-12	$3^1/4$	$4^3/8$	$11/16$	$2^1/4$	$1^1/4$	$1^1/2$	3	134,000
AC1750	$1^3/4$-12	$3^1/4$	$4^3/8$	$11/16$	$2^1/4$	$1^1/4$	$1^1/2$	3	134,000

Fig. 1.2.44 *Catalog excerpt from TRD Manufacturing, Inc., giving the dimensions for alignment couplers of various sizes.*

The alignment coupler should be installed so that it has an engagement of 1.25″. In other words, it extends 1.25″ into the end of the hydraulic cylinder shaft. Begin your reconstruction of the alignment coupler by establishing this distance to locate the right end of the coupler.

1. Offset the left end of the hydraulic cylinder shaft to the left by 1.25″.
2. Now that the rightmost edge of the coupler has been established, use it to offset the other vertical lines as specified in Figs. 1.2.43 and 1.2.44. You will need to compare the figures closely and do a certain amount of math to create the lines at the correct offset distances.

3. Offset the main centerline up and down by half the largest radius of the coupler. Extend the vertical lines to meet them. Then change the offset lines to the **Object** layer and trim them to finish the outline of this part of the coupler.
4. Repeat step 3 for each of the other radii according to the specifications given in the table in Fig. 1.2.44.

Note: If you become confused by the large number of vertical lines in steps 2 and 3, you may wish to use an alternate method. Another way to accomplish the same task is to add each component of the coupler in sequence from right to left. This method requires more back-and-forth referencing to find the required dimensions, but it may be less confusing for some people to complete each section independently.

Attach the Beak and Bell

The next step is to attach the beak to the face of the bell. You created the holes for the screws in earlier exercises. Now the screws must be added to both the plate at the face of the bell and the flange at the back of it. You will also install the setscrew in the needle at this time.

The screws should be installed as shown in Fig. 1.2.45. Notice that at the back of the beak, only the upper screw is installed. The lower hole is left empty to accommodate a note on the final assembly drawing. Follow the steps in Practice Exercise 1.2K to install the four screws. Refer to Fig. 1.2.45 as necessary.

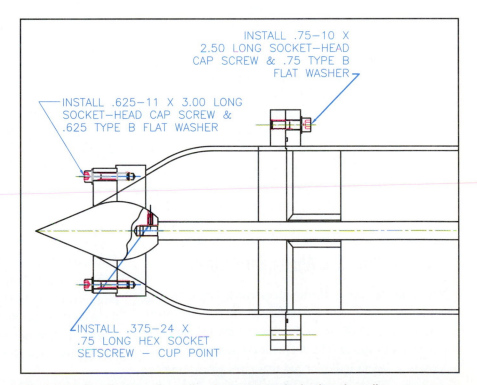

INSTALL .75–10 X 2.50 LONG SOCKET–HEAD CAP SCREW & .75 TYPE B FLAT WASHER

INSTALL .625–11 X 3.00 LONG SOCKET–HEAD CAP SCREW & .625 TYPE B FLAT WASHER

INSTALL .375–24 X .75 LONG HEX SOCKET SETSCREW – CUP POINT

Fig. 1.2.45 Specifications for adding the screws to the beak and needle.

These and all of the other screws in the nozzle assembly drawing do not need to be drawn exactly to scale because they are standard parts that are purchased separately. They will not be manufactured by the people who manufacture the nozzle assembly. However, you must create reasonable facsimiles of the screws at their approximate sizes to include in the assembly drawing. *Note:* Exact dimensions can be found in the latest edition of *Machinery's Handbook*.

1. First create the screws to attach the beak to the face of the bell. The engineers have specified .625–11 × 3.0 socket-head cap screws. A closeup view of the cap screw is shown in Fig. 1.2.46A. Create the screw in a blank space on the screen. Then make a block of it and delete the original. Insert the new block twice: in the hole on the upper plate/beak flange, and in the hole on the lower plate/beak flange. Refer again to Fig. 1.2.45 for proper placement.

2. Again using a blank spot on the screen, create the .75–10 × 2.50 socket-head cap screw and its associated washer. A closeup of this screw is shown in Fig. 1.2.46B. Make a block

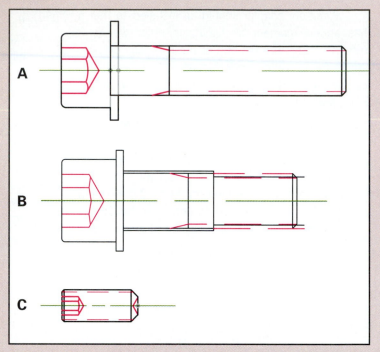

Fig. 1.2.46 *Closeup of the screws to be added in Practice Exercise 1.2K. The screws are shown here at their relative sizes. You should be able to find the sizes by reading the names of the screws in the text and referring to a table of standard sizes.*

of it, delete the original, and install the new block as shown in Fig. 1.2.45.

3. Create the setscrew as shown in Fig. 1.2.46C. Make a block of it, delete the original, and then insert it vertically as shown in Fig. 1.2.45.

Attach the Shaft and Alignment Coupler

Return to the back of the nozzle assembly and turn your attention once again to the cylinder spacer. The main shaft runs from the back of the needle, through the 14″ pipe, and into the cylinder spacer. The slot in the cylinder spacer allows onlookers to see an indicator rod that gives a visual indication of where the shaft and needle are positioned when the actuator strokes.

The detail in Fig. 1.2.47 shows the dimensions and specifications for attaching the main shaft to the alignment coupler. Notice that the shaft has a 1.0–14 tapped end to engage the end of the alignment coupler. Follow the steps in Practice Exercise 1.2L to draw the connection between the shaft and the alignment coupler. Refer to Fig. 1.2.47 as necessary.

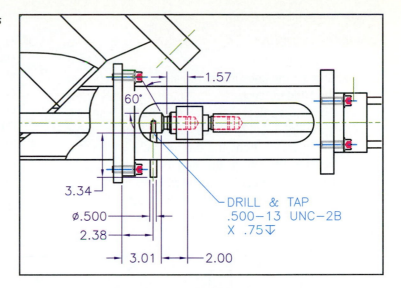

Fig. 1.2.47 Dimensions and specifications for attaching the main shaft to the alignment coupler.

Practice Exercise 1.2L

1. Extend the main shaft into the right end of the cylinder spacer **3.01** from the end of the lower nozzle housing. Do this by offsetting the appropriate line, extending the main shaft lines, and then trimming as necessary.

2. Step down the shaft diameter to 1″ at a 60° taper. To do this, enter the **LINE** command and use the **Endpoint** object snap to select the right end of the lower shaft line. Use the polar coordinates **@1<60** to create a line at 60° to the lower shaft line. Then mirror this line using the main centerline as the mirror line. Offset the main centerline up and down by half of the new diameter and change the lines to the **Object** layer. Trim the 60° lines to the edges of the stepped-down shaft diameter, and trim the 1″ shaft diameter to the right end of the 60° lines. Draw a line from the upper 60° line to the lower one to create the right side of the stepdown.

3. Offset the left side of the stepdown to the right by **2.00** to extend the shaft another 2.00″ into the cylinder spacer. Trim the shaft lines to this line. Trim the offset line to the shaft diameter.

4. Offset the right end of the shaft back to the left by **1.57** to establish the tapped section. Trim the offset line back to the shaft diameter.

Notice that part of the tapped section is hidden from view by the alignment coupler. To show the hidden lines correctly, you will need to break the lines at the point at which they become hidden.

5. Enter the **BREAK** command. At the Select object prompt, use the **Intersection** object snap to select the intersection of the shaft line with the vertical line as shown in the detail in Fig. 1.2.48. (You will probably have to zoom in very closely.) At the prompt for the second point, enter the @ symbol to break the line at the intersection. Repeat this procedure for the bottom shaft line.

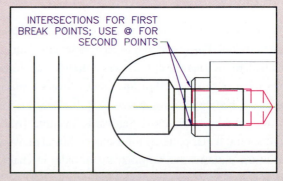

Fig. 1.2.48 Intersections for breaking the lines in step 5.

6. Change the lines inside the alignment coupler to the **Hidden** layer.

7. Add the vertical indicator rod. As you can see from Fig. 1.2.47, the centerline of the indicator rod is 2.38″ from the end of the lower nozzle housing, so offset the end of the housing to locate the centerline of the rod. Then offset the rod centerline by half of the diameter of the rod. Change the centerline to the **Center** layer.

8. The indicator extends 3.34″ below the lower edge of the Ø1.5″ shaft, so offset the lower edge of the shaft to establish the lower end of the indicator rod. Trim the lines to finish the lower end of the rod.

9. Create a **.5–13 × .75** tapped hole in the shaft to accommodate the rod indicator. Refer again to Fig. 1.2.47 for placement.

This completes the attachment of the main shaft to the alignment coupler. This part of your drawing should now look like the one in Fig. 1.2.49.

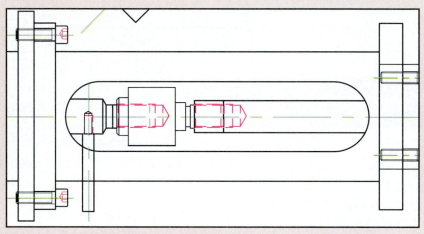

Fig. 1.2.49 **The completed cylinder spacer.**

Hatch the Sections

The nozzle assembly drawing incorporates several broken-out sections. Solid pieces shown in section are crosshatched, or hatched, to indicate the cut surfaces. A **hatch** is a standard pattern of lines called **section lines** that conveys a specific meaning to people reading a technical drawing. Prior to 1992, the American National Standards Institute (ANSI) defined several hatch patterns for use in technical drawings. A few of the patterns that you may see in older mechanical drafting are shown in Fig. 1.2.50. However, in 1992, ANSI simplified the standard so that all hatches are now created using the old general-purpose pattern, ANSI 31.

The ANSI 31 pattern is one of several patterns that are predefined in AutoCAD. This makes the hatching process fairly simple. You should already be familiar with the basic hatching process in AutoCAD. If you need to review, refer to the *CAD Reference* at the end of this textbook

According to drafting convention, neighboring pieces in an assembly should be hatched using lines that run in different directions so that it is evident that they are separate pieces. Therefore, AutoCAD allows you to change the default angle of the lines. By default, the ANSI 31 hatch pattern runs at 45°, so this is considered angle "0." To change the hatch angle, simply enter a different Angle value in the Boundary Hatch dialog box. See Fig. 1.2.51.

ANSI 31
CAST AND MALLEABLE IRON; GENERAL-PURPOSE SYMBOL

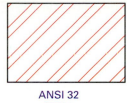

ANSI 32
STEEL

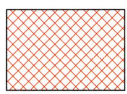

ANSI 33
BRONZE, BRASS, COPPER, AND COMPOSITIONS

Fig. 1.2.50 Hatch symbols commonly found on older technical drawings. ANSI 31 is the only pattern recognized by the current ANSI standard (ANSI Y14.2M-1992).

ANSI 34
RUBBER, PLASTIC, ELECTRICAL INSULATION

ANSI 37
WHITE METAL, ZINC, LEAD, BABBITT, ALLOYS

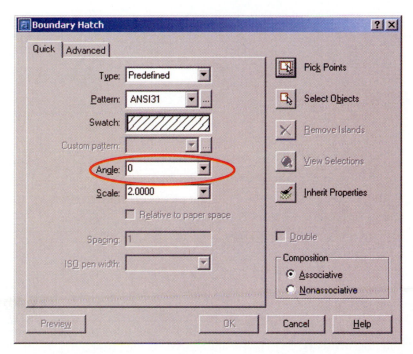

Fig. 1.2.51 The Boundary Hatch window allows you to change various characteristics of a hatch, including the angle at which the lines run.

You can also change the scale of the hatch, which determines how far apart the section lines are spaced. Line spacing for a hatched area is specified in ANSI Y14. In general, the line spacing should be appropriate for the size of the area and the individual drawing.

Follow the steps in Practice Exercise 1.2M to add hatching to the nozzle assembly drawing. Refer to Fig. 1.2.52 as necessary for the general locations of the hatches.

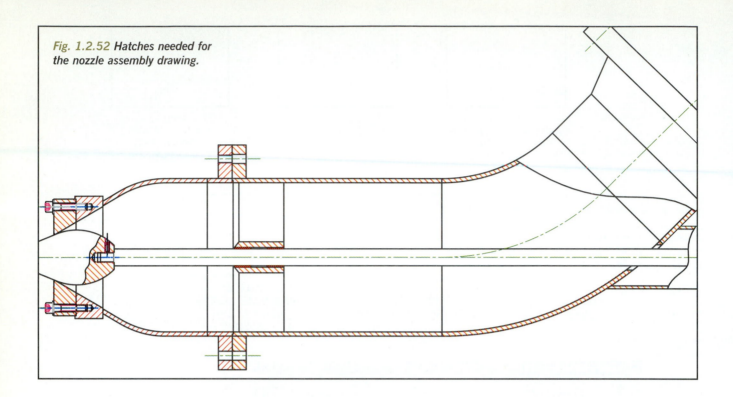

Fig. 1.2.52 Hatches needed for the nozzle assembly drawing.

Practice Exercise 1.2M

Notice in Fig. 1.2.52 that adjoining parts have been hatched using different line angles, but the different areas of a single part are always hatched using the same line angle. For example, on the bell, all of the section lines on both the top and the bottom pipe thickness and their flanges run at the same angle. We will start with the hatch for the bell because in this part, the lines run at their default angle of 0 (45°).

1. Zoom in on the bell portion of the nozzle and make **Hatch** the active layer. Enter the **BHATCH** command (or pick the Hatch button) to display the Boundary Hatch window. Click the **Pick Points** button. The Boundary Hatch window temporarily disappears. Pick with the mouse in each area indicated in the detail in Fig. 1.2.53. As you pick each area, the outline of the area becomes highlighted by dashed lines. Press **Enter** to end the selection process.

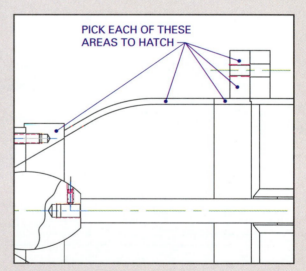

Fig. 1.2.53 Pick points for the first hatch.

2. In the Boundary Hatch window, change the hatch scale to **2.0**.
3. Pick the **Preview** button. If the preview looks similar to that in Fig. 1.2.52, pick the **OK** button to finish the hatch.

4. Repeat steps 1 through 3 for the bottom half of the sectioned bell. Leave the hatch angle at 0.

5. Zoom in on the beak and needle assembly. The plate to which the bell attaches and the inside of the needle will be hatched as a single component. Enter the **BHATCH** command and pick the points shown in Fig. 1.2.54 to hatch. Leave the scale at 2.0, and enter a new hatch angle of **270**. This creates a hatch with lines that are perpendicular to those in the bell hatch. Preview the hatch, and if everything looks okay, pick the **OK** button to finish the hatch.

6. Use the **Dynamic** option of the **ZOOM** command to move the zoom window to the center part of the nozzle assembly drawing and focus on the lower nozzle housing.

The section lines in this piece of the assembly will also run at an angle of 270. This distinguishes the 14" pipe of the lower nozzle housing from the bell to which it attaches. You could add the hatch using the same procedure you used in previous steps. However, since you have already hatched an area using this hatch angle, AutoCAD allows you to inherit the properties of the hatch you have already created. This feature saves time especially when you have set up a hatch with several options, because you don't have to set them up again for the new hatch.

7. Check to be sure that at least a small portion of the hatch on the needle assembly is visible on the screen. Enter the **BHATCH** command and pick the **Inherit Properties** button. The Boundary Hatch window temporarily disappears, and the cursor changes to a pickbox with a paintbrush beside it. Pick a point anywhere in the hatched area of the needle assembly to indicate which hatch properties you want to inherit. Then pick the points shown in Fig. 1.2.55 to hatch the lower nozzle housing. Preview the result and then pick the **OK** button.

8. The round stock and bearings at the center of the star support also show in section. Zoom in on this area and hatch it as shown in Fig. 1.2.56

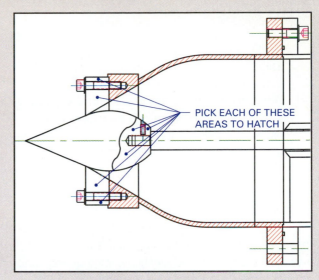

Fig. 1.2.54 *Pick points for the needle assembly.*

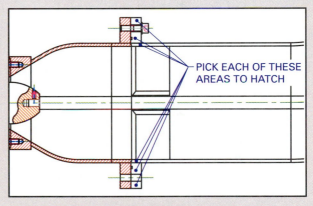

Fig. 1.2.55 *Pick points for the lower nozzle housing.*

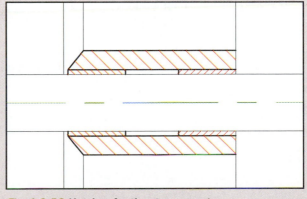

Fig. 1.2.56 *Hatches for the star support.*

using any of the techniques discussed in this exercise. Notice the difference in hatch scale. The round stock has line spacing of **2.0**, but the smaller bearings have a line spacing of **1.0**.

9. Move the zoom window to the 45° elbow and the rear housing assembly. Finish the hatching as shown in Fig. 1.2.57.

When you finish this practice exercise, your drawing should look very similar to the one in Fig. 1.2.58. At this point, most of the actual drawing work has been done. All you need to do is add the weld representations and the specifications, dimensions, and details to finish the assembly drawing.

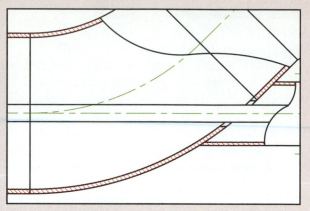

Fig. 1.2.57 Hatches for the 45° elbow and rear housing.

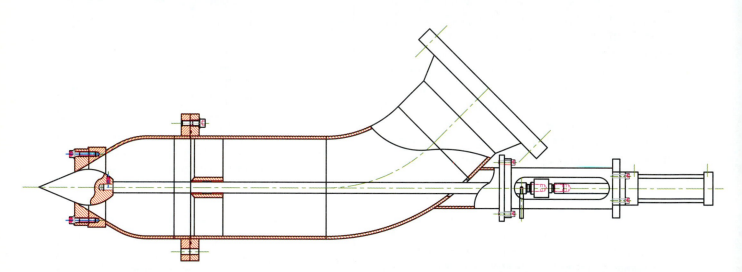

Fig. 1.2.58 Status of the nozzle assembly drawing after Practice Exercise 1.2M.

Add the Weld Representations

Drafters at some companies draw weld representations for reference at each place on the assembly drawing where a weld is specified. Welds are shown on a drawing as solid shapes that approximate the shape of the actual weld. They are not drawn to scale, but they should be of a size to "look right" in the drawing.

There are several ways to accomplish welds in AutoCAD. The SOLID command is sometimes used because it contains an automatic fill, which makes it appear solid. However, curves are difficult to accomplish with the SOLID command, and many welds have curved surfaces. The PLINE command can be used with a nonzero width and the Arc option to create a weld representation, but this, too, can be a time-consuming process of trial and error. Therefore, most weld representations are created using a combination of the ARC and HATCH commands. The drafter creates an arc to enclose the space where the weld will appear and then applies a hatch with a very small line spacing.

The disadvantages of using the arc-and-hatch method are as follows:

- Hatches with such a small line spacing require many individual lines, so they take a relatively long time to generate on the screen.
- In spite of the fact that a hatch is considered a single composite object in AutoCAD, the individual lines that make up the hatch have to be tracked in the file's database. A hatch with small line spacing results in a large number of lines that need to be kept in the file's database, which in turn increases the size of the file.

A few hatches of this type do not noticeably affect the drawing, but in drawings that have many weld representations, both regeneration time and file size can be affected.

The nozzle assembly drawing requires a fairly small number of weld representations. You can therefore use the arc-and-hatch method for this drawing. Follow the steps in Practice Exercise 1.2N to add the weld representations to the nozzle assembly drawing.

Practice Exercise

It is a good idea to place the weld representations on a separate layer. This will allow you to control the line width of the weld hatches independently from the line width of other hatches in the drawing. As you know, hatch lines are usually thin. However, in the weld representations, the hatch lines are used to create a solid-looking object. By using thicker lines, you can reduce the number of lines needed to produce the solid appearance.

1. Create a new layer named **Welds**. Assign the color **White**, and make it the current layer.
2. Enter the **ARC** command to create curved lines to form the outline of the welds at the bottom of the bell housing, as shown in Fig. 1.2.59. Use object snaps as necessary to place the weld against the flange and housing.
3. Enter the **HATCH** command. Select the **Pick Points** button, and pick the weld areas. Use an

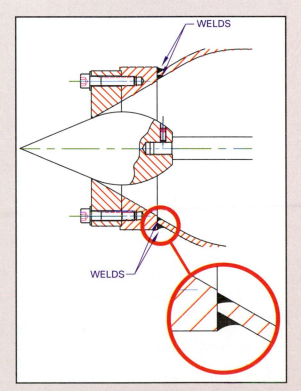

Fig. 1.2.59 *Weld representations.*

angle of **0**, but change the hatch scale to **.2** to space the lines very closely together. Preview the hatch. The lines should be close enough together to form an almost solid surface. If this is the case, pick the **OK** button. When you zoom out to see the entire drawing, the surface should appear to be solid. *Note:* Some versions of AutoCAD include a Solid Fill hatch pattern. If this pattern is available to you, you can use it instead of the ANSI 31 pattern to create a true solid fill.

4. Repeat steps 2 and 3 to form the welds at the top of the bell housing. Refer again to Fig. 1.2.59.

5. Using **ZOOM Dynamic**, move the viewing window to the right side of the bell housing and focus your attention on the upper portion of the joint between the bell housing and the 14″ pipe segment. Using a procedure similar to the one you used in steps 2 and 3, add welds to both sides of both of the flanges at this joint. See Fig. 1.2.60.

6. Add welds at both sides of the bar in the star support, as shown in Fig. 1.2.60.

7. Repeat steps 5 and 6 to create welds on the bottom of the flanges and the welds on the bar of the star support.

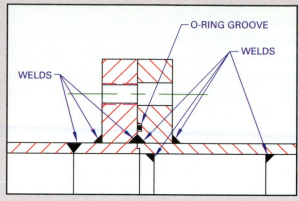

Fig. 1.2.60 *Positions of the welds and O-ring groove for the top flanges on the joint between the bell housing and the pipe.*

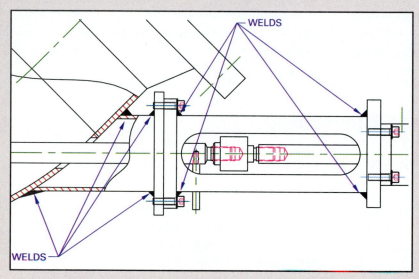

Fig. 1.2.61 *Welds for the back of the nozzle and the cylinder spacer.*

8. While the drawing is at this zoom magnification, use the **CIRCLE** command to add a representation of the O-ring in the O-ring groove. Refer again to Fig. 1.2.60. Hatch the O-ring using the same properties you are using for the welds so that the O-ring appears solid.

9. Repeat steps 5 through 8 for the welds and O-ring representation at the bottom of the joint.

10. Use **ZOOM Dynamic** to change the viewing window to the back of the nozzle assembly and the cylinder spacer. Use the techniques discussed in previous steps to add the eight welds as shown in Fig. 1.2.61. This finishes the weld representations for this drawing.

Add Dimensions and Notes

Because this is an assembly drawing, only the most important overall dimensions are needed. Most of the linear dimensions will be located using the turbine housing interface. This is an important location face because it interfaces with the turbine housing assembly.

Before you can add the dimensions and notes, you must set up an appropriate dimension style for the drawing.

AutoCAD's default dimension style, named Standard, uses the Standard text font and gives decimal values to a precision of four decimal places. You will need to change the font to ROMANS and the precision to two decimal places. You will also need to change the size of the dimension text, arrowheads, and several other settings.

Follow the steps in Practice Exercise 1.2O to set up a dimension style and add dimensions and notes to the drawing. Refer to Fig. 1.2.62 for placement and overall guidance.

Practice Exercise 1.2O

ANSI has set specific standards for the height of text on technical drawings. In general, the text has to be large enough to be readable, but not so large as to be incompatible with the drawing. For general text on a drawing, ANSI specifies a minimum of $1/8''$ text. Other sizes are allowed for drawing titles and title block text. See ASME Y14.2M for more specific information about these text sizes. For the nozzle assembly drawing, the sizes will be specified for you in the following steps.

1. If the Dimension toolbar is not displayed on your screen, display it now. You may want to dock it to the top or side of the screen.

2. Pick the **Dimension Style** button to display the Dimension Style Manager.

There are two ways to proceed from here. You can either modify the default Standard dimension style, or you can create an entirely new dimension style. For this drawing, modify the Standard style.

3. Make sure **Standard** is highlighted in the box on the left side of the Dimension Style Manager. Then pick the **Modify...** button to display the Modify Dimension Style window.

Notice that this window has several tabs. You will need to make several changes to settings in the first four tabs. (The last two tabs are not applicable to this drawing, so you do not need to change their settings.)

4. Pick the **Lines and Arrows** tab to display its contents. In the Dimension Lines area (top left), change the color of dimension lines to **ByLayer**. In Extension Lines area, change the color of extension lines to **ByLayer**. Also change the following settings:
Extend beyond dim lines: **.50**
Offset from origin: **.40**
Finally, in the Arrowheads area, change the arrow size to **.50**. Leave the remaining options at their default values.

5. Pick the **Text** tab to display its contents. Change the following settings:
Text style: **ROMANS**
Text color: **ByLayer**
Text height: **.5**
Offset from dim line: **.30**

6. Pick the **Fit** tab to display its contents. In the Fit Options area, pick the second radio button to specify that if the arrows and dimensions won't both fit, the arrows should be the first items to move outside the extension lines.

7. Pick the **Primary Units** tab to display its contents. Change the following settings:
Unit format: **Decimal**
Precision: **0.00**
Scale factor: **1.00**
Also, select zero suppression for both linear and angular dimensions.

8. Pick **OK** and **Close** to close the window. The changes you made are saved automatically.

Now you can begin dimensioning the drawing. Follow these steps:

9. The turbine housing interface is labeled in Fig. 1.2.62. Locate this interface now, because you will use it as a basis for most of the horizontal dimensions.

10. Prepare to dimension the drawing by making **Dimensions** the current layer. If you have not already done so, set up **Endpoint** and **Intersection** as running object snaps. This will help you dimension the drawing accurately.

11. Pick the **Linear Dimension** button on the Dimension toolbar. Pick the turbine housing interface as the first extension line origin. Pick the top left corner of the plate on the beak as the second extension line origin. Place the dimension approximately as shown in Fig. 1.2.62.

12. Create another linear dimension from the turbine housing interface to the left end of the needle. Place it as shown in Fig. 1.2.62.

13. Dimension the horizontal distance from the turbine housing interface to the centerline at the right (upper) end of the nozzle.

14. Dimension the horizontal distance from the turbine housing interface to the joint between the rear housing and the cylinder spacer.

15. Dimension the horizontal distance from the turbine housing interface to the right end of the hydraulic cylinder.

16. Add a vertical dimension to show the distance between the centerline at the right (upper) end of the nozzle and the main centerline that runs through the hydraulic cylinder.

The last dimension to be placed is a reference dimension for the 4.063 beak diameter. Notice that this dimension is given to three decimal places. You will need to override the dimension style to:

- change the number of decimal places from two to three
- add the diameter symbol
- add the parentheses to indicate that this is a reference dimension

You could set up a formal dimension style override to accomplish these changes, but that really isn't worth the effort for an isolated dimension. Formal dimension style overrides are appropriate for cases in which you need to apply the same overrides to several different dimensions. You will use that method to dimension the bell housing detail drawing in the next section of this project. To override the values for a single dimension, as in the present case, it is more efficient to use the Properties window to change the values.

17. Use the **Linear Dimension** button to create the dimension. The default dimension value should be 4.07. Doubleclick the dimension to show its properties in the Properties window. If the options in the Text category are not displayed, pick the + in the box next to Text to display them. Near the bottom of the list of options, find **Text Override** and click in the blank box next to it. You can make all three of the necessary changes simply by entering the required text here. For the diameter symbol, use AutoCAD's code of %%c. Specifically, enter **(%%c4.063)** in the box. Pick anywhere in the drawing area to return to the drawing. Press the Escape key to remove the grips from the dimension. *Note:* To retain AutoCAD's dimension text, you can enter **(%%c<>)** instead of the actual number. If you do this, however, be sure to change the precision to three decimal places.

18. If you have not yet trimmed the construction line that forms the main centerline, the centerline probably runs through the reference dimension you created in step 17. Trim the main centerline now so that it extends only a short distance at either end of the nozzle assembly.

Several other important dimensions need to be described in notes. These notes can either be placed on the Dimensions layer or on a separate Text (or Notes) layer. For this drawing, you will place them on the separate Text layer you created when you set up the drawing. *Note:* The general notes that pertain to the entire drawing should not yet be completed. You will create them in a later exercise.

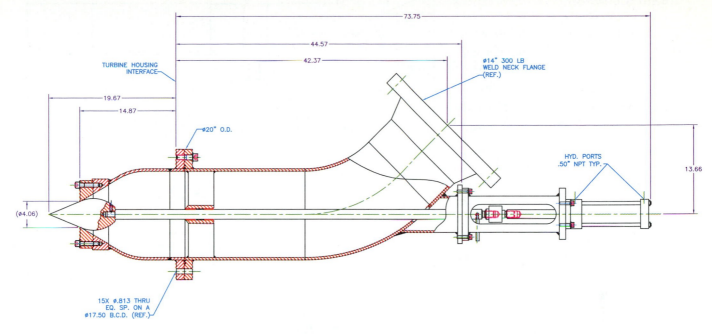

Fig. 1.2.62 Overall dimensions and specifications for the nozzle assembly drawing.

19. Make **Text** the current layer.

20. Zoom in on the front of the nozzle assembly and note the .813 through holes. There are 15 holes on a B.C.D. of 17.500. To create the note, pick the **Quick Leader** button on the Dimension toolbar and specify a first point at the intersection of the centerline of the hole on the bottom flange of the bell housing. Pick a second point down and to the left of the first point to establish the other end of the leader. Right-click twice to end the leader and accept the default text width. At the prompt for annotation text, enter the following note text:

15X Ø.813 THRU
EQ. SP. ON A
Ø17.500 B.C.D. (REF)

Press the **Enter** key at the end of each line to continue the text. After you have finished the third line, press **Enter** a second time to complete the leader. *Hint:* You can move the text and leader to a better position by picking the grip on the text and moving it. The leader will follow the text automatically.

21. Use the same technique to note that the flange that interfaces with the turbine housing has a 20.0 outside diameter.

22. Note the turbine housing interface.

23. Note the 14″ outside diameter of the pipe. Make this a reference note by placing the entire note in parentheses.

24. Note the 14″–300 weld-neck flange. Make this a reference note by adding **(REF)** at the end of the note.

25. Note the hydraulic port sizes on the cylinder assembly. They are .5″ NPT, typical bolt ports.

Create the Detail

Because the nozzle assembly drawing will be used for part identification and product assembly, the individual parts must be identified on the drawing. This will be accomplished by including part identification numbers that are keyed to a parts list.

For this method to be effective, all of the parts must be clearly visible on the drawing. A quick review of the drawing reveals that the only part users may have trouble seeing is the O-ring groove on the flange that interfaces with the bell housing. To make the O-ring groove easier to see and identify, we will create a detail at a scale of 2:1.

There is no need to redraw the geometry to create the detail in AutoCAD. Instead, you can simply copy the items to be included in the detail, trim the larger objects to the edge of the bounding circle, and use the SCALE command to increase their size.

Follow the steps in Practice Exercise 1.2P to create the detail. Refer to Fig. 1.2.63 as necessary for placement.

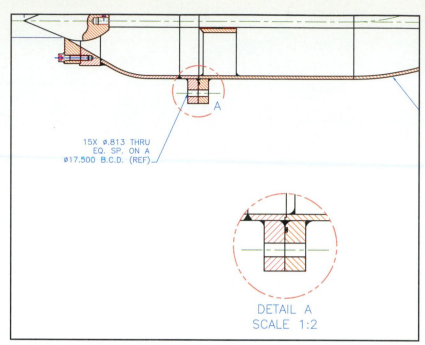

15X ∅.813 THRU
EQ. SP. ON A
∅17.500 B.C.D. (REF)

DETAIL A
SCALE 1:2

Fig. 1.2.63 *Creation and placement of Detail A.*

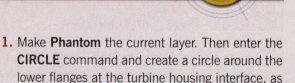

Practice Exercise 1.2P

1. Make **Phantom** the current layer. Then enter the **CIRCLE** command and create a circle around the lower flanges at the turbine housing interface, as shown in Fig. 1.2.63.
2. Enter the **COPY** command and use a crossing window to select the circle and its contents. Some of the objects selected will extend beyond the circle. Place the copy approximately as shown in Fig. 1.2.63.
3. Zoom in on the copy of the circle.
4. Enter the **TRIM** command and trim all of the lines from the outside of the circle.

5. Use **EXPLODE** to explode the hatches that extend outside of the circle. Erase and trim the portions of the hatch that are outside the circle.
6. Zoom in on the original circle on the lower part of the main drawing. Break it as shown in Fig. 1.2.63 and place an A in the break to show that this portion of the drawing is shown in detail A.
7. Now enlarge the scale to 2:1. Enter the **SCALE** command and use a crossing window to select the entire detail. Select a base point anywhere inside the circle, and specify a scale factor of **2**.
8. Below the detail, add the following text:
   DETAIL A
   SCALE 1:2
   Center the lines of text under the detail for a neater appearance.

Add Identification Numbers

The part identification numbers show where each part listed in the parts list is used in the assembly. To make them easy to identify, the part identification numbers are shown in circles called *bubbles*. The number in the bubble matches the item number in the parts list, which you will add later. Follow the steps in Practice Exercise 1.2Q to add the identification numbers to the drawing. Refer to Fig. 1.2.64 for the numbers and their placement on the drawing.

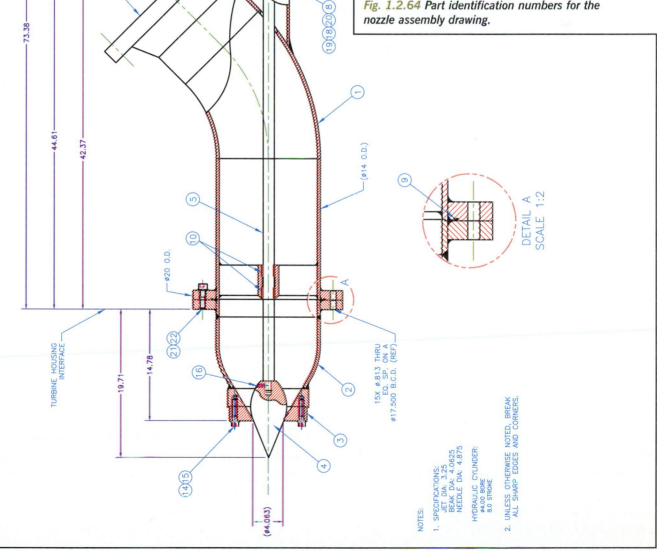

Fig. 1.2.64 *Part identification numbers for the nozzle assembly drawing.*

Practice Exercise

Notice in Fig. 1.2.64 that each bubble is connected to a leader, but there is no "arm" with text at the end of the leader. The bubble and number take its place. The bubble can be added directly to the leader using the Settings option of the QLEADER command, but doing so limits the angles at which you can place the leader. Therefore, we will use the Settings option to remove the arm, but we will not put anything in its place. We will then add the bubbles and identification numbers in a separate step.

1. Begin by entering the bubble labeled "1" in Fig. 1.2.64. Enter the **QLEADER** command, but instead of picking a starting point, press **Enter** or the **S** key to enter the Settings option. Pick the **Annotation** tab to display its contents. Set the annotation type to **None** and select **Reuse Next**. Then use the **Nearest** object snap to select a point on the lower nozzle housing, as shown in Fig. 1.2.64. Select a second point and press **Enter** to end the command.

2. Enter the **CIRCLE** command and create a circle (bubble) with a diameter of **1.75**. Place the center of the circle at the endpoint of the leader. With Object Snap on, trim the leader to the intersection of the leader and the rim of the circle.

3. Use the **DTEXT** command to place a **1** inside the bubble.

4. Continue around the drawing, creating all of the bubbles with their leaders and part identification numbers. There are several ways to approach this task. You can simply repeat steps 1 through 3 again and again, or you can analyze the drawing in Fig. 1.2.64 to find leaders that run in the same direction. After you have created one, you can copy it to create others that run in a similar direction and then just change the text in the bubble. Another approach is to create all of the leaders and then go back to add the circles and numbers. When you finish, check to make sure that you have inserted an information number for all 22 items shown in Fig. 1.2.63. Note that some item numbers are shown more than once on the drawing. Be sure to include the part identification number that refers to detail A.

Create the Parts List

All of the work you have completed so far on the nozzle assembly drawing has been done in AutoCAD's **model space**, as indicated by the MODEL tab at the bottom of the drawing area. Model space is the area, black by default, in which you create the geometry and do most of the basic drawing work. To display the drawing on paper, AutoCAD provides layouts that exist in **paper space**. In paper space, you can view the drawing as it will appear on the printed page. By default, paper space has a white background.

When you use one of AutoCAD's predefined drawing templates to create a new file, an appropriate border and title block are included automatically in the first layout tab. When you created the PROJECT1.dwg file, you used a predefined ANSI D template. If you look at the tabs at the bottom of the drawing area, you will see that AutoCAD created a single layout tab called ANSI D Title Block.

It is important to keep in mind that the company you work for may have its own custom border and/or title block that you will be required to use. Most companies that use this approach save the border and title block as a block in AutoCAD that you can simply insert into the drawing. For Project 1, however, you are using the ANSI template.

A **parts list** is a list of all the parts needed for the assembly, along with an indication of how many of each part are needed. Other names for a parts list include "list of materials" and "bill of materials," but we will use *parts list*—the official ANSI name. The parts list is created directly on the ANSI D layout to improve control over its size and placement relative to the border and title block. However, before you can work on the ANSI D template, you must create a paper-space viewport to contain the drawing. All of the geometry that you created in model space will appear in this viewport. Follow the steps in Practice Exercise 1.2R to add the viewport, customize the layout, and create the parts list.

Practice Exercise 1.2R

1. To activate paper space and display the ANSI D border, click the **ANSI D Title Block** tab at the bottom of the drawing area. A blank border and title block appear, as shown in Fig. 1.2.65.
2. Make **Viewport** the current layer.
3. If the geometry appears automatically in the layout, delete the existing viewport. From the **View** pulldown menu, select **Viewports** and then **1 Viewport**. Use the cursor to create a new viewport with boundaries as shown in blue in Fig. 1.2.65. (Blue is used for illustration purposes only. The viewport you create will be the color you chose in Section 1 for the Viewport layer.) For best results, use the **Endpoint** object snap to select the points to define the viewport.

When you pick the second point to define the viewport window, the nozzle assembly drawing appears automatically in the new viewport. It will not be at the correct size, however.

4. Doubleclick the viewport you created in step 3 to display its properties. In the **Misc** section of the Properties window, notice that the value for Standard Scale is Custom. Click the word **Custom** to display a dropdown menu of scales. Select **1:4** to scale the drawing properly for the ANSI D layout. The drawing should now look like the one in Fig. 1.2.66.

The next step is to add the parts list. According to ANSI, the parts list can be a separate page in a drawing set or included on a sheet that also contains geometry. Because the parts list for this drawing identifies the parts in the assembly, we will include it on the same sheet. The parts list should be located just above the title block. Ideally, its left edge should align with the edge of the title block. The preferred format for a parts list is to place the title "PARTS LIST" and the column heads at the bottom of the list. The numbers run from bottom to top.

5. Create a new layer called **Parts List** and make it the current layer. Make its color the same as the color of the Viewport layer.

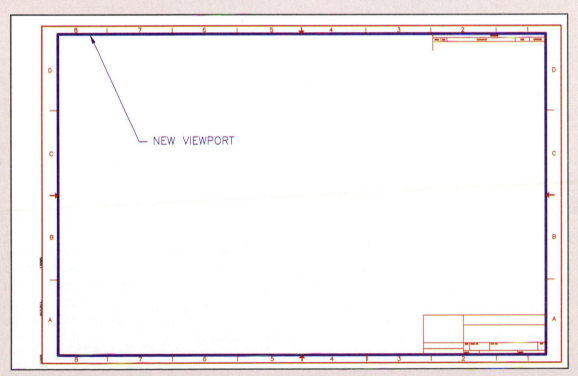

Fig. 1.2.65 AutoCAD's ANSI D template.

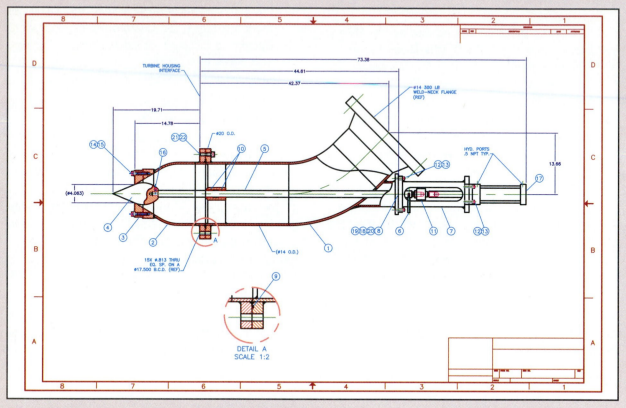

Fig. 1.2.66 *The nozzle assembly drawing layout at a scale of 1:4.*

6. Zoom in on the lower right corner of the layout.

7. Prepare the structure of the parts list by creating the lines that separate the entries. To do this, first use the **LINE** command to create the bottom horizontal line just above the top of the title block. End the line a little to the left of the right side of the viewport line.

8. Use the **ARRAY** command to create evenly spaced copies of the line. Pick the line as the object to array, and set the options as follows:

Rows: **25**

Columns: **1**

Row offset: **.25**

Angle of array: **0**

Pick the **Preview** button to view the array. It may interfere with the bubbles at the lower right edge of the drawing, but that's okay for now. You can change the position of the geometry later. If the lines are positioned correctly above the title block, accept the array.

9. Add the vertical lines at the sides of the parts list to finish the basic structure of the parts list.

10. Zoom out to see the entire layout. You will need to move the drawing within the viewport to place it in such a way that it does not interfere with the parts list. To do this, pick the **PAPER** button at the bottom of the AutoCAD screen to change into model space. Then enter the **PAN** command and use it to position the drawing correctly. Be sure that the drawing doesn't interfere with the revision block at the top of the layout. Pick the **MODEL** button to return to paper space.

11. Freeze the Viewport layer so that the viewport lines will not display or plot.

The parts list for the nozzle assembly drawing is shown in Fig. 1.2.67. Notice that two different sizes are used in the text: one for items that take up only one line, and another, smaller size for those that take up two lines. It is acceptable to use different text sizes, but you should limit the number of sizes to two.

12. Use the **MTEXT** command to create the text for the first column (ITEM). One line of text will fit comfortably into each line of the structure at a height of **.11**. *Important:* MTEXT imposes a limit on how small text can be created, but after it has been created, you can change it to any size you want. If you get an error message, enter the text at the smallest size allowed. After you have entered the text for the first column, end the MTEXT command and proceed to the next step.

13. The text object you created in step 12 will need some tweaking to fit properly in the parts list structure. Select the text object to display its properties in the Properties window. Change the text height to **.11** and the line space factor to **1.36**. (If these values don't work exactly, experiment to find values that do.) Use the **MOVE** command to position the text correctly.

ITEM	QTY	DESCRIPTION	DWG/PART NO.	VENDOR
22	15	WASHER, FLAT ANSI B18.221 NARROW	.75"	STANDARD
21	15	HEX−HEAD CAP SCREW ANSI B18.3	.75−10 X 2.50	STANDARD
20	1	O−RING	2−231	STANDARD
19	1	BACKUP, O−RING	8−222	STANDARD
18	1	O−RING	2−222	STANDARD
17	1	HYDRAULIC CYLINDER 4" BORE X 8" STROKE	4−ADN−F1N− 8−2−3−1000	CUNNINGHAM
16	1	HEX SOCKET SETSCREW, 316 STAINLESS STEEL, CUP POINT − ANSI B18.3	.375−24 X .75	STANDARD
15	8	WASHER, FLAT NARROW, 316 STAINLESS STEEL	.625"	STANDARD
14	8	HEX−HEAD CAP SCREW, 18−8 STAINLESS STEEL, PER MS 16995	.625−11 X 3.0	STANDARD
13	10	WASHER, FLAT ANSI B18.221 NARROW	.50"	STANDARD
12	10	HEX−HEAD CAP SCREW ANSI B18.3	.50−13 X 1.75	STANDARD
11	1	ALIGNMENT COUPLER	TRD−AC−1000	TRD
10	2	BUSHING, BRONZE	⌀1.5 I.D. X ⌀1.75 O.D. X 1.50	STANDARD
9	1	O−RING	2−384	STANDARD
8	1	REAR BUSHING	N00008	CANYON IND.
7	1	SPACER, CYLINDER	N00007	CANYON IND.
6	1	INDICATOR ROD	N00006	CANYON IND.
5	1	SHAFT, NEEDLE	N00005	CANYON IND.
4	1	NEEDLE	N00004	CANYON IND.
3	1	BEAK	N00003	CANYON IND.
2	1	BELL HOUSING	N00002	CANYON IND.
1	1	LOWER NOZZLE HOUSING	N00001	CANYON IND.
ITEM	QTY	DESCRIPTION	DWG/PART NO.	VENDOR
		PARTS LIST		

Fig. 1.2.67 **Parts list for the nozzle assembly drawing.**

14. Enter **MTEXT** again to create the QTY column. Use the procedure described in steps 12 and 13 to create and tweak the text. Make the text the same size as the text in the ITEM column.

15. Proceed to the DESCRIPTION column. You will need more than one text object for this column. In general, begin a new text object each time the text size changes from the one-line size to the two-line size. Use a height of **.07** for the two-line text. *Hint:* To align the text in the different text boxes, create the first text box, which has .11 text in it. Then, with Ortho on, copy the entire text object to the next line. Change the text, and then change the text size to .07. Now you have one box with .11 text and one with .07 text. Copy one of these two boxes, depending on the text size needed, to create the rest of the boxes. Be sure to keep Ortho on during each copy. The text will be perfectly centered on the column.

16. Repeat the procedure outlined above for the remaining columns.

17. In the last row, add the title: PARTS LIST. Center it horizontally across the whole parts list, and trim any vertical lines that extend through this bottom row.

18. Add the vertical lines that separate the columns.

Now you can add the general notes at the lower left corner of the drawing. By creating these notes in paper space (directly on the layout), you can better control their placement and size in relation to other paper-space objects such as the parts list.

19. Add the general notes and specifications at the bottom left corner of the drawing. The text should be as follows:

NOTES:

1. SPECIFICATIONS:
   JET DIA: 3.25
   BEAK DIA: 4.0625
   NEEDLE DIA: 4.875

HYDRAULIC CYLINDER:
   Ø4.00 BORE
   8.0 STROKE

2. UNLESS OTHERWISE NOTED, BREAK ALL SHARP EDGES AND CORNERS.

AutoCAD does not provide a simple way to align text columns evenly. If you create the text using MTEXT, use spaces to align the text as much as possible. If you use DTEXT, you can align the indents by moving the individual lines, but you can't easily control the vertical line spacing. For this application, both commands have both advantages and disadvantages. For example, you can move text created in MTEXT as a single unit, whereas you must select each individual DTEXT line to move the text without altering the line spacing. However, you can see the text appear in place on the drawing if you use DTEXT. The command you use is a matter of personal preference. Use whichever you prefer to create the notes and specifications. Your drawing should now look similar to the one in Fig. 1.2.68.

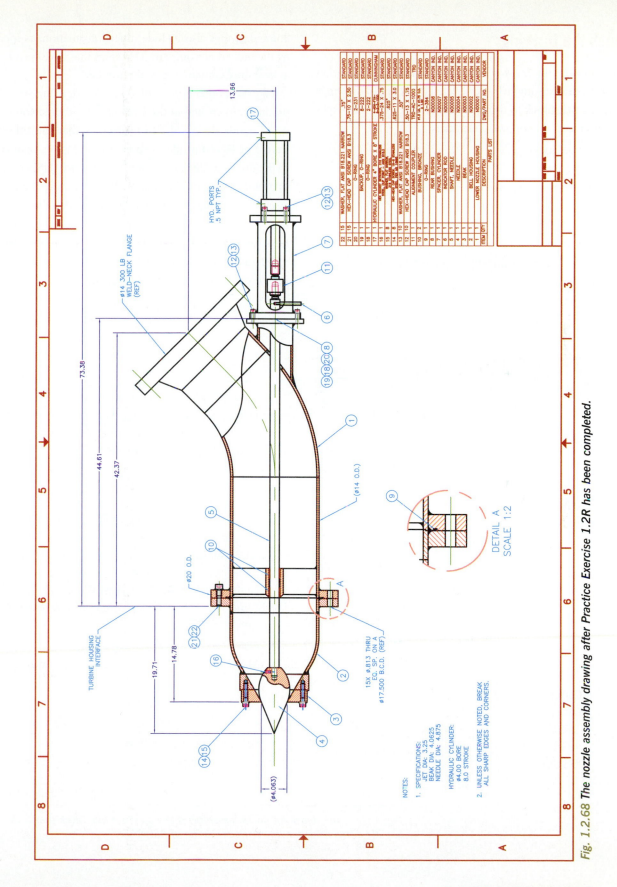

Fig. 1.2.68 The nozzle assembly drawing after Practice Exercise 1.2R has been completed.

Finish the Drawing

In industry, no drawing is complete without information about the drawing and who completed it. Technical drawings contain a **title block** to deliver this information. Title blocks include at least the following elements:

- title of the drawing
- drafter's name
- date the drawing was completed
- name or initials of checker
- name or initials of person responsible for final approval
- scale to which the drawing has been prepared
- sheet number of the drawing and the number of sheets in the set of drawings

Some drawings also include other information, as specified by the individual company. Drawings that have been revised at least once contain a revision number.

Like most other aspects of a technical drawing, title blocks on drawings created in the United States should conform to ANSI standards. ANSI Y14.1 specifies that the title block should be placed inside the border at the lower right corner of the drawing sheet. AutoCAD inserted an ANSI-compliant title block form automatically when you created the nozzle assembly drawing because you based the drawing on one of AutoCAD's ANSI D templates. The form is not complete, however. It is merely intended to get you started within the ANSI guidelines. Complete Practice Exercise 1.2S to finish both the title block and the drawing.

Practice Exercise 1.2S

1. Zoom in on the title block form in the lower right corner of the drawing, and make **Text** the current layer.
2. Use **DTEXT** to insert the name of the drawing as shown in Fig. 1.2.69. Use a text height of **.24** for **LOWER NOZZLE ASSEMBLY**. Then reenter the **DTEXT** command and use a text height of **.15** for **HYDROELECTRIC TURBINE**.
3. In the rectangular area at the upper left of the title block, use **MTEXT** to create the text. Specify a height of **.09** and a line-space factor of **2.50**. Enter the following text.

   FILE NAME
   CONTRACT NO.
   DRAWN
   CHECK
   APPR.
   ISSUED

4. Using **DTEXT** and a text height of **.125**, add the file name (**PROJECT 1**). Also add your name after the DRAWN entry. None of the other items in this group need to be specified at this time.
5. After the word SCALE at the bottom of the title block, enter the scale at which you inserted the drawing onto the layout, which was **1:4**. After the word SHEET, enter **1 OF 2**. Use **DTEXT** with a text size of **.09**.
6. **ZOOM All** to see the entire layout. Your drawing should look like the one in Fig. 1.2.70.
7. Right-click the **ANSI D Title Block** tab at the bottom of the drawing area and change its name to **NOZZLE**.

Fig. 1.2.69 Title block for nozzle assembly drawing. (Your title block may look slightly different depending on your version of AutoCAD.)

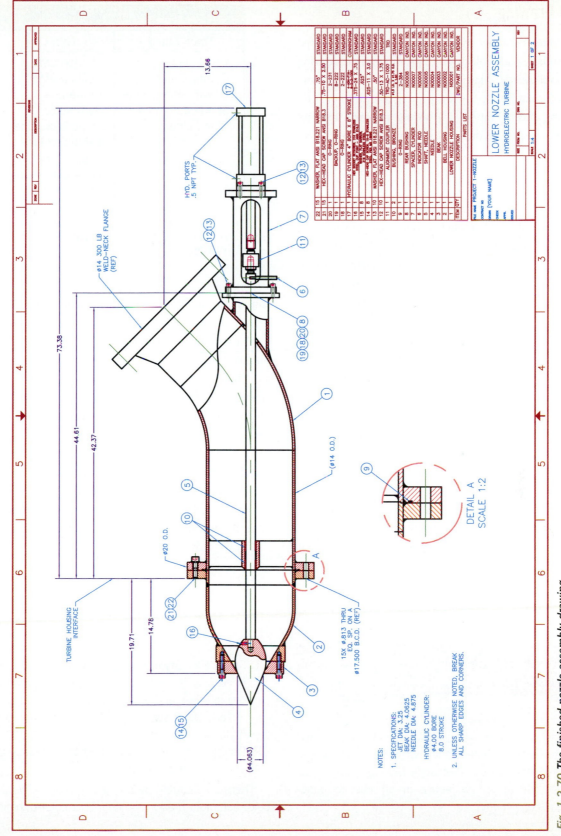

Fig. 1.2.70 *The finished nozzle assembly drawing.*

Plot the Drawing

At one time or another, most CAD files need to be plotted on paper. In fact, the whole point of positioning the drawing correctly on the ANSI D layout in the previous practice exercise was to format it properly for printing (plotting) on a D-size drawing sheet. Plotting a drawing properly requires thinking ahead to create the proper settings.

Plot Styles

Before you can actually print the drawing, however, you should set up a **plot style table** to ensure that your drawing meets ANSI standards. ANSI Y14.2M specifies two line widths for use on drawings. The thinner line is half the width of the thicker line. The table in Fig. 1.2.71 shows which types of lines should be thick and which should be thin. The standard says that it is acceptable to use a single line width on CAD-generated drawings. However, technical drawings that use two line widths are easier to read when plotted, so best practice is to use two line widths unless the company you work for specifically requests a single line width.

By using plot styles, you can plot using two line widths while keeping all of your lines in the electronic file at the default width of 0 for use with CAD/CAM equipment. Plot styles do not change the width of the lines in the drawing. They change only the appearance of the *plotted* lines.

Plot styles also allow you to control colors, linetypes, and other properties that are associated with layers and objects. In general, these items default to the values you specified when

ANSI Line Weights	
Thick Lines	**Thin Lines**
Visible Line	Hidden Line
Cutting-Plane Line	Section Line
Viewing-Plane Line	Centerline
Short Break Line	Symmetry Line
Chain Line	Dimension Line
	Extension Line
	Leader Line
	Long Break Line
	Phantom Line
	Sewing/Stitch Line

Fig. 1.2.71 Line weights for various types of lines according to ANSI Y14.2M.

you set up the layer, so you can usually leave them at the default values.

Plot style tables are administered through the Plot Style Manager. AutoCAD stores each plot style table as a file with an .stb extension. For the nozzle assembly drawing, you will create a new plot style table that contains two plot styles: THICK and THIN. After you have created the plot style table, you can select it for use and assign plot styles to individual layers using the Layer Properties Manager. Note that once you have set up a plot style table, you can assign the file to any drawing. It is not specific to the current drawing file.

In Practice Exercise 1.2T, you will set up a new plot style table. You will then assign the new plot styles to the layers in PROJECT 1.dwg.

Practice Exercise 1.2T

The easiest way to set up a plot style table is to use the Add-a-Plot Style Table wizard. This wizard, accessed through the Plot Style Manager, takes you through the setup procedure in an easy-to-follow step-by-step manner.

1. Select **Plot Style Manager...** from the File menu to display the Plot Styles window. This window displays all of the plot style tables that have already been set up, along with the Add-a-Plot Style Table wizard.
2. Doubleclick on the **Add-a-Plot Style Table** wizard and read the opening screen.

There are actually two types of plot style tables: color-dependent and named. The color-dependent tables assign properties (such as line width) according to a layer's assigned color. This approach works well unless you have two layers the same color but need to assign different line widths.

To guard against this problem, many drafters use named plot styles instead. In this type of plot style table, you can name individual plot styles and then assign them to layers regardless of layer color. This is the type of plot style table you will use for the nozzle assembly drawing.

3. Pick the **Next** button to proceed to the next step. AutoCAD allows you to build a plot style table from scratch, from an existing plot style table, from an R14 CFG file, or from a PCP or PC2 file. Choose the **Start from scratch** option and pick the **Next** button again.

4. Select **Named Plot Style Table** and pick the **Next** button.

5. AutoCAD prompts you to enter a name for the plot style table. Enter the name **Projects**. AutoCAD creates a new file called Projects.stb to contain your plot style table. Pick the **Next** button.

The Finish screen appears, but do not pick the Finish button yet. So far, AutoCAD has only established the file for the plot style table. You must add the individual plot styles.

6. Pick the **Plot Style Table Editor...** button. Notice that the table currently has one plot style called Normal. Pick the **Add Style** button near the bottom of the window.

7. A new style named Style1 appears. If the name Style1 is not highlighted, highlight it now and change its name to **THICK**. Skim down to the Line weight column, which defaults to the object's line weight. Pick in the white box that says **Use object line weight** to display a list of line weights. Scroll down and select **.35** mm. Leave everything else at its default value.

8. Pick the **Add Style** button again to create another plot style. Name this one **THIN** and select a line weight of **.18**. Pick the **Save & Close** button to finish the plot style table.

It is important to display the NOZZLE layout on the screen before you select plot styles because AutoCAD allows you to select a different plot style table for each layout in a drawing file. You want to attach this table to NOZZLE.

9. Display the NOZZLE layout. Pick the **Layers** button to display the Layer Properties Manager. Near the right side of the window, notice the Plot Style column. All of the layers use the Normal plot style by default. You may leave layer 0 at its default setting, but you should change all of the other layers.

10. Pick **Normal** in the Plot Style column for the Center layer to display the Select Plot Style window. Because this is the first plot style you are specifying, you must indicate which plot style table you want to use. In the Select Plot Style dialog box, click in the white box for **Active plot style table** to display a list of plot styles. Choose **Projects.stb** from the list. The plot styles you set up in Projects.stb display as a list in the upper half of the window. Centerlines are thin lines, so select **THIN** and then pick the **OK** button to return to the Layer Properties Manager.

11. Pick **Normal** next to each of the other layers in the drawing, assigning the plot styles as shown in Fig. 1.2.72.

Assignment of Plot Styles	
Layer	**Plot Style**
0	Normal
Center	THIN
Dimensions	THIN
Hatch	THIN
Hidden	THIN
Object	THICK
Parts List	THIN
Phantom	THICK
Text	THIN
Title Block	THIN
Viewport	THICK
Welds	THICK

Fig. 1.2.72 Assign these plot styles to the layers in the nozzle assembly drawing.

Plot the NOZZLE Layout

The NOZZLE layout is now ready to be plotted. Before you plot it for the first time, you will need to select a printer and set up the plotting parameters. *Note:* The terms *plot* and *print* are used interchangeably in this discussion.

The drawing was set up for plotting on D-size paper (hence the ANSI D drawing template). This size is commonly used instead of smaller sizes for this type of drawing because it allows the drawing to be shown at a scale that makes detail easier to see and understand. Therefore, the plotting instructions in this project assume that you have access to a printer that can print a D-size drawing.

If you do not have access to a D-size printer, you can plot to a smaller size sheet by displaying the drawing at a smaller scale. To do so, you should create a new layout using a smaller sheet template. Do not attempt to repurpose the ANSI D layout. You will need to change the scale at which the drawing appears on the layout. Be sure to note the scale on the title block. However, the drawing will be difficult to read because it was set up for a D-size plot. At a smaller scale, much of the text will be unreadable. To allow the drawing to be plotted at a slightly larger scale, you may place both the parts list and the detail on separate pages. Refer to ANSI Y14.34M for guidelines on including a parts list as a separate page in a set of technical drawings.

Practice Exercise 1.2U steps you through the procedure for plotting a D-size drawing. If your printer is not set up for use with AutoCAD or if you need help with adding a new printer, refer to the *CAD Reference* at the end of this book.

Practice Exercise 1.2U

This exercise assumes that you have already set up a printer that will plot a D-size drawing. If you have not done so, do it before continuing. Then follow the steps below.

1. With the NOZZLE layout visible on the screen, enter the **PLOT** command to display the Plot window. Check to be sure the Layout name in the upper left corner says NOZZLE.

2. Activate the **Plot Device** tab if it is not already displayed. In the Plotter configuration area, pick the box next to **Name** to show a list of available plotters. Pick the plotter you will use to plot the D-size drawing.

3. Check to be sure that Projects.stb appears as the name of the plot style table. If not, pick the box to display a list of plot style tables and select **Projects.stb**. Leave all of the other settings at their default values.

4. Pick the **Plot Settings** tab to display its contents. Check to be sure that the plotter you have selected is shown as the plot device and that the paper size is ANSI D. Set the drawing orientation to **Landscape**, and set the Plot area to **Extents**.

In the Plot scale area, notice that the plot is set at a scale of 1:1. This is correct, even though the drawing will appear at a scale of 1:4. The scale in this dialog box is for the paper-space layout, which should always be printed at 1:1. You specified the 1:4 scale for the drawing when you imported it into the paper space viewport, so no further scaling is necessary.

5. In the Plot offset area, click the box next to **Center the plot**. AutoCAD will automatically do the math to center the plot exactly on the paper, although it gives you the opportunity to change the vertical and horizontal offsets. Leave them at their default values.

6. In the Plot options area, pick to select **Plot with plot styles**. This ensures that plot style table you have set up is used in the plot.

7. Pick the **Full Preview...** button to check the appearance of the drawing. Make any necessary changes.

8. Pick **OK** to plot the drawing.

Review Questions

1. What should you set up in an AutoCAD drawing file before you begin the actual drawing work?

2. Why do many CAD operators set up text styles with a height of 0?

3. Think about how layers are used in the PROJECT 1.dwg file. Why is it a good idea to set up a layer for each different type of line?

4. What is a pipe schedule, and why should you consult one before creating a scheduled pipe on a technical drawing?

5. If you needed to create lines to represent a 12″ pipe and the centerline for the pipe already existed in the drawing, what AutoCAD command would you use to create the lines? What further step(s) would then be necessary?

6. What types of objects in a technical drawing might you block using AutoCAD's BLOCK or WBLOCK command? Why?

7. Explain the effect of the MIRROR command. How is a copy made by mirroring different from a copy made using the COPY command?

8. According to ANSI, what is the preferred hatch pattern for adding section lines to drawings? How do you create this pattern in AutoCAD?

9. What is the purpose of detail A in the nozzle assembly drawing?

10. What is a parts list? In the nozzle assembly drawing, what relation does it have to the identification numbers?

11. Why is the parts list done in paper space rather than model space?

12. At a minimum, what information should the title block of a CAD drawing contain?

13. Briefly describe the two kinds of plot style tables. What is the advantage of using the kind that was specified in the nozzle assembly drawing?

14. How do you assign plot styles to geometry in a drawing?

15. What changes must you make in the NOZZLE layout to print it at a smaller size if you do not have access to a printer that can plot D-size sheets?

Portfolio Project

Needle Valve (continued)

Refer to your notes from the Needle Valve project at the end of Section 1.1. Create a new drawing file called PORTFOLIO 1.dwg using one of AutoCAD's ANSI templates. *Note:* Be sure to use one of the "named plot style" templates. The color-dependent templates use a different method of handling line thicknesses on plots. Plan to include all of the needed drawings in the PORTFOLIO 1.dwg file. Set up the units and limits for the drawing, being sure to take into consideration the total space needed for all of the drawings. Set up layers and any other necessary drawing characteristics. Create at least one of the drawings to be included in the portfolio project. Create a parts list if required. Plot the drawing(s).

Practice Problems

1. In the nozzle assembly drawing project, suppose the engineers had specified Schedule 40 pipe instead of Schedule 30 for the 14″ pipe. Consult a table of pipe schedules to determine how this change would affect the nozzle assembly drawing. Then use the SAVEAS command to save the PROJECT 1.dwg file with a new name of Practice Problem 1.2.3.dwg. *Do not* work in your original PROJECT 1.dwg file. Make all necessary changes to the new Practice Problem 1.2.3.dwg file so that the drawing reflects the Schedule 40 pipe. Add the revision number and date to the revision block at the top right corner of the layout.

2. Refer again to Fig. 1.2.1. Calculate the limits that would be needed for the two drawings in Project 1 if you were to place them in separate drawing files.

3. One company's set of technical drawings for a cell phone includes:
   • reference assembly drawing that identifies all the parts in a parts list
   • fully dimensioned working drawing of the case in the open and closed positions

   The company's policy is to include all of the drawings in a single AutoCAD file, using a different layout for each drawing. The space required for each drawing has been blocked out by the chief drafter and is shown in Fig. 1.2.73. Calculate the limits for this drawing file. In your answer, include both your calculations and a roughly dimensioned sketch to show how you plan to arrange the drawings in the file.

4. ANSI C drawing sheets will be used for the set of drawings discussed in the previous problem. At what scale should the working drawing be plotted?

5. Create a new drawing file using the ANSI A - Named Plot Styles template (landscape) and name it Practice Problem 1.2.5.dwg. Use the AutoCAD commands and procedures discussed in this section to draw and dimension the 90° pipe elbow and the connecting pipes shown in Fig. 1.2.74. Use Ø6″ Schedule 80 pipe. The horizontal straight pipe is 7″ long, and the vertical pipe is 5″ long. Consult a table of pipe schedules to determine the inner diameter of the pipe. Document your work by keeping a log of the commands you used and the order in which you used them.

6. Set up the ANSI A layout for the pipe drawing you created in Practice Problem 5. Determine the best scale at which to display the drawing on this size sheet, and complete the title block as discussed in this section. Name the layout PIPE ASSEMBLY. Assume that this is the only sheet in the drawing set. Do not add a parts list.

7. Assign the Projects.stb plot style table to the drawing and specify the correct line weights for each layer you used. Plot the PIPE ASSEMBLY layout.

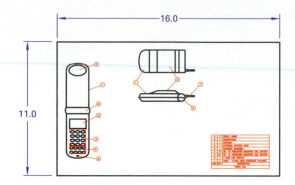

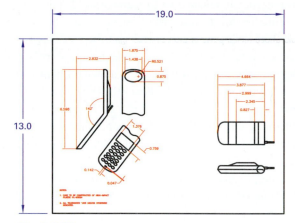

Fig. 1.2.73

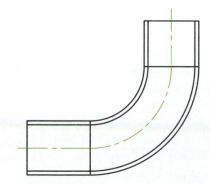

Fig. 1.2.74

Bell Housing Detail Drawing

Manufacturing departments often need more precise information than can easily be shown on an assembly drawing. Closely interconnecting parts require very specific geometric tolerances to function correctly. Welds must also be specified in many cases. In addition, some parts of an assembly are complex enough to warrant their own detail drawings.

The bell housing in the lower nozzle assembly is such a part. To describe it adequately, a detail drawing showing the bolt holes and their tolerances is required. An associated offset section taken through both sets of bolt holes is needed to provide precise locational and positional tolerances for all parts of the bell housing, as well as welding specifications.

Section Objectives
- Create an offset section.
- Use polar arrays to add concentric details to a drawing.
- Dimension a drawing using symmetrical and deviation tolerances.
- Specify geometric tolerances using standard GD&T techniques.
- Specify welds correctly on a detail drawing.

Key Terms
- cutting-plane line
- datum
- geometric dimensioning and tolerancing (GD&T)
- offset section
- polar array
- tolerance

Fig. 1.3.1 *A cutaway solid model of the bell housing shows its basic form.*

Drawing Strategy and Techniques

While you are working on a detail drawing, it often helps to keep in mind a three-dimensional image of the piece you are detailing. The computer-generated solid model in Fig. 1.3.1 shows a cutaway view of the bell housing. Notice the spacing of the bolt holes and the alternating bolt holes in the large flange.

Two views will be needed in the detail drawing, as explained in the introduction above. The front view will provide information about the position and size of all three sets of bolt holes. The side view is a sectional view through the bolt holes. To create the sectional view, you need to understand the concept behind an offset section. To simplify the task of creating the bolt holes in the front view, you need to know how to use AutoCAD's ARRAY command.

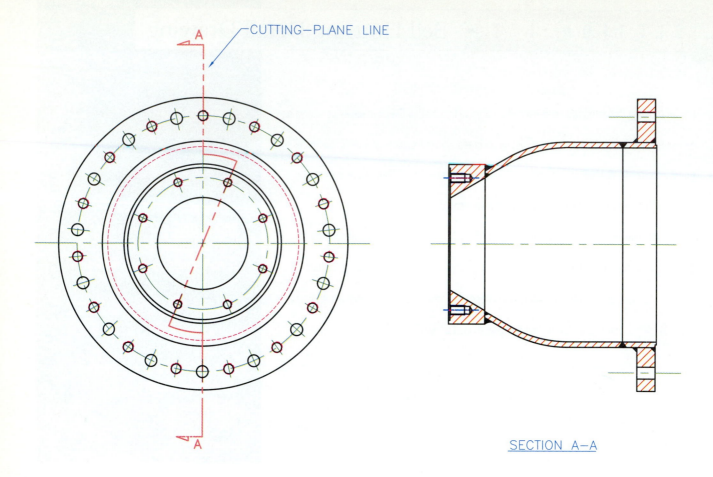

CUTTING—PLANE LINE

SECTION A—A

Fig. 1.3.2 *Offset cutting-plane line A-A and the resulting Section A-A.*

Offset Section

The side view consists of an offset section taken through all three sets of bolt holes. An **offset section** is a view that shows the cut face of an object along a nonlinear path in order to include features that would not ordinarily be visible in a full or half section. The face of the section is defined by the **cutting-plane line** in another view. Sections are numbered A-A, B-B, etc., and the cutting-plane lines that define them are identified using the same letters.

For the bell housing detail drawing, the section will be called Section A-A. The cutting-plane line will be chosen so that all three different sizes of bolt holes can be displayed in the section. The cutting-plane line is shown in the front view as a phantom line. See Fig. 1.3.2.

Polar Array

The bolt holes can be added to the front view quickly and easily using the ARRAY command. You created a rectangular array in the previous section while working on the nozzle assembly drawing. In this section, you will use a polar array to create the bolt holes. A **polar array** is one in which the arrayed objects are arranged in a circular pattern around a center point. A simple polar array is shown in Fig. 1.3.3.

To specify a polar array in AutoCAD, you must include the following information:

- the item(s) to be arrayed
- total number of times the item(s) appear in the array
- number of degrees through which the array will pass
- whether the items should be rotated as they are arrayed

Using this information, AutoCAD calculates the exact position of each item in the array. Figure 1.3.4 shows the result of various array specifications. Notice that if you specify less than 360°, the arrayed items are calculated to fit within the specified arc. The result of picking a center point on the object to be arrayed is shown in Fig. 1.3.4A. In Fig. 1.3.4B, notice the negative degree specification. This directs AutoCAD to create the array in a clockwise direction. By default, arrays are constructed in a counterclockwise direction. Specifying a large number of copies in the array results in a geometric overlap, as shown in Fig. 1.3.4C. As you can see, a polar array can be used to create many different types of patterns.

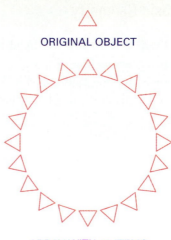

ORIGINAL OBJECT

ARRAY WITH 20 ITEMS
ROTATED THROUGH 360°

Fig. 1.3.3 *A polar array of a simple triangle through 360°.*

Fig. 1.3.4 *Examples of polar arrays with a variety of specification patterns.*

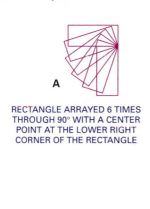

A

RECTANGLE ARRAYED 6 TIMES
THROUGH 90° WITH A CENTER
POINT AT THE LOWER RIGHT
CORNER OF THE RECTANGLE

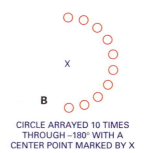

B

CIRCLE ARRAYED 10 TIMES
THROUGH −180° WITH A
CENTER POINT MARKED BY X

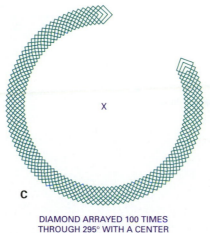

C

DIAMOND ARRAYED 100 TIMES
THROUGH 295° WITH A CENTER
POINT MARKED BY X

Build the Basic Geometry

In Practice Exercise 1.3A, you will build the two basic views of the drawing. Because the sectional view is similar to the nozzle assembly drawing you created in Section 2, you can copy and paste parts of the nozzle assembly to start the bell housing detail. After trimming away the unneeded parts, you can use the sectional view as a basis for the front view. In the front view, you will use three separate polar arrays to create the three sets of bolts. Refer to Fig. 1.3.5 for specifications and dimensions as you complete the basic geometry and specifications for the two views.

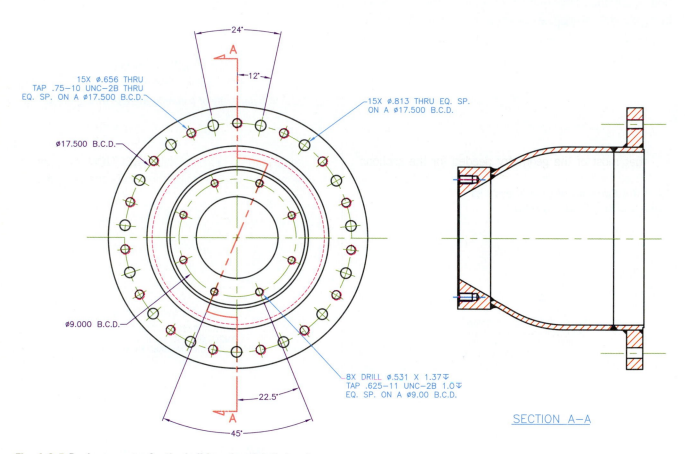

*Fig. 1.3.5 **Basic geometry for the bell housing detail drawing.***

Recall that we have planned to include both the nozzle assembly drawing and the bell housing detail in the PROJECT 1.dwg file. If necessary, refer again to Fig. 1.2.1 and your notes to refresh your memory about where you planned to place the detail drawing.

1. Open the **PROJECT 1.dwg** file and enter **ZOOM All** to see the entire drawing area. The nozzle assembly drawing should be complete, and there should be plenty of space available for the bell housing detail drawing.

Before you begin, compare the geometry needed for the detail drawing in Fig. 1.3.5 with the nozzle assembly drawing you created in Section 1.2. Notice that most of the geometry needed for the sectional side view in the detail is already present in the nozzle drawing. You can take advantage of this by copying the appropriate parts to start the detail drawing.

2. Zoom in on the left part of the nozzle assembly drawing. Before you copy the basic geometry of the bell housing for use in sectional side view, you can simplify your work by freezing layers that contain items you do not need to copy. Freeze the following layers:

Dimensions
Phantom
Text

3. Enter the **COPY** command and use a crossing box to select the bell housing. Be sure to include both flanges. Enter **'ZOOM All** and place the copy in its approximate position according to your plan. Notice the apostrophe before the ZOOM command. This makes the command transparent, so it does not interrupt the COPY command.

4. Unfreeze the frozen layers.

5. Zoom in on the copy. Use the **ERASE** and **TRIM** commands as necessary to remove all of the unneeded parts that were included in the copy. When you finish, your copy should look like the right side of Fig. 1.3.5. *Note:* You may need to delete and redefine some of the hatches.

6. Make Text the current layer. Use the **DTEXT** command to add the label **SECTION A-A**, as shown in Fig. 1.3.5. To underscore the text, use AutoCAD's %%u code:

%%uSECTION A-A

Now you can use the section to project the lines for the front view. Although you already know the diameters of most of the circles, it is often faster to create construction lines to guide their creation.

7. Zoom out to the approximate finished area of the detail drawing. With Ortho on, extend the centerline of the side view to the left to create the centerline for the front view.

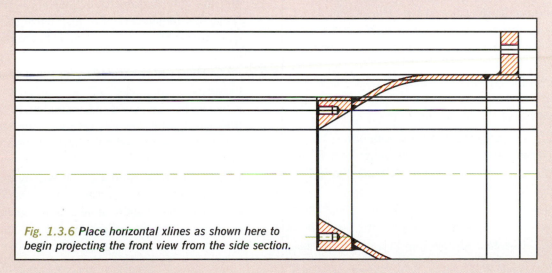

Fig. 1.3.6 Place horizontal xlines as shown here to begin projecting the front view from the side section.

8. Enter the **XLINE** command. Enter **H** to select the Horizontal option. Use object snaps to place horizontal construction lines at key points, as shown in Fig. 1.3.6. Note that you only need to place lines on the upper half of the drawing.

9. Make sure **Perpendicular** is set as a running object snap and that Ortho is on. Then enter the **CIRCLE** command. Use the **Nearest** object snap to select a point on the centerline. This point will become the center point of the front view. When prompted for the radius or diameter, move the cursor up to the first construction line. When the perpendicular symbol appears, click to establish the circle.

10. Repeat step 9 for all of the remaining construction lines to set up the concentric circles of the front view.

11. Two of the circles you just created are bolt circle diameters, so they should appear as centerlines. Follow the two construction lines that lead from the centers of the bolt holes to find the bolt circle diameters, and change these two circles to the **Center** layer.

12. The inside diameter of the 14″ pipe shows as a hidden line in the front view. Change this circle to the **Hidden** layer.

13. Zoom out to look critically at the two views you have started for the detail drawing. If the front view seems too close to the side section, select all of the circles and move them to the left. Position the views approximately as shown in Fig. 1.3.5.

14. Erase the construction lines. Trim or extend the main centerline if necessary.

15. Add the vertical centerline to the front view.

Now it's time to add the three sets of bolt holes. All three sets can be created by drawing one hole per set and then using a polar array to place the rest of the holes correctly.

16. Focus first on the two sets of bolt holes that alternate on the larger bolt circle diameter. Begin with the Ø.656 tapped hole that falls on the vertical

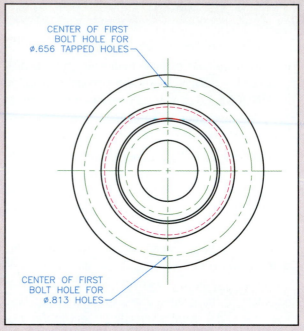

CENTER OF FIRST
BOLT HOLE FOR
ø.656 TAPPED HOLES

CENTER OF FIRST
BOLT HOLE FOR
ø.813 HOLES

*Fig. 1.3.7 **Place the center of the first bolt hole at the upper intersection of the vertical centerline and the larger bolt hole diameter.***

centerline of the front view. Enter the **CIRCLE** command. Use the **Intersection** object snap to select the intersection of the vertical centerline and the larger of the two bolt circle diameters, as shown in Fig. 1.3.7. Specify a diameter of **.656**.

17. Add the tapped diameter by creating a second circle, concentric with the first circle, with a diameter of **.75**. Change this circle to the **Hidden** layer.

18. Add a short vertical centerline for the hole. Even though this short line will be masked by the longer vertical centerline, it is important to create the short line before creating the array. Otherwise, the arrayed holes will not have vertical centerlines.

19. Enter the **ARRAY** command and specify a polar array. Select the two circles that represent the bolt hole and the short centerline as the objects to array. Choose the center point of the front view (where the horizontal and vertical centerlines

intersect) as the center point of the array. Specify a total of **15** items and an angle to fill of **360**. Preview the array, and if the holes are positioned correctly, pick **OK** to finish the array.

20. Now concentrate on the other set of bolt holes on the larger bolt circle diameter. Enter the **CIRCLE** command, and use **Intersection** to snap to the lower intersection of the vertical centerline and the bolt circle diameter, as shown in Fig. 1.3.7. Specify a diameter of **.813**. Again, add a vertical centerline.

21. Enter the **ARRAY** command and create a polar array of the Ø.813 hole and its centerline using the same center point, **15** items, and an angle to fill of **360**.

22. Create the bolt holes on the smaller flange. This process is a little trickier because none of the holes fall neatly on a centerline. Notice on Fig. 1.3.5 that one of the holes has a center point 22.5° to the right of the lower vertical centerline. To locate the center point of that hole, create a temporary 18″ line using polar coordinates. You know that radial distances in AutoCAD increase by degrees counterclockwise starting at 0° on the positive X axis. Calculate the number of degrees to specify as shown in Fig. 1.3.8. The intersection of this line and the bolt circle diameter marks the center point of the first hole.

23. Enter the **CIRCLE** command and create a **Ø.531** hole with the center point you determined in step 20. Add a **Ø.625** tap and change it to the **Hidden** layer. Add a short centerline and change it to the **Center** layer.

24. Array the tapped hole and its centerline around the center of the front view, specifying a total of **8** holes and an angle to fill of **360**. Check the array to make sure it formed correctly. Then delete the temporary line.

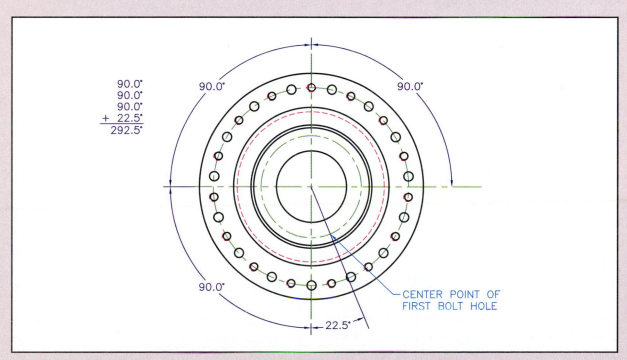

Fig. 1.3.8 Add the degrees in the three quadrants, moving counterclockwise from 0°; then add the 22.5° to find the angle at which to specify the temporary line. Specify the center point of the front view as the first point on the line.

25. Make **Phantom** the current layer and construct the cutting-plane line. The easiest way to do this is to create a circle first to define the curved parts of the line. Then use the **LINE** command to add the straight portions. *Hint:* For the offset portion of the line, you may want to change the center-line of the .531 tapped hole to the **Phantom** layer and extend as necessary instead of trying to duplicate the exact angle of the offset line through the hole centerlines. Add the arrows at the ends of the cutting-plane line. Use **DTEXT** to add the letter **A** near the top and bottom arrows.

26. The dimensions and notes in Fig. 1.3.5 that you used to construct the front view are needed by the manufacturer also. Make **Dimensions** the current layer. Pick the **Angular Dimension** button on the Dimension toolbar to add the four angular dimensions. (The dimension style should still be set up from your work on the nozzle assembly drawing.) Pick the **Quick Leader** button on the Dimension toolbar to specify the two bolt circle diameters. Then make **Text** the current layer and add the three notes for the bolt holes. When you finish, your drawing should look like the one in Fig. 1.3.5.

Limits and Tolerances

For manufacturing purposes, dimensions must be stated in very specific ways. This does not always mean showing a dimension with a very high precision. The precision that can be achieved using machinery—even computer-controlled machinery—has finite limits. Drawings that specify a high precision where it is not necessary may either make a piece completely unmanufacturable or drive up the cost of manufacture to the point that the product is not feasible.

Some dimensions can vary slightly without causing difficulty in assembling or using the product. Others, however, must be tightly controlled. Drafters and engineers specify a **tolerance** to allow for variability introduced by machining. The tolerance tells the manufacturer how much the part can vary from the stated dimension.

All of the dimensions on a working drawing must include a tolerance. To improve readability, the tolerances most commonly used on a specific drawing are often stated in a note, as shown in Fig. 1.3.9. For dimensions that need a different tolerance, the tolerances are included with the dimension. These tolerances are stated using either limit dimensioning or plus-and-minus tolerancing according to the rules set forth in ANSI Y14.5M.

Limit Dimensioning

In the limit method of dimensioning, the theoretical, or basic, dimension is not shown at all. Instead, the minimum and maximum values for the dimension are given, as shown in Fig. 1.3.10. By convention, the minimum value is placed above the maximum value.

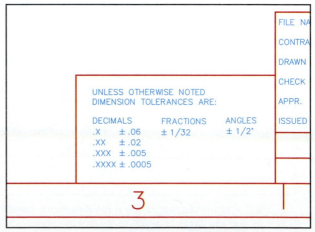

Fig. 1.3.9 *Tolerance stated as a note on the drawing. In this case, it has been placed near the title block for high visibility.*

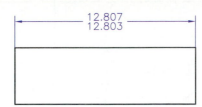

Fig. 1.3.10 *In limit dimensioning, the basic dimension is not shown. In this case, the basic dimension is 12.805.*

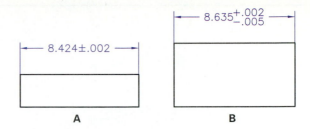

A　　　**B**

Fig. 1.3.11 *Plus-and-minus tolerancing: (A) symmetrical tolerances; (B) nonsymmetrical deviations.*

Plus-and-Minus Tolerancing

The disadvantage of the limit dimensioning method is that it is difficult to read at a glance because the basic dimension is not given. In plus-and-minus tolerancing, the basic dimension is specified, followed by the allowable positive and negative deviations, as shown in Fig. 1.3.11. When the upper and lower tolerances are the same, they can be combined with a ± symbol (Fig. 1.3.11A). When the tolerances are different, the positive tolerance is placed above the negative tolerance (Fig. 1.3.11B).

AutoCAD's Implementation

When we set up the dimension style for the nozzle assembly drawing, we simply modified the Standard dimension style. We ignored the Tolerances tab on the Modify Dimension Style window because we did not need to specify tolerances. Tolerances are necessary for some of the dimensions on the sectional view of the bell housing detail, however. The drawing does not incorporate limit dimensions, but it does need both symmetrical and nonsymmetrical plus-and-minus tolerances.

The Tolerances tab is shown in Fig. 1.3.12 with the Method drop-down box displayed. Notice that AutoCAD breaks the plus-and-minus tolerancing method into two types so that symmetrical tolerances are listed separately from nonsymmetrical, or deviation, tolerances. For the bell housing detail, we will use both symmetrical and deviation tolerances.

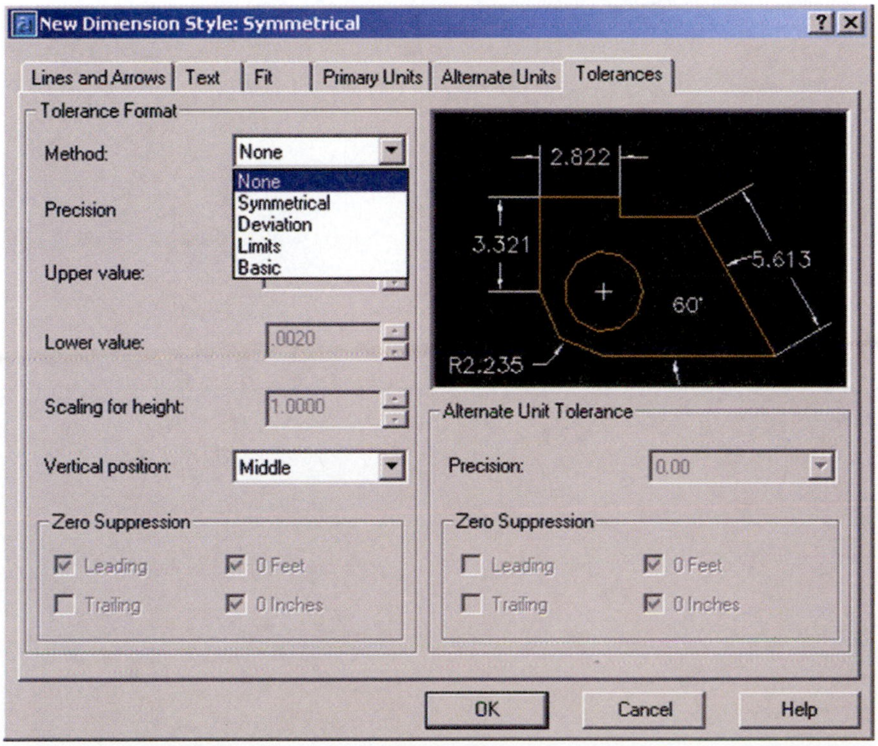

Dimension Section A-A

In Practice Exercise 1.3B, you will create two new dimension styles and use all three styles as you dimension Section A-A. As you work, refer to Fig. 1.3.13 for the location of dimensions and notes.

Fig. 1.3.12 *Types of tolerances available in AutoCAD.*

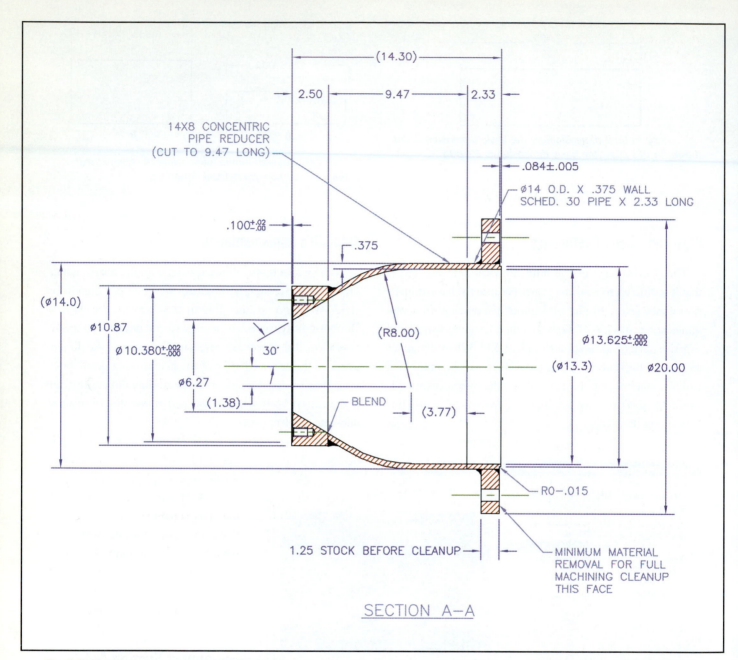

Fig. 1.3.13 Dimensions and notes for Section A-A.

Practice Exercise

1.3B

By looking carefully at Fig. 1.3.13, you can see that five different types of dimensions are used:

- one-place decimals without tolerances
- two-place decimals without tolerances
- three-place decimals without tolerances
- three-place decimals with symmetrical tolerances
- three-place decimals with nonsymmetrical deviations

Recall that in the assembly drawing, you used two-place decimals. Three-place decimals are used only when the precision needed for the dimension requires it. Also, according to ANSI, the same number of decimal places should be used in the basic dimension as is used in the tolerance values. Therefore, for example, a dimension with a tolerance of ±.002 must be specified to three decimal places: 3.005±.002.

For the nozzle assembly drawing, you modified the Standard dimension style for use with two decimal places and used overrides for the few dimensions that required different characteristics. For the bell housing detail, you will set up a separate style for each different kind of dimension.

1. Open the **PROJECT 1.dwg** file and zoom in on Section A-A. Display the **Dimension** toolbar if it is not already displayed. *Note:* If you dock the Dimension toolbar to the side of the drawing area, the dropdown box that lists the defined dimension styles will not display. Either leave the toolbar floating or dock it to the top of the drawing area so that you can change dimension styles with one click.

2. Pick the **Dimension Style** button on the Dimension toolbar to display the Dimension Style Manager. The Standard style should be set up for the two-decimal dimensions you used on the nozzle assembly drawing. Pick the **Modify...** button to check the settings. Then pick **OK** to return to the Dimension Style Manager. You will use the Standard style for all two-decimal dimensions.

3. Pick the **New...** button to begin a new dimension style. Name the new style **3 Decimal Places**. Make sure that **Standard** appears in the Start With box and that **All dimensions** appears in the Use for box. The style will be used for all dimensions. Pick the **Continue** button.

4. Because you are starting with a copy of the Standard style, you do not need to reenter any of the settings you have already set up. Instead, all you have to do is change the number of decimal places. To do so, pick the **Primary Units** tab. Pick the dropdown box next to Precision and select **0.000**. Pick **OK** to return to the Dimension Style Manager. The 3 Decimal Places style should now appear in the Styles box above the Standard style.

5. Pick the **New...** button again to set up the style for one decimal place. Name it **1 Decimal Place**, and select **3 Decimal Places** as the style to start with. Change the precision to **0.0**, and pick **OK** to return to the Dimension Style Manager.

6. Pick the **New...** button again to set up the next dimension style. Name this style **Symmetrical**, and select **3 Decimal Places** as the style to start with. Pick the **Tolerances** tab to display the tolerance options. Pick the dropdown box next to **Method** and select **Symmetrical**. The precision should already be set at 0.000 because you are basing this style on the 3 Decimal Places style. For the upper tolerance value, enter **0.005**. Pick the box next to **Leading** in the Zero Suppression area. Then pick **OK** to return to the Dimension Style Manager.

7. Pick the **New...** button again to set up the last dimension style. Name this style **Deviation**, and select **Symmetrical** as the style to start with. The Tolerances tab should already be displayed. Pick the dropdown box next to **Method** and select **Deviation** as the method of tolerancing. Set the upper tolerance value to **0.000**, and set the lower value to **0.002**. Pick **OK** and check to be sure that all four styles now appear in the Styles window of the Dimension Style Manager. Pick **Close** to return to the drawing area.

8. Set the current dimension style to **Symmetrical**.

9. Start by placing the **.084±.005** dimension, as shown in Fig. 1.3.14. Notice that the tolerance appears automatically. Leave extra space between this dimension and the next set of three dimensions. You will be adding further information in this space in a future practice exercise.

10. Pick the dropdown box in the Dimension toolbar and change the dimension style to **Standard**. Use the **Linear Dimension** button to add the **2.33** dimension. Then pick the **Continue Dimension** button to place the other two dimensions on that tier.

11. Place the overall reference dimension as the third dimension tier. Doubleclick the dimension to display its properties and override the text to add the parentheses to indicate that this is a reference dimension.

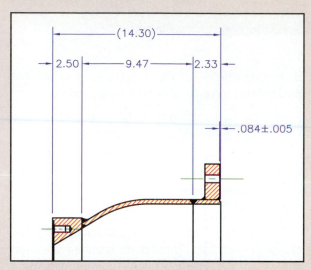

Fig. 1.3.14 *Placement of the horizontal dimensions for Section A-A.*

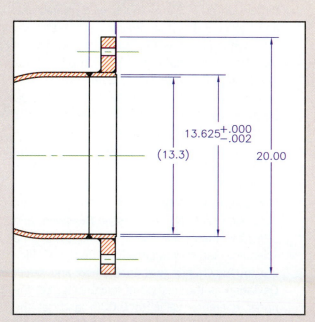

Fig. 1.3.15 *Vertical dimensions on the right side of Section A-A.*

12. Move on to the vertical dimensions at the right side of the section, as shown in Fig. 1.3.15. Change the dimension style to **1 Decimal Place** before setting the innermost reference dimension. Double-click the dimension to see its properties, and add the parentheses to indicate that this is a reference dimension.

13. Change the dimension style to **Deviation** and add the next dimension as shown. By default, the text of this dimension will be placed next to the Ø13.3 reference dimension, and the two will probably interfere with each other. Pick the grip on the text for the toleranced dimension and move it up so that the text does not interfere.

14. Change the dimension style to **1 Decimal Place** and add the last vertical dimension on this side of the view.

15. Zoom in on the left side of Section A-A to create the dimensions there, as shown in Fig. 1.3.16. Use the **BREAK** command to break the centerline between the front and sectional views to make room for these dimensions.

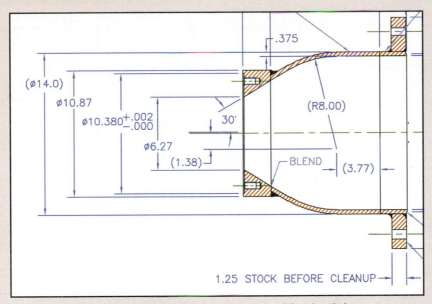

Fig. 1.3.16 *Dimensions on the left side and bottom of Section A-A.*

16. Change the dimension style to **3 Decimal Places** and use the **Linear Dimension** button to add the (.375) reference dimension for the pipe width.

17. Change the dimension style to **Standard,** pick the **Radius Dimension** button, and add the (R8.00) dimension as a reference dimension.

18. Add the vertical and horizontal dimensions that locate the center point of the R8.00 curve. Make both of these reference dimensions.

19. Pick the **Angular Dimension** button and add the 30° dimension at the beak.

20. Use **Linear Dimension** to add the vertical Ø6.27 dimension. Note that you will need to override the text to add the diameter symbol. AutoCAD's text code for the diameter symbol if **%%c**.

21. Change the dimension style to **Deviation** and add the next vertical dimension. Doubleclick the new dimension and change the values of the upper and lower deviations to **+.002** and **–.000**, respectively. Override the text to add the diameter symbol. Move the text up if necessary to avoid interfering with the previous dimension.

Note: If the tolerances disappear when you override the text, undo the override and add the diameter symbol as a separate DTEXT object and position it before the dimension.

22. Change the dimension style to **2 Decimal Places** and add the next vertical dimension. Override the text to add the diameter symbol.

23. Change the dimension style to **1 Decimal Place** and add the (Ø14.0) reference dimension.

24. Change the dimension style to **Deviation** and add the .100 dimension. Doubleclick the dimension and change the precision of the text and tolerances to **0.00**. Change the tolerance values to **+.02** and **–.00**, respectively.

25. Move to the bottom of the sectional view and add a horizontal dimension. Override the text to say **1.25 STOCK BEFORE CLEANUP**.

26. Make **TEXT** the current layer.

27. Zoom out to see the entire sectional view. Pick the **Quick Leader** button and add each of the five notes shown in Fig. 1.3.13 on page 82. Position them as shown.

Geometric Dimensioning & Tolerancing (GD&T)

Drawings that will be used in the manufacturing process require information that goes beyond the linear dimensions and notes you added to Section A-A in Practice Exercise 1.3B. This information, known as **geometric dimensioning and tolerancing (GD&T)**, is considered a type of tolerancing and is added in accordance with ANSI Y14.5M.

Keep in mind that not every dimension on a given drawing needs this extra information. In many cases, it is enough if the part falls within its specified tolerance. The individual manufacturing processes and the equipment to be used usually determine what GD&T information should be included. This information should be verified prior to completing the GD&T dimensioning process.

Geometric Characteristic Symbols

There are several types of geometric tolerances. Some control the form of individual features such as lines or surfaces; others control the profile of the feature. Still others control the orientation, location, or runout of related features. For example, material information such as the required flatness of a surface after machining may be needed, or the concentricity of parts must be called out. A chart showing the various types of geometric tolerances and their symbols, known as *geometric characteristic symbols*, is shown in Fig. 1.3.17.

Type of Tolerance	Characteristic	Symbol
Form	Straightness	—
	Flatness	▱
	Circularity	○
	Cylindricity	⌀
Profile	Profile of a Line	⌒
	Profile of a Surface	⌓
Orientation	Angularity	∠
	Perpendicularity	⊥
	Parallelism	//
Location	Position	⊕
	Concentricity	◎
	Symmetry	⹀
Runout	Circular Runout	↗
	Total Runout	↗↗

Fig. 1.3.17 *Geometric characteristic symbols.*

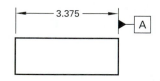

Fig. 1.3.18 *Method of identifying a datum.*

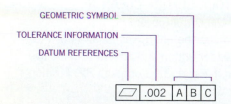

Fig. 1.3.19 *Example of a feature control frame.*

Datum Identifiers

In addition to characteristic symbols, one or more datums may need to be defined on a drawing. A **datum** is a theoretically exact surface area or line from which all geometric dimensions are taken. Datum identifiers are placed in a box connected to a surface or extension line by a small triangle, as shown in Fig. 1.3.18.

Feature Control Frames

On technical drawings, geometric characteristic symbols are placed in boxes known as *feature control frames*. These boxes are divided into several compartments depending on the information needed. An example of a typical feature control frame is shown in Fig. 1.3.19. The order of the items

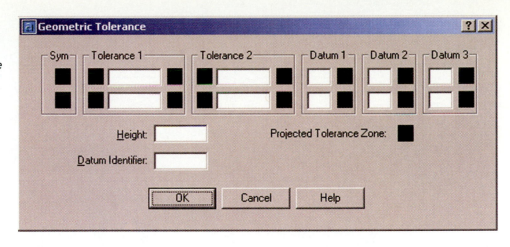

Fig. 1.3.20 AutoCAD's Geometric Tolerance window allows you to build a feature control frame easily using the correct "sentence structure."

must always be the same. The first compartment on the left contains the geometric characteristic symbol. Any necessary tolerance information is placed in the next compartment, followed by one or more datum references, if applicable. The information in the feature control frame shown in Fig. 1.3.19 is read as: "The flatness of the feature must be within two thousandths relative to datum feature A."

According to ANSI Y14.5M, feature control frames can be placed in the following positions:

- immediately below a callout or dimension pertaining to the feature
- at the end of a leader that points to the feature
- attached to an extension line that runs from the feature
- attached to an extension of the dimension line pertaining to a feature

GD&T in AutoCAD

AutoCAD makes the addition of feature control frames to a drawing easy by providing a window in which you can build them. Entering the TOLERANCE command produces the window shown in Fig. 1.3.20. In this window, you can click on the various black boxes to see appropriate choices. For example, clicking the black box under Sym presents a collection of geometric characteristic symbols from which you can select.

To create a datum identifier instead of a feature control frame, leave all of the boxes blank except the white box next to Datum Identifier. Place the letter of the datum in this box. AutoCAD does not provide an automatic method for creating the triangle at the end of the datum identifier. Many drafters who use GD&T in AutoCAD create a block of the triangle so that it can be inserted easily into the drawing.

Add GD&T to the Bell Housing Detail

In the bell housing detail drawing, both the front view and Section A-A require GD&T specifications as follows:

- The concentricity of the bolt holes needs to be controlled.
- The inside diameter of the main 14″ pipe will become datum feature A.
- The locating shoulder on the 14″ pipe will become datum feature B.
- The edge of the flange next to the locating shoulder requires a combined feature control frame for both flatness and perpendicularity controls.
- The concentricity of the beak opening must be controlled.

In Practice Exercise 1.3C, you will add the necessary datums and geometric characteristic symbols to the bell housing detail drawing. Refer to Fig. 1.3.21 as necessary for placement.

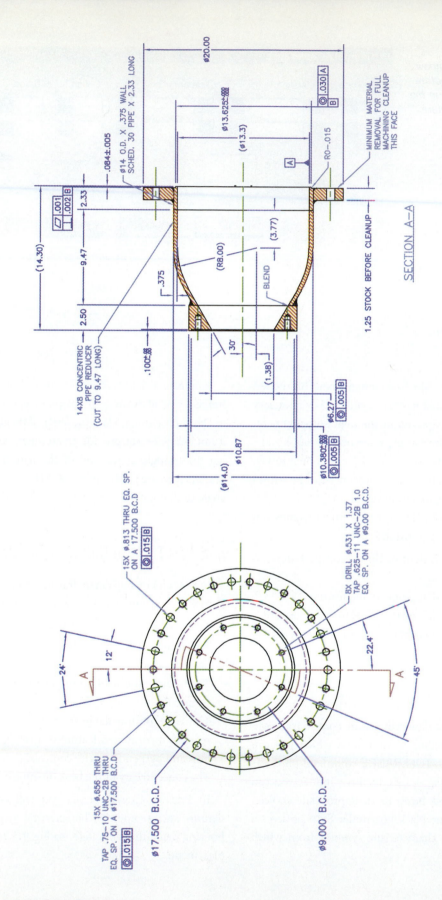

Fig. 1.3.21 *Addition of datums and feature control frames to the bell housing detail drawing.*

Practice Exercise

Start the geometric dimensioning and tolerancing of the bell housing detail by adding the callout for datum feature A. Zoom in on the right side of Section A-A as shown in Fig. 1.3.22 before you begin.

1. Make **Dimensions** the current layer.
2. Enter the **TOLERANCE** command to display the Geometric Tolerance window.
3. Enter the capital letter **A** in the box labeled Datum Identifier. Pick **OK**. The Geometric Tolerance window disappears and the datum identifier becomes attached to the cursor. Move it to a point above and to the right of the dimension line of the Ø13.3 reference dimension and pick a point to insert it into the drawing.
4. Use the **LINE** command to extend a line straight up from the top of the dimension line and then over to the datum identifier, as shown in Fig. 1.3.22.
5. Create the filled triangle at the base of the line. *Hint:* You can do this using the same technique you used to fill the weld representations in Practice Exercise 1.2N. Place the hatch on the Dimensions layer, however. This finishes the datum A callout.
6. Datum feature B can be attached to the associated feature control frame, so you can create both items in the same operation. To do so, enter the **TOLERANCE** command. Pick the first black box under Sym to display a window of geometric characteristic symbols. Pick the symbol for concentricity. Working across the first row, skip the first black box under Tolerance 1. (Picking in this box adds a diameter symbol, which is not needed in this case.) In the next white box, enter the tolerance **.030**. Skip over to the white box under Datum 1 and enter a capital **A**. In the white box next to Datum Identifier, enter a capital **B**. Pick **OK** and place the feature control frame below and to the right of the dimension line for the locating shoulder. There is no need to insert a triangle for this datum because it is attached to the feature control frame.

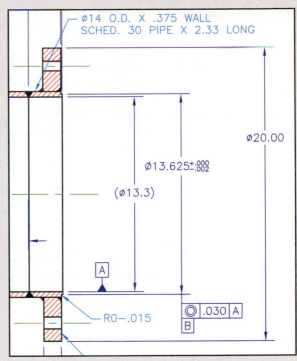

Fig. 1.3.22 *Datum identifiers A and B and the first feature control frame.*

7. Use the **LINE** command to extend a line straight down from the bottom of the dimension line. Enter the **MOVE** command and select the feature control frame. For the base point, select the top left endpoint of the frame. For the second point, use the Nearest object snap to snap to the line you just created. Refer again to Fig. 1.3.22.
8. Move the viewing window up to the top of Section A-A. The locating shoulder requires a flatness of .001 and must be perpendicular to datum B within .002. You can specify both of these requirements using a combined feature control frame, as shown in Fig. 1.3.23. Enter the **TOLERANCE** command and pick the first black box under Sym. Select the symbol for flatness, and enter **.001** in the white box under Tolerance 1. Then move down to the second line and insert the symbol for perpendicularity. Enter **.002** in the white box under Tolerance 1, and place a capital **B** in the white box under Datum 1. Pick **OK** and place the feature control frame on the extension line as shown in Fig. 1.3.23.

Two concentricity requirements must be noted at the front of the nozzle. Notice in Fig. 1.3.24 that the dimensions with feature control frames have been moved down. AutoCAD has no way of interrupting the dimension line for the feature control frame without losing the associativity of the dimension. If you placed the frame under the dimension in its previous position, the dimension line would run through it, making it difficult to read. Moving the dimensions is the easiest way to solve this problem.

9. Zoom in on the front of the nozzle and add the two feature control frames to show concentricity within **.015** to datum **B**. Move the dimensions as shown in Fig. 1.3.24. To extend the dimension line between the arrows, doubleclick the dimensions. In the **Fit** area of the Properties window, change **Dim line forced** to **On**.

Now move to the front view of the nozzle to insert the remaining requirements. Both sets of holes on the larger bolt circle diameter must be concentric with datum B to within .015.

10. Enter the **TOLERANCE** command and create the two feature control frames. Place them as shown in Fig. 1.3.25.

The geometric dimensioning and tolerancing of the bell housing detail is now complete. Your drawing should look like the one in Fig. 1.3.21.

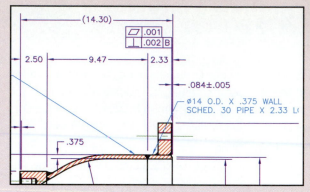

Fig. 1.3.23 Placement of the compound feature control frame for the locating shoulder.

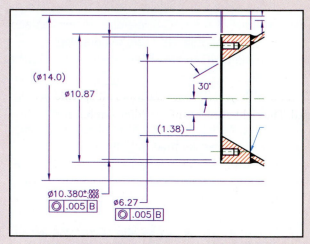

Fig. 1.3.24 Placement of the feature control frames for the nozzle, showing movement of the dimensions.

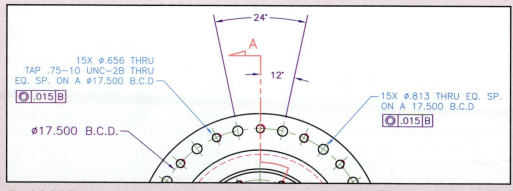

Fig. 1.3.25 Feature control frames for the front view of the bell housing detail.

Weld Symbology

Because some pieces of the bell housing require welds, welding specifications must be included on the drawing. Instead of writing out the instructions, drafters use a set of symbols established by the American Welding Society (AWS). These symbols, like those used in GD&T, are arranged in a standard order and pattern so that the welder can read them and understand their intent at a glance.

Elements of a Welding Symbol

Note that there is a difference between a welding symbol and a weld symbol. A *welding symbol* includes the entire weld specification and can include up to eight elements, depending on the specifications needed. One of those elements is the *weld symbol*, which describes the specific weld to be used. The eight elements of a welding symbol are:

- **reference line**—the line around which all of the other elements are oriented.
- **arrow**—at the end of the leader, the arrow points to one side of the joint (for groove, fillet, flange, and flash welds) or to the outer surface of one of the members of the joint at the centerline of the weld. The side or surface to which the arrow points is called the *arrow side*. The other side or surface is known as the *other side* or *other side member*.
- **basic weld symbols**—describe the type of weld to be formed. Basic weld symbols are shown in Fig. 1.3.26.
- **supplementary symbols**—symbols that further define the weld by giving more information, such as where the welding is to be performed. "Weld-all-around" and "field weld" symbols are also included in this category. Supplementary symbols are shown in Fig. 1.3.27.
- **dimensions**—include the size, length, etc. of the weld to be formed, as well as other supplementary information. These dimensions are placed in specific places on the reference line so that they cannot be confused. For example, the size of the weld is always on the left side of the basic weld symbol, regardless of which side of the reference line the arrow is placed. The length is always on the right of the basic weld symbol.

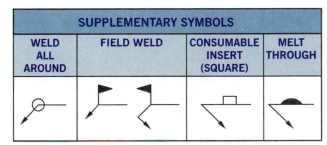

BASIC WELD SYMBOLS						
FILLET	PLUG OR SLOT	STUD	SPOT OR PROJECTION	SEAM	BACK OR BACKING	SURFACING

Fig. 1.3.26 *Basic weld symbols.*

FLANGE		GROOVE							
EDGE	CORNER	SQUARE	SCARF	V	BEVEL	U	J	FLARE-V	FLARE-BEVEL

SUPPLEMENTARY SYMBOLS			
WELD ALL AROUND	FIELD WELD	CONSUMABLE INSERT (SQUARE)	MELT THROUGH

CONTOUR			BACKING OR SPACER (RECTANGULAR)
FLAT	CONVEX	CONCAVE	

Fig. 1.3.27 *Supplementary symbols.*

FINISH SYMBOLS	
LETTER SYMBOL	METHOD
C	Chipping
G	Grinding
M	Machining
R	Rolling
H	Hammering

Fig. 1.3.28 *Finish symbols.*

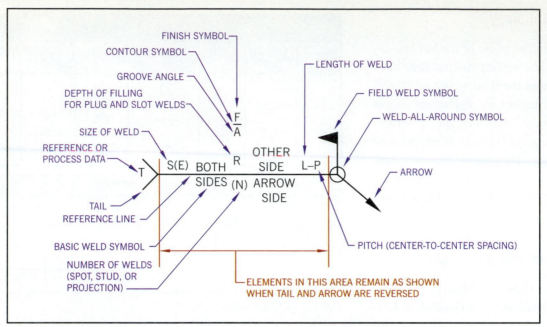

Fig. 1.3.29 Positions of the various elements of a weld symbol.

- **finish symbols**—letters that indicate the finish method and contour of the weld. See Fig. 1.3.28 for typical letters and their meaning.
- **tail**—appears on the opposite end of the reference line from the arrow; used only when a specification, process, or other reference is needed in the welding symbol. The tail is omitted when these are not needed.
- **process or reference**—appears within the tail to provide information about specifications, processes, or other instructions; this designation is made by letter symbols approved by ANSI. These letter designations are listed in Appendix B.

The placement of these elements on the weld symbol is shown in Fig. 1.3.29. Note that not all of these eight elements are necessarily required on a given weld specification.

Weld Symbols in AutoCAD

The DesignCenter in AutoCAD 2002 and later includes a symbol library of basic weld symbols. The library file is located in the DesignCenter folder, which is located in the Sample folder, and is named Welding.dwg. To use them, open DesignCenter and select the Welding.dwg file. Pick Blocks to display the symbols, and drag the symbols you want to use into your drawing file.

If your version of AutoCAD does not contain a symbol library of welds, you should either find a commercial one or create your own, especially if you expect to need them frequently. Several companies make third-party symbol libraries for welding, and some of them are more extensive than the one provided in AutoCAD 2002. In fact, software is available that works within AutoCAD to build the entire welding symbol, including all of the elements discussed in this section, instead of just providing the basic weld symbol.

Practice Exercise 1.3D steps you through the procedure to add the welding symbols needed on Section A-A of the bell housing detail drawing. This exercise uses the basic weld symbols provided with AutoCAD 2002. If you do not have access to these symbols, or if your institution uses a different or more advanced symbol library or program, you may need to alter the steps. Whatever method you use, your welding symbols should look like those in Fig. 1.3.30 when you finish.

Fig. 1.3.30 *Section A-A with welding symbols in place.*

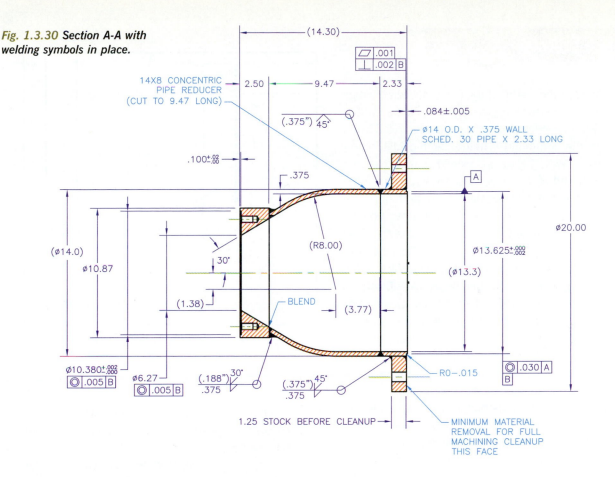

14X8 CONCENTRIC
PIPE REDUCER
(CUT TO 9.47 LONG)

Ø14 O.D. X .375 WALL
SCHED. 30 PIPE X 2.33 LONG

1.25 STOCK BEFORE CLEANUP

MINIMUM MATERIAL
REMOVAL FOR FULL
MACHINING CLEANUP
THIS FACE

Practice Exercise 1.3D

AutoCAD provides only the basic weld symbols in its symbol library. You must therefore construct the entire welding symbol using other AutoCAD commands as well as the blocks in the symbol library.

The welding specification you are about to insert is for a V weld between the 14″ pipe and the concentric pipe reducer. This weld will extend through the thickness of the pipe, or .375″.

1. Zoom in on the top wall of the 14″ pipe, as shown in Fig. 1.3.31, and make **Dimensions** the current layer.

2. Pick the **Quick Leader** button on the Dimension toolbar and press **Enter** to view the annotation settings. Set the annotation to **None**. Pick **OK**

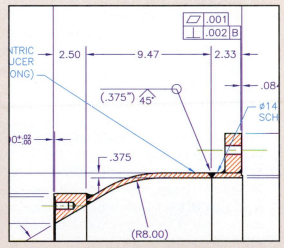

Fig. 1.3.31 *Add the welding symbol for the V weld between the pipe and pipe reducer.*

and start the leader at the joint between the pipe and the concentric pipe reducer. Pick a second point for the leader as shown in Fig. 1.3.31 and end the command. Then enter the **LINE** command. With Ortho on, extend a line from the upper end of the leader a short way to the left.

3. Pick the **AutoCAD DesignCenter** button on the Standard toolbar to open DesignCenter. Double-click the **Sample** folder (located in the AutoCAD 2002 folder) and then click the **DesignCenter** folder to display its contents. Doubleclick the **Welding.dwg** file and then doubleclick **Blocks** to see the blocks included in the file. Doubleclick the block named **Partial V Weld** to display its insertion properties. To scale the block correctly for use in the bell housing detail, specify a uniform scale of **2.00**. Notice from the thumbnail in the DesignCenter window that the block is upside-down from its orientation in the drawing. Specify a rotation angle of **180** to orient it correctly. Then pick **OK** to close the window. The block will be attached to the cursor. Use the **Endpoint** object snap to position the block at the end of the short line that extends from the leader.

4. Enter the **DTEXT** command and specify the same text height you used for dimensions. Enter **.375** (the size of the weld) as shown in Fig. 1.3.31. Before you end the DTEXT command, pick a point beneath the V symbol and enter the text **45°**. You may need to adjust the position of this text using grips.

5. Enter the **LINE** command and snap to the left end of the horizontal line of the weld symbol. With Ortho on, create a line to the left so that it extends over the text.

6. Enter the **CIRCLE** command and add the weld-all-around symbol. The center of the circle should be at the intersection of the leader and the right side of the weld symbol. Use a radius of **.35** for the circle.

Next you will add the welding symbols for the weld at the joint between the 14″ pipe and its flange, as shown in Fig. 1.3.32.

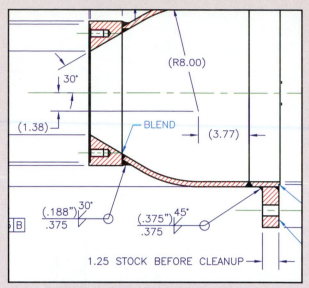

Fig. 1.3.32 *Welding symbols created in steps 8 and 9.*

7. Zoom in on the lower part of the 14″ pipe.

8. Follow the procedure outlined in steps 2 through 6 to create the weld at the joint between the 14″ pipe and the flange. Note that this weld calls for a fillet weld on the arrow side and a 45° bevel weld on the other side. Use the **Fillet Weld - Left** block for the fillet weld, and use the **Partial Bevel Weld** block for the bevel weld. You will not need to rotate the blocks before you insert them, but make the scale **2.00**. Notice that the fillet weld supplied by AutoCAD includes a tail, which you do not need. You will have to explode the block to remove the tail. Be sure to change the remaining pieces of the block to the Dimensions layer.

9. Create the welding symbol for the weld between the concentric pipe reducer and its flange. You can follow steps 2 through 6 again if you wish, but notice how similar this welding symbol is to the one you created in step 8. It would be faster simply to copy the weld you created in step 8 and edit the text to create this weld. If you do this, however, it is very important to remember to change the text, because the specifications are different.

Finish the Layout

All of the actual drawing work has now been completed for the bell housing detail. It is time to do all of the final tasks such as placing the drawing on a layout, adding the general notes and tolerances, and finishing the title block.

To do this, you must add a layout to the PROJECT 1.dwg file. Practice Exercise 1.3E steps you through the process of creating a new layout, placing the bell housing detail on the layout, and adding the final touches such as general notes and the title block. This will finish the drawing as well as your work in Project 1.

Practice Exercise 1.3E

Since you have already set up the NOZZLE layout, the simplest way to create a new layout is simply to copy the existing one. By making a copy, you ensure that all of the settings and the title block text that you entered for the NOZZLE layout are transferred automatically—you won't have to reenter information that already exists.

1. Pick the **NOZZLE** tab at the bottom of the drawing area to display the nozzle assembly layout. Right-click the tab and pick **Move or Copy...** from the shortcut menu that appears.

2. At the bottom of the Move or Copy window, pick to select **Create a copy**. In the Before layout window, select **(move to end)** and then pick **OK**.

A new tab appears at the bottom of the screen. By default, it is named NOZZLE (2).

3. Pick the new **NOZZLE (2)** tab, right-click to display the shortcut menu, and select **Rename**. Rename it **BELL HOUSING DETAIL**.

4. Use the **ERASE** command to delete the parts list and the general notes for the nozzle assembly.

5. If the Viewport layer is still frozen, thaw it now and make it current. Select the existing viewport and delete it from the layout.

6. From the **View** menu, select **Viewports** and **1 Viewport** to create a new viewport to hold the bell housing detail. Position it as you did for the nozzle assembly drawing.

7. Doubleclick one of the viewport lines to see its properties and change the standard scale to **1:2**.

8. Click **PAPER** to enter model space and use the **PAN** command to position. *Note:* If the drawing is too large for the space on the drawing sheet, you may need to return to the Model tab and adjust the spacing between dimensions and between the two views to make it fit.

9. In the layout view, add the general notes in the lower left corner as follows:

NOTES:

1. **SPECIFICATIONS:**
   BEAK DIA: 4.0625

2. **MATERIAL = CARBON STEEL**

3. **UNLESS OTHERWISE NOTED, BREAK ALL SHARP EDGES AND CORNERS**

4. **PAINT OUTSIDE**

10. Check the title block and change any information necessary. Title this drawing **BELL HOUSING DETAIL**. Note the **1:2** scale in the space provided, and number this sheet **2 of 2**.

11. On the left side of the title block, add the general tolerance notes as follows:

UNLESS OTHERWISE NOTED DIMENSION TOLERANCES ARE:

DECIMALS		ANGLES
.X	±.06	±.5°
.XX	±.02	
.XXX	±.005	
.XXXX	±.0005	

12. If required by your instructor, plot the layout.

When you finish this practice exercise, Project 1 is complete. Your bell housing detail drawing should look like the one in Fig. 1.3.33.

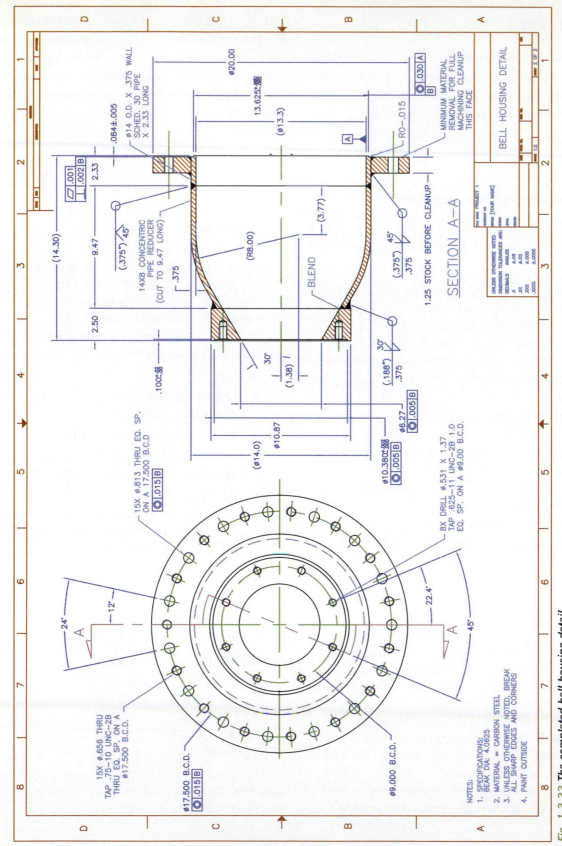

Fig. 1.3.33 The completed bell housing detail.

Review Questions

1. What is an offset section, and why is it used in the bell housing detail?

2. What information does AutoCAD need to form a polar array? To form a rectangular array?

3. How would an array with 20 objects rotated through 180° differ in appearance from an array with the same 20 objects rotated through 360°?

4. What was the advantage of using the XLINE command when you developed the front view from the geometry in Section A-A?

5. Why are tolerances needed on working drawings?

6. According to AutoCAD, what is the difference between a symmetrical tolerance and a deviation tolerance?

7. What term is used by ANSI to include both symmetrical and deviation tolerances?

8. On a working drawing, some dimensions may not have tolerances stated with the dimension text. Do these dimensions have a specified tolerance? Explain.

9. The width of a part is 3.125. Using the limit method, how should its tolerance be stated if the upper tolerance is +.002 and the lower tolerance is −.005?

10. Explain why more than one dimension style may be needed for a working drawing created in AutoCAD.

11. What is the purpose of GD&T?

12. What are the five basic types of tolerances defined by geometric characteristic symbols?

13. What is a datum identifier, and how is it related to the geometric characteristic information provided in a feature control frame?

14. What command in AutoCAD is used to build feature control frames?

Portfolio Project

Needle Valve (continued)

Refer to your notes from the Needle Valve project at the end of Sections 1.1 and 1.2. Open your PORTFOLIO 1.dwg file and create any drawings not yet completed to finish the set of working drawings. Be sure to place each drawing on a separate layout. Name the layouts so that the drawings are easy to recognize. Add any necessary GD&T or welding information. Plot the entire set of drawings and submit it for approval. When the drawings have been checked and approved, place them in your portfolio for safekeeping.

15. What is the difference between a weld symbol and a welding symbol? Which is included with AutoCAD 2002 and later?

16. If your version of AutoCAD does not include the Welding.dwg symbol library, what other options do you have?

17. Why might you need to make last-minute adjustments to your drawing when you place it on the layout?

Practice Problems

1. The top view of an alignment bracket is shown in Fig. 1.3.34. Create a new drawing in AutoCAD and name it Practice Problem 1.3.1.dwg. Draw the top view of the bracket and then create the front view as an offset section based on the cutting-plane line shown in the top view. The overall height of the bracket is .50″, and the height of the boss is 1.50″.

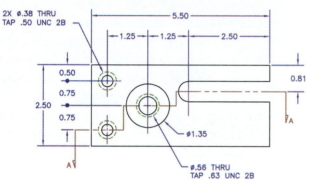

Fig. 1.3.34

2. The linked plates in the assembly shown in Fig. 1.3.35 are regular pentagons inscribed in a Ø1.00″ circle. The bolt holes are Ø.124″, and the B.C.D. is 1.50″. Use one or more polar arrays in AutoCAD to create the assembly. Dimension the drawing. Name the file Practice Problem 1.3.2.dwg.

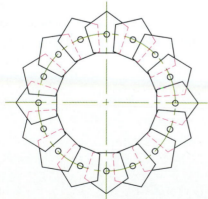

Fig. 1.3.35

3. Refer to Fig. 1.3.36. What are the dimensional requirements for the cylinder?

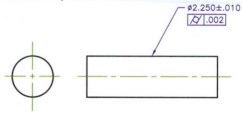

Fig. 1.3.36

4. Refer to Fig. 1.3.37. What constraints are placed on the rightmost vertical surface?

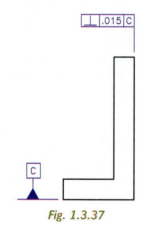

Fig. 1.3.37

5. Create a new drawing named Practice Problem 1.3.5. Draw the block in Fig. 1.3.38. Make the bottom horizontal surface datum A and add a feature control frame to restrict the top horizontal surface to be parallel within .010 of datum A and flat within .004.

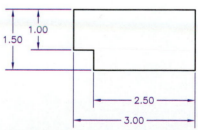

Fig. 1.3.38

6. Create a new drawing named Practice Problem 1.3.6. Draw the two views of the cylindrical spacer shown in Fig. 1.3.39. Then add the following changes and additions to the dimensions, using plus-and-minus tolerances where necessary:

- Change the precision of the Ø.51 holes to three decimal places.
- Change the tolerance of the Ø.51 holes to ±.002.
- In the front view, make the inner surface of the spacer datum A.
- In the side view, make the right surface datum B.
- Require the Ø.51 holes to be positioned within .014 of their stated dimension with respect to datums A and B.
- Change the precision of the outer diameter of the spacer to three decimal places.
- Change the tolerance of the outer diameter to +.005 and −.002.

7. Create a new drawing named Practice Problem 1.3.7. Draw two steel bars at right angles to each other, as shown in Fig. 1.3.40. Add a welding symbol to specify fillet welds on both sides with a depth of .125.

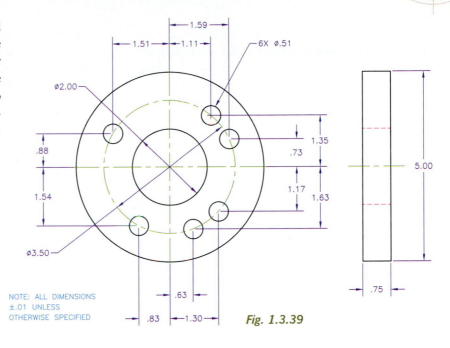

NOTE: ALL DIMENSIONS
±.01 UNLESS
OTHERWISE SPECIFIED

Fig. 1.3.39

Fig. 1.3.40

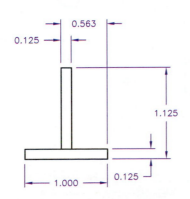

In Practice Problems 8 through 11, you will design each device to your own specifications. Each problem will require the following:

- **Assembly Drawing:** A pictorial view of the entire assembly. Label all components and state all critical assembly dimensions.
- **Detail Drawings:** A complete detail drawing for each component in the assembly. Draw them to ANSI drafting standards. Include at least one detail for every part.

The designs should adhere to the following guidelines:

- **Commercially Viable:** Design should be producible using common techniques such as molding, casting, machining, stamping, forming, etc.
- **Multiple Components:** The design should contain at least three unique components.
- **Fit:** Design each component to fit and operate with the other components in the assembly. Use generous tolerances and pay close attention to interferences.

8. Develop drawings for a yo-yo. Examine existing designs at a local toy shop before beginning. Resesarch materials from which to produce the components, manufacturing techniques, and cost associated with producing the product. The design goal is to produce the toy for less than $1. Refer to Figs. 1.3.41 and 1.3.42 for ideas, but do not copy the drawings exactly. Rather, develop your own ideas and use them to produce your finished drawings.

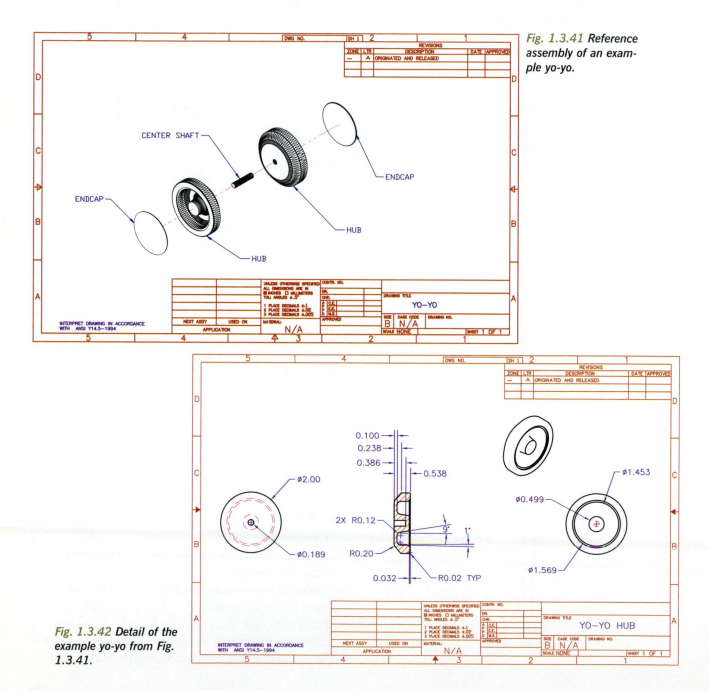

Fig. 1.3.41 *Reference assembly of an example yo-yo.*

Fig. 1.3.42 *Detail of the example yo-yo from Fig. 1.3.41.*

9. Develop drawings for a tape dispenser. You should research existing designs at a local office supply store before beginning your design. Resesarch materials from which to produce the components, manufacturing techniques, and cost associated with producing the product.

The design goal is to produce the tape dispenser for less than $2. Refer to Figs. 1.3.43 and 1.3.44 for ideas on what your final output should look like, but do not copy the drawings exactly. Rather, develop your own ideas and use them to produce your finished drawings.

Fig. 1.3.43 Reference assembly of an example tape dispenser.

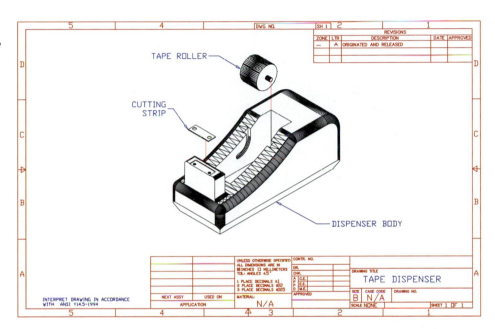

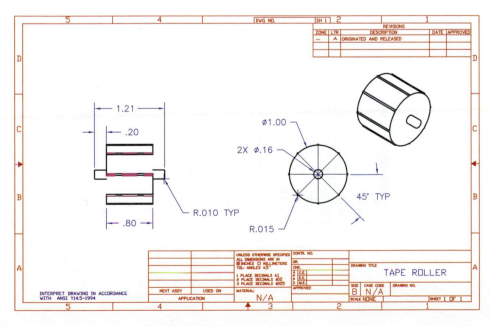

Fig. 1.3.44 Detail of the example tape dispenser from Fig. 1.3.43.

10. Develop drawings for a picture frame. You should research existing designs at a local craft store or specialty shop before beginning your design. Resesarch materials from which to produce the components, manufacturing techniques, and cost associated with producing the prod- uct. The design goal is to produce the picture frame for less than $0.50. Refer to Figs. 1.3.45 and 1.3.46 for ideas on what your final output should look like, but do not copy the drawings exactly. Rather, develop your own ideas and use them to produce your finished drawings.

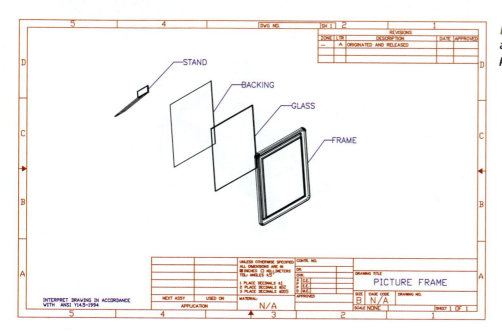

Fig. 1.3.45 *Reference assembly of an example picture frame.*

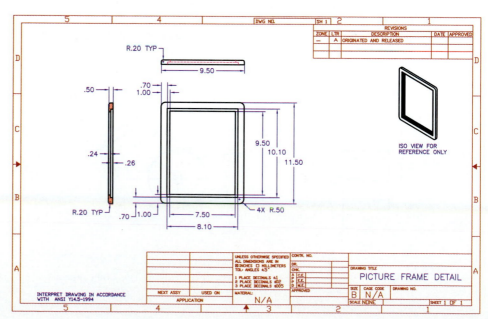

Fig. 1.3.46 *Detail of the example picture frame from Fig. 1.3.45.*

11. Develop drawings for a pair of eyeglasses. You should research existing designs at local eyewear shops before beginning your design. Resesarch materials from which to produce the components, manufacturing techniques, and cost associated with producing the product. The design goal is to produce the eyeglasses for less than $10. Refer to Figs. 1.3.47 and 1.3.48 for ideas on what your final output should look like, but do not copy the drawings exactly. Rather, develop your own ideas and use them to produce your finished drawings.

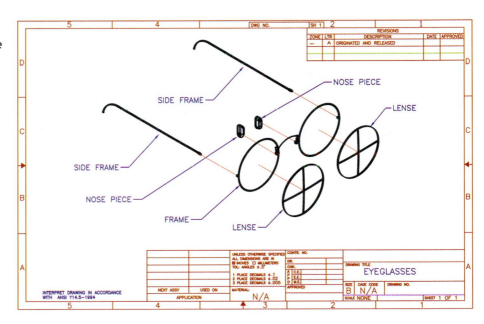

*Fig. 1.3.47 **Reference assembly of an example pair of eyeglasses.***

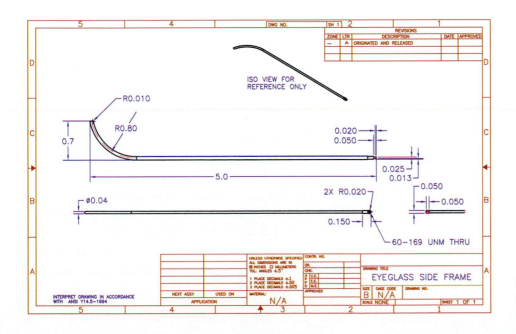

*Fig. 1.3.48 **Detail of the example eyeglasses from Fig. 1.3.47.***

Summary

- Before the actual drawing work for a design/drafting project begins, the design parameters should be set and background material should be obtained about the project.
- Setting up a drawing file in AutoCAD includes calculating and setting drawing limits, units, text styles, and layers.
- In AutoCAD, all of the drawings in a set of working drawings can be placed in the same file and then displayed and printed using different custom layouts.
- Line widths on plotted AutoCAD drawings can be controlled by using color-dependent or named plot styles.

- All dimensions on working drawings must be toleranced.
- AutoCAD can be set up to create any type of tolerance automatically when you dimension the drawing.
- Geometric dimensioning and tolerancing information is added to drawings that will be used in the manufacturing process to control the form, profile, orientation, location, or runout of a surface or feature.
- AutoCAD's TOLERANCE command allows the drafter to build feature control frames in accordance with ANSI Y14 standards.

Project 1 Analysis

1. Discuss the relative advantages and disadvantages of placing all of the drawings in a set of working drawings into a single AutoCAD file.
2. Why might a drafter prefer to create a new drawing file in AutoCAD using one of the standard drawing templates supplied with the software?
3. What is the purpose of a reference assembly drawing? In this project, why was a detail drawing also required?
4. Why should drafters use a symbol library to add welding symbols to a drawing? When might the library supplied with the AutoCAD software not be sufficient?
5. Explain why geometric dimensioning and tolerancing is sometimes required on working drawings in addition to strict linear tolerancing.

Synthesis Projects

Apply the skills and techniques you have practiced in this project to these special follow-up projects.

1. Analyze the drawings you created in this project. Use the information they contain to create a cost estimate for building the lower nozzle assembly. Do whatever research is necessary to build the estimate, and keep a record of the sources you use.

2. A client has asked you to design and create the working drawings for an external recordable CD unit for a computer. Determine the design parameters and do the necessary research to plan the project.

3. Set up a drawing file for use in the project described in the previous problem. Show your limits calculations and include a sketch to show where you intend to place each drawing in the file.

4. If time permits, develop the working drawings for the recordable CD unit, plot them, and add them to your portfolio.

Structural Field Assembly of Tire Conveyor

When you have completed this project, you should be able to:

- Identify the roles of the drafter in a field assembly project.
- Explain the differences between traditional technical documentation and documentation that is produced as a prototype is developed.
- Document a field assembly project thoroughly and accurately.
- Increase efficiency by reusing geometry among the files in a set of working drawings.
- Represent standard parts in an assembly.

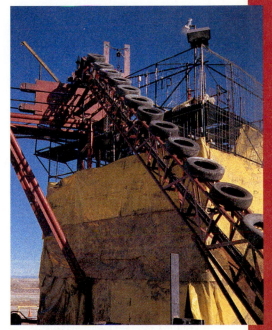

Fig. 2.1 **Tire conveyor for the prototype tire gasification system.**

Introduction

During the development of a new technology, drafters sometimes need to work interactively with engineers and construction supervisors. Instead of providing a finished set of plans, the drafter goes on-site and works with others to provide documentation as the project develops. The documentation consists of working drawings for creating the finished assembly.

In this project, you will work through the steps required to provide the documentation for a tire conveyor for use with a prototype tire gasification system. The prototype was erected by Emery Energy Company, LLC, as part of its effort to develop a workable gasification technology. Before you begin the actual drawings, you will need to become familiar with the concept of gasification and Emery's original ideas for implementation. You must understand the function of tire conveyor and how it fits into the overall gasification system. Then you will create a set of working drawings to describe the tire conveyor for future manufacture.

Project Parameters

The drafter's role in a field project is somewhat different from the traditional role in which the drafter presents a set of carefully developed working drawings according to which an item is built or manufactured. In a prototype project such as the tire gasification system built by Emery Energy Company, LLC, the drafter's primary responsibility is to document the construction of the prototype and the decisions made in the field so that the project can be duplicated later.

There is no less need for accuracy in a field project, however. Inaccurate documentation can mean major delays and rework if the prototype is accepted and taken into production. Therefore, the drafter in a field project must be flexible during the project and should doublecheck actual measurements after each component has been finalized to be sure the drawings are accurate.

Section Objectives

- Explore the concept of tire gasification as it relates to recycling technology.
- Identify the role of the tire conveyor in the gasification system.
- Identify the roles of the drafter in a field assembly project.
- Plan the drawings necessary to document the tire conveyor.

Key Terms

- gasification
- reverse engineering

The Tire Gasification System

The tire conveyor that forms the basis for this project was built as part of Emery Energy Company's prototype tire gasification system in the mid-1990s. **Gasification** is the transformation of solid and/or liquid materials into a gaseous form. Emery Energy Company's idea was to take a waste product—discarded tires—and convert it to gases that can be used for other purposes. The first-generation prototype that forms the basis for this project was operational from 1996 to 1998. Emery used the prototype in developing its patented gasification process. The original prototype still stands and may one day be reopened to supply energy through the use of biomass gasification.

In this prototype, scrap tires were fed into an egg-shaped chamber called a *geodesic reactor* for processing. See Fig. 2.1.1. The reactor (nicknamed "the Egg") gasified the tires and fed the gases into attached pipelines for distribution.

The reactor was already in place before the tire conveyor was designed. Until the conveyor was added, the tires had to be hand-fed into it. The drafting firm worked with designers and field construction personnel to develop the tire conveyor.

In cases such as this one, in which part or all of the system is already in place, the drafter may have two independent tasks at the construction site:

- document existing equipment by **reverse engineering**. In other words, the drafter measures equipment that has already been installed and then creates documentation—working drawings—to match it.
- help retrofit new equipment to existing construction.

Both of these tasks may seem quite different from the traditional role of creating the technical drawings from which equipment can be manufactured. In reality, the result is similar: the drawings provide documentation so that plans can be made to manufacture the product commercially.

Requirements for the Tire Conveyor

The sole purpose of the tire conveyor was to transport the tires from ground level to the tire feeder at the top of the gasification system, as shown in the diagram in Fig. 2.1.2. Because the tire conveyor was designed and assembled in the field, Emery chose to incorporate common materials and standard components that could easily be delivered to the site. The components would then be readily available if parts later needed to be repaired or replaced.

Because the conveyor had to be long enough to reach from the ground to the top of the four-story gasifier, the framework needed to be rigidly constructed. The framework also had to be designed to be strong enough to support its own weight as well as the weight of the tires being transported.

Fig. 2.1.1 Geodesic reactor for the tire gasification system.

Fig. 2.1.2 Assembly drawing of the tire gasification system showing the function and placement of the tire conveyor developed in this project.

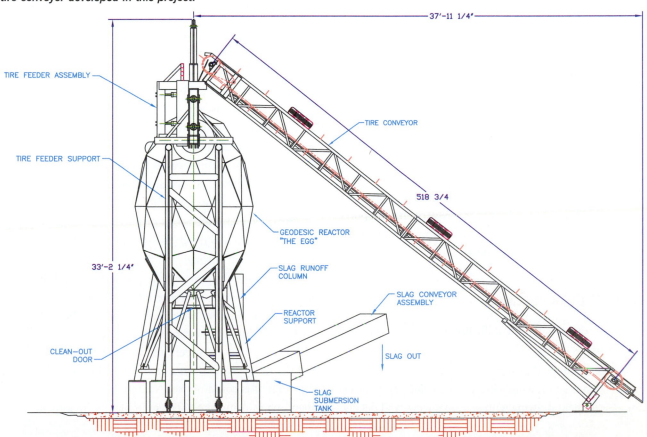

Fig. 2.1.3 Tire conveyor for the tire gasification system. The main framework was welded from 2^1/$_2''$ Schedule 40 pipe.

To perform its function of transporting tires to the tire feeder at the top of the gasifier, the tire conveyor needed the following components:

- **roller chain**—to convey tires continually to the tire feeder. Hooks added to this chain assured that tires were held securely on the conveyor.
- **sprockets**—to locate the roller chain and link it to the hydraulic motor.
- **roller bearing units**—to support the sprockets and roller chain.
- **hydraulic motor**—to drive the roller chain.

As required on uphill inclined conveyors, the hydraulic motor was located at the upper end. To keep the conveyor synchronized with the tire feeder, the hydraulic motor and other hydraulic equipment on the gasification system were controlled by computer.

Because this was a field assembly, the designers chose to use 2^1/$_2''$ Schedule 40 pipe for the framework that supports the conveyor, as shown in Fig. 2.1.3. This pipe is readily available and is suitable for welded construction.

Technical Documentation

Complete documentation of the tire conveyor includes the following items:

- reference assembly drawing
- upper weldment drawing with sectional views and details
- lower weldment drawing with sectional views and details
- shaft detail
- drive chain detail
- cut sheet to show the dimensions to which flat (plate) and bar stock should be cut

In the following sections, you will recreate the drawings as they were determined by drafters and field personnel during construction of the tire conveyor. Keep in mind as you work through this project that the drafter interacted on a daily basis with others at the construction site. The drafter had to stay up-to-date on all changes as they were made in the field.

Drawing Files

In Project 1, you created both of the drawings for the nozzle assembly in a single drawing file. In Project 2, you will take a different approach because of the large number of drawing sheets involved and because we will use a different method of creating the parts lists. Each drawing will be created in a different drawing file, and the drawing files will be numbered in sequence for easy reference.

Drawing Numbers

Drafters generally assign a drawing number to each individual drawing in a set of working drawings. When a single drawing file is used, as was the case in Project 1, the drawing numbers are often used to name the layout tabs. In projects such as the tire conveyor, which incorporates several drawing files, the AutoCAD files themselves may be named according to the drawing numbers.

This practice makes it easier to refer to a specific drawing within the entire set of working drawings. For example, the parts lists in this project will refer to drawing numbers instead of actual dimensions or part numbers for complex details that are described on separate drawings. Instead of having to search through the drawings for the shaft detail, anyone reading the drawings simply needs to find the drawing numbered 0023 for more information about the shaft. The drawing number for each drawing must be recorded as part of the title block information.

It is important to understand that the drawing number is not the same as the sheet number. Within a set of working drawings, it is common to have drawings that occupy more than one drawing sheet. For example, the weldment drawings for the tire conveyor require two sheets each. The numbers we will use in Project 2 and the original drawing numbers from Emery Energy Company, LLC, are shown in Fig. 2.1.4. Notice that the drawing number is the same for each sheet of a multiple-sheet drawing, and the sheet number is appended to the drawing number for each additional sheet required. If the upper weldment had required a third sheet, its sheet number would have been 00052-3.

Emery Drawing No.	Project 2 Drawing No.	Description
00050	0020	REFERENCE ASSEMBLY DRAWING
00051	0021	LOWER WELDMENT DRAWING (SHEET 1)
00051-2	0021-2	LOWER WELDMENT DRAWING (SHEET 2)
00052	0022	UPPER WELDMENT DRAWING (SHEET 1)
00052-2	0022-2	UPPER WELDMENT DRAWING (SHEET 2)
00053	0023	SHAFT DETAIL
00054	0024	DRIVE CHAIN DETAIL
00055	0025	CUT SHEET

Fig. 2.1.4 *Drawing numbers for the tire conveyor drawing in Project 2.*

Practice Exercise 2.1A

In this section, you will create all of the drawing files needed for Project 2. As you work on the drawings throughout this project, you will use AutoCAD's COPYCLIP and PASTECLIP commands to copy AutoCAD geometry back and forth among the drawing files. COPYCLIP allows you to copy information from an AutoCAD file to the Windows Clipboard, and PASTECLIP allows you to paste the contents of the Clipboard into another AutoCAD file.

1. Create a new drawing file using AutoCAD's **ANSI D - Named Plot Styles.dwt** drawing template.

   It is not necessary to calculate the drawing limits for all of the drawings at this time. For now, you can set the limits for all of the drawings to the size needed for one of the largest drawings. Later, you can adjust the limits of individual files if necessary.

2. Enter the **LIMITS** command. Leave the lower limit at 0.0000,0.0000. Set the upper limit to **340,220**. Then **ZOOM All**.

3. Use the **UNITS** command to set up fractional inches with a precision of **0 1/16**.

4. Set up the same layers you used in Project 1. To do this quickly, open **DesignCenter** and find your **PROJECT 1.dwg** file. Single-click the file name to see a list of transferrable items in the view window. Click **Layers** to see a list of the layers in PROJECT 1.dwg. Highlight all of the layers and drag them into the drawing area to add them to your current drawing file.

5. Follow the procedure in step 4 to add the text styles and linetypes you set up for Project 1. You will create new dimension styles later.

6. Save the file as **PROJECT 2-0020.dwg**.

7. Close the file and exit AutoCAD.

8. In Windows, navigate to the folder in which you store your AutoCAD files. Highlight the **PROJECT 2-0020.dwg** file and right-click. Select **Copy** from the menu. Deselect the drawing file. Then right-click again and choose **Paste**. A copy of the file appears in the list of files. Highlight the copy and pick the name to edit the name of the file. Rename it **PROJECT 2-0021.dwg**.

Review Questions

1. What is gasification?
2. Briefly describe the purpose of the tire gasification system for which the tire conveyor in this project was constructed.
3. How does the role of the drafter at a field assembly project differ from his or her traditional role?
4. Other than the supporting framework, what four main parts are needed for the tire conveyor?
5. Why was 2½″ Schedule 40 pipe chosen for the supporting framework?
6. Why was the hydraulic motor attached to the top of the tire conveyor rather than at the bottom?
7. Briefly describe the procedure for using AutoCAD's DesignCenter to transfer complex settings such as layers, linetypes, and text styles from one drawing to another.

Practice Problems

1. Create a new AutoCAD template file that you can use for other mechanical drawings to be displayed on ANSI D-size sheets. Begin by creating a new drawing based on AutoCAD's ANSI D - Named Plot Styles.dwt template. Then use DesignCenter to add layers, linetypes, and other settings that you think would be useful. Finally, save the drawing as a template file named ANSI D with Settings.dwt.
2. Emery's gasification process has now been further refined to include biomass as one of the sources of gasification. Research further background material about gasification in general and tire and biomass gasification in particular. Prepare a PowerPoint® or other electronic presentation to show the basic gasification process. Include a description of the necessary inputs, any specific materials or processes needed, and the types of gases released as products. Are there any waste products?
3. Investigate commercially available conveyors. What types are generally available? Why wouldn't these types work for the tire conveyor in this project?

Portfolio Project

Biomass Conveyor

The gasification system for which you developed the tire conveyor in Project 2 played an important part in research by Emery Energy Company, LLC, into fuel production from recyclable materials. They discovered that tires alone were not an efficient source of energy, so they are now working with combinations of tires and various types of biomass.

For this portfolio project, you will design and create working drawings for a conveyor that feeds various types of biomass into a geodesic reactor. You may assume that the reactor is the same four-story, egg-shaped structure that was used in the tire gasification prototype.

1. To begin, research biomass. What is it, and what forms can it take? Have researchers yet discovered what forms of biomass are the most efficient fuel producers?
2. Design a conveyor that will transport as many of the feasible forms of biomass as possible to the feeder at the top of the geodesic reactor. Note that you are not limited to the structural framework used in the tire conveyor, but you should design the conveyor to be constructed of readily available materials.
3. Sketch out your design ideas and submit them to your instructor for approval.
4. Upon approval, plan the necessary drawings and set up a separate AutoCAD file for each drawing to be created. Name the drawings **PORTFOLIO 2-0010.dwg**, **PORTFOLIO 2-0011.dwg**, etc.

The Preliminary Views

In a field assembly project, the drafter usually doesn't have enough information at the beginning of the project to complete any one drawing in a set of working drawings. Instead, he or she works on several drawings concurrently, completing information in each drawing as it becomes available in the field. In this project, you will begin by creating preliminary views for the reference assembly drawing because variations of these views will be needed in several of the drawings. Instead of recreating the views for each drawing, you can copy these preliminary views and use them as a basis for each drawing in which normal views are needed.

Fractional inches are used instead of decimal inches in this project. Fractional-inch drawings are common in field assembly projects in the United States because many of the standard parts are specified using fractional inches. Although ASME prefers the use of decimal inches, it does allow the use of fractional inches, and you are likely to encounter this practice in the field.

Section Objectives
- Develop preliminary normal views of the tire conveyor.
- Use alignment paths to transfer points and distances among the normal views.
- Estimate dimensions not provided by manufacturers of standard parts.

Key Terms
- alignment path
- normal views
- object snap tracking

Required Views

The purpose of the reference assembly drawing in a field assembly project is to identify all of the parts and show them in their correct positions in the assembly. The tire conveyor is a long, narrow structure that includes relatively small parts at both ends. It cannot be shown in a single view with one detail, as was the case in Project 1. Four views are needed, as well as a section and two details.

Normal Views

Working drawings provide a way to describe a three-dimensional (3D) object completely on two-dimensional (2D) paper. Most 3D objects require views of more than one side of the object to accomplish this complete description. The most common views are the top, front, and right-side

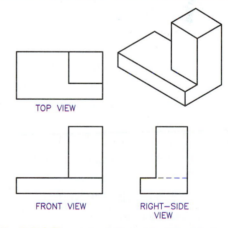

Fig. 2.2.1 *The normal views of a 3D object.*

views. These are known as the **normal views** of an object. See Fig. 2.2.1. All three of the normal views will be required for the reference assembly drawing of the tire conveyor.

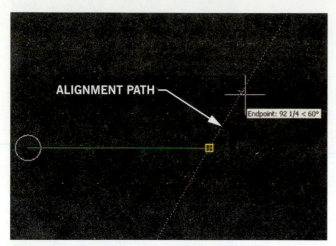

Fig. 2.2.2 *An alignment path from the right endpoint of the line with object snap tracking set to track polar angles at 15° intervals.*

Other Required Views

Because the detailing differs at the two ends of the tire conveyor, a left-side view will be needed as well as the right-side view. Also, neither side view shows clearly the infrastructure of the conveyor framework. A section taken through the main part of the framework is necessary to identify its various parts. Finally, two of the parts—the upper and lower wear pads—require details to show their exact construction.

Object Snap Tracking

In the United States, the normal views of a part are set up so that the top view is above and aligned with the front view. The side view is located to the side and aligned with the front view. Because the views must be aligned precisely, it is reasonable to use construction lines (XLINE) to transfer points from one view to another. For example, after establishing the length of a part in the front view, you could create vertical construction lines at the ends of the front view to establish the endpoints in the top view.

However, AutoCAD's AutoTrack™ feature provides **object snap tracking**—a less cumbersome way to transfer some distances and points. AutoTrack works with running object snaps to set up alignment paths from the active object snaps. An **alignment path** is a temporary line that appears to help you place new geometry at a specified angle and distance

from an acquired object snap. See Fig. 2.2.2. In this project, you will use alignment paths to align the normal views properly. As you practice using them, you will find this to be an efficient method of aligning views and transferring points.

To acquire an object snap at a specific point, simply move the cursor over that point. For example, to acquire the endpoint of a line, make sure the Endpoint object snap is active. Run the cursor over the endpoint you want to acquire. Then, as you move the cursor around the drawing, various alignment paths appear, along with the current distance from the acquired point and the angle of the line. *Note:* Running the cursor over the same object snap a second time unacquires the point.

By default, object snap tracking produces alignment paths at 0°, 90°, 180°, and 270° angles, producing vertical or horizontal tracking only. However, you can set it to track at other polar angles and intervals. To set up object snap tracking, use the Polar Tracking tab of the Drafting Settings window (DSETTINGS). See Fig. 2.2.3. Notice that, in addition to the angles available in the Increment Angle dropdown box, you can add other angles that meet your needs for a specific drawing. To do this, check the Additional angles box and pick New. Add the angle in the white display box. To delete custom angles, highlight the angle and pick Delete.

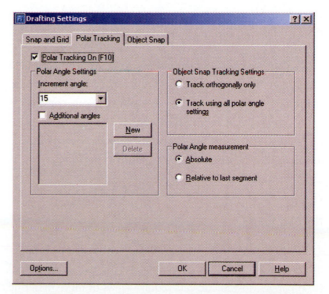

Fig. 2.2.3 *The Polar Tracking tab of the Drafting Settings window allows you to adjust the angles and increments of alignment paths.*

Fig. 2.2.4 **Overall working dimensions of the tire conveyor.**

Project File Backups

Because you will be working on more than one drawing concurrently, some of the sections in Project 2 encompass multiple drawings or views. Not all of the drawings or views will be completed within a given section. You may sometimes need to go back to drawings done in a previous section to make adjustments. The reference assembly drawing will be drawn on an ongoing basis throughout the project.

Because you will be working in several different drawing files at once, it is very important to back up your work often. In fact, you may want to keep a series of backup files in case you accidentally work on the wrong view at the wrong time. One good strategy is to use the SAVEAS command to save the file periodically. For each new file, append to the file name the date and time of you saved the file. For example, you might have files named:

- PROJECT 2-0020 8-6-03 10AM.dwg
- PROJECT 2-0020 8-7-03 10:30AM.dwg
- PROJECT 2-0020 8-7-03 11:30AM.dwg

This method makes it easy to see at a glance which is the latest file. Also, keeping older versions of a drawing in progress gives you additional flexibility. If you decide to retrieve something that you have since erased, you can copy it from the older version of the file.

Store your backup files in a safe place in case you need them. You may want to delete files older than one or two weeks to make space for newer backups. Note: When using this approach, it is extremely important to be certain the correct drawing file is active each time you resume work on a Practice Exercise.

Construct the Preliminary Views

In Practice Exercise 2.2A, you will begin work on the front, top, and left-side views for the reference assembly drawing. The overall dimensions for the front and top views are shown in Fig. 2.2.4. Refer to Fig. 2.2.4 as necessary as you complete Practice Exercise 2.2A.

As stated earlier, the reference assembly drawing will need a top view, front view, and both side views. The right-side view will show the lower end of the conveyor, and the left-side view will show the upper end. Until the lower weldment is developed, you will not have enough information to create the right-side view. For now, concentrate on the front, top, and left-side views.

1. Open the **PROJECT 2-0020.dwg** file that you created in Section 1. Make **Object** the current layer, and **ZOOM All**.

Begin by drawing a circle to represent the upper (left) sprocket in the front view. The sprockets chosen to drive the chain for this conveyor have a diameter of Ø18 1/16″.

2. In the lower left part of the drawing area, draw a **Ø18 1/16″** circle to represent the outside diameter of the upper sprocket, as shown in Fig. 2.2.4. *Hint:* To enter a fractional value in AutoCAD, insert a hyphen between the whole number and the numerator of the fraction, as follows: 18-1/16. Add the vertical and horizontal centerlines for the sprocket.

Recall that the tire conveyor must transport tires to the top of a four-story reactor assembly. Although the chain will be fitted with hooks to help hold the tires in place, the risk of accident due to falling tires must be minimized. Engineers have calculated that the sprocket-to-sprocket length of the drive chain must be 518 3/4″ to transport the tires at an angle at which the tires will not fall off.

3. Offset the vertical centerline of the upper sprocket **518 3/4″** to the right to determine the center-to-center distance of the sprockets.

4. Use grips to extend the horizontal centerline of the upper sprocket to the right so that it passes through the lower (right) sprocket, as shown in Fig. 2.2.5.

5. Enter the **LTSCALE** command and set the linetype scale to **10** so that the center linetype is visible on the drawing.

6. Copy the upper sprocket, using its center point as the base point. Place the copy at the intersection of the horizontal and vertical centerlines for the lower sprocket.

7. Note from Fig. 2.2.4 the distance from the center of the sprocket to the assembly split line. Offset the vertical centerline of the upper sprocket to the right by this distance to establish the weldment split line, as shown in Fig. 2.2.6.

Fig. 2.2.5 Centerlines in place for the lower (right) sprocket.

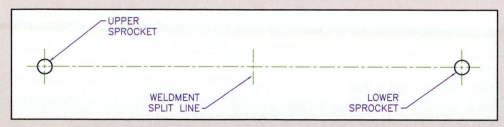

Fig. 2.2.6 Add the lower sprocket and the weldment split line.

Now you can add the 2 1/2″ Schedule 40 pipe that forms the framework of the conveyor. The framework consists of four long pieces of pipe, including two upper and two lower pieces. One upper and one lower piece are visible in the front view.

8. Offset the centerline of the sprocket up to create the centerline of the upper pipe.

9. Offset the centerline of the upper pipe to create the centerline of the lower pipe. Refer to Fig. 2.2.4 for offset distances.

10. Check a pipe schedule to find the outer diameter of the Schedule 40 pipe. Offset the centerlines to create the walls of the upper and lower pipes in the front view. Move the pipe walls to the **Object** layer.

11. Establish the left end of the upper and lower pipes by offsetting the vertical centerline of the sprocket by the distances shown in Fig. 2.2.4. Trim the horizontal pipe lines to position the left ends accurately. Move the offset lines to the object layer and trim them to form the left ends of the pipes.

12. To establish the right end of the upper pipe, off-set the left end to the right. The length of the upper pipe is **520″**. Zoom in on the lower sprocket and trim the horizontal pipe lines to the right end line.

The right end of the lower pipe is flush with the right end of the upper pipe. It is most easily added using object snap tracking. Before you can use this feature, you must set it up for the current drawing purpose, as described in step 13. This setup is a one-time procedure, unless you need to change the settings later for a different drawing purpose.

13. Enter the **DSETTINGS** command and select the **Object Snap** tab. Set up running object snaps for the object snaps you use most. Suggested running object snaps include Endpoint, Midpoint, Center, Quadrant, Intersection, and Perpendicular. Then switch to the **Polar Tracking** tab to set up object snap tracking. Pick the box next to **Polar Tracking On** and set the increment angle to

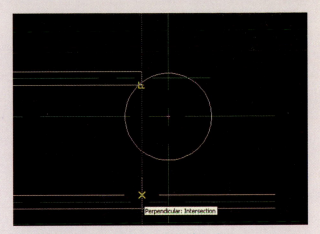

Fig. 2.2.7 Acquire the lower endpoint of the upper pipe and use the vertical alignment path to start the new line perpendicular to the top line of the lower pipe.

45. On the right side of the box, pick **Track using all polar angle settings**. For polar angle measurement, select **Absolute**. Pick **OK** to save these settings and return to the drawing area.

14. Enter the **LINE** command. Acquire the lower endpoint of the right end of the upper pipe line by running the cursor over it. Then move the cursor straight down, following the alignment path that appears. When the cursor passes over the lower pipe line, notice that AutoCAD displays the perpendicular object snap symbol. Position the cursor so that the perpendicular object snap symbol appears on the upper line of the lower pipe, as shown in Fig. 2.2.7. Click to select the first endpoint of the line. Then move the cursor straight down again until the perpendicular symbol appears on the lower line and click again to end the line. End the LINE command.

15. Trim the lower pipe lines to the right end line.

Now begin the top view of the conveyor. Remember that it must be aligned exactly with the front view.

16. Zoom out slightly and position the front view near the bottom of the drawing area. Offset the centerline of the sprocket upward by 128″ to establish the centerline of the top view.

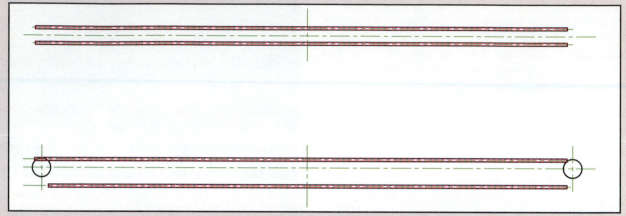

Fig. 2.2.8 Preliminary front and top views of the conveyor.

17. Create the two 2 1/2″ Schedule 40 pipes and their centerlines in the top view. The easiest method is to copy them from the front view. Refer again to Fig. 2.2.4 for their exact locations.

18. Add the end lines to complete the pipes in the top view. Since the pipe length in the top view is the same as the length of the upper pipe in the front view, use object snap tracking to place the end lines accurately. Follow the alignment paths up vertically from the endpoints in the front view.

19. Create the assembly split line in the top view. Use object snap tracking to align it accurately with the split line in the front view.

20. Referring to a pipe schedule, add the inner pipe diameter for all of the pipes in the front and top views. Place these lines on the **Hidden** layer, as shown in Fig. 2.2.8.

Now begin the left-side view to show the upper end of the conveyor. In this view, all four of the pipes that make up the framework are shown in their end view. Refer to Fig. 2.2.9.

21. Offset the vertical sprocket centerline in the front view **120″** to the left to create the vertical centerline for the left-side view. Then create the horizontal centerline. Use object snap tracking to line it up exactly with the horizontal sprocket centerline in the front view. If these centerlines are not aligned, the views will not align properly.

22. Using the dimensions in Fig. 2.2.9, create the circles to represent the inner and outer diameters of the four pipes in the end view and add their centerlines. The centers of the pipes should be exactly in line with those in the front view. *Hint:* First offset the sprocket centerlines to establish the center points for all of the circles. Then create one set of circles and copy it to the other three locations. Trim the centerlines.

To drive the main roller chain, the engineers have decided to use a #120R36 steel sprocket with split taper bushings manufactured by Browning/Emerson Power Transmission. So far on your drawing, the sprocket is represented by the **Ø18 1/16″** circle you created in the front view. To finish the front view representation and add the top and side views, you need more information about the sprocket.

In this project, as in the real world, you will need to look up the information using the manufacturer's data sheets. Appendix C contains pages from the Browning catalog that give technical specifications for Browning roller chain sprockets.

23. Look up sprocket #120R36 on the appropriate catalog page in Appendix C. Use the dimensions given in the table to complete the front view of the sprocket. Do not show the individual teeth. Instead, use a centerline to show the pitch diameter. Use a bore diameter of **Ø2 3/16″**.

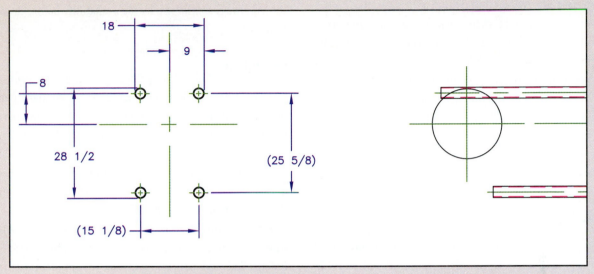

Fig. 2.2.9 *Placement of the pipes for the left-side view.*

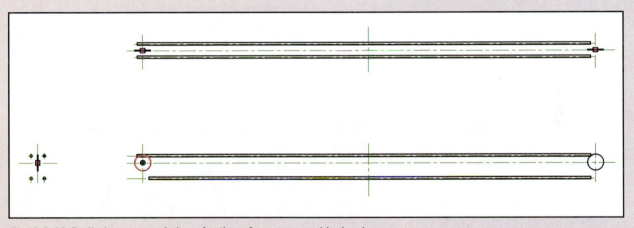

Fig. 2.2.10 *Preliminary normal views for the reference assembly drawing.*

24. Complete the left-side view of the #120R36 sprocket. Transfer the appropriate points and distances using construction lines or object snap tracking. Refer to Appendix C for dimensions as necessary.

25. Complete the top view of the #120R36 sprocket. Transfer the appropriate points and dis-

tances using construction lines or object snap tracking. Refer to Appendix C for dimensions as necessary. When you finish, the three views of the #120 sprocket should look similar to those in Fig. 2.2.10.

26. When you finish creating the sprocket in all three views, be sure to save the drawing file.

Drawing Practices for Standard Parts

When standard parts such as sprockets, roller bearings, and fasteners are incorporated into a project, the drafter must represent them in their correct positions and sizes. Technical data sheets such as those reproduced in Appendix C can provide most of the information needed to draw the parts. However, data sheets generally only include the dimensions that are critical to the intended use of the part; they are not intended to describe the part fully.

The unavailability of complete dimensions for standard parts is a fairly common problem for drafters. They solve the problem in one of two ways.

Part File Downloads

Many manufacturers now allow drafters to download precise CAD files of their standard parts for use in working drawings. The drafter accesses the manufacturer's Web site and chooses the necessary part. The file is downloaded in a common file format such as DWG or DXF. The drafter can then insert the entire file into the working drawings as a block reference.

When it is available, this is the preferred method of illustrating standard parts. In addition to ensuring accuracy, it can save the drafter an enormous amount of time. You should always check for part file availability before drawing a standard part.

Estimation

Unfortunately, some manufacturers of standard parts do not offer this option. However, because standard parts are purchased, there is really no need for the drafter to reproduce every detail accurately on working drawings. Those dimensions that affect the assembly must be correct. Nonessential dimensions that are not provided on the manufacturer's data sheets are often estimated.

Although this method is not as precise as using CAD files from the manufacturer, it is acceptable drafting practice. In Practice Exercise 2.2B, you will have an opportunity to practice estimating dimensions as you continue to work on the upper weldment drawings. Keep in mind that it is necessary to draw the important dimensions accurately. In general, if the dimension is stated on the data sheet, you should present it accurately on the working drawings.

Practice Exercise 2.2B

In this exercise, you will continue working on the three-view drawing you created in Practice Exercise 2.2A. The next step is to add the shaft and bearings that support the sprocket.

Specifically, the sprocket is supported by a 2 3/16″ shaft and two 2 3/16″ tapered roller bearing units with two-bolt bases. The bearings the engineers specified are Browning part number PBE920 pillow blocks.

1. Zoom in on the front view of the sprocket.
2. Obtain the specifications for the bearings from the Browning data sheet in Appendix C. Add the bearing to the front view of the detail drawing, as shown in Fig. 2.2.11. Estimate all dimensions

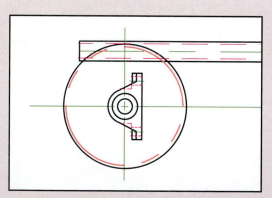

Fig. 2.2.11 *Front view of the roller bearing.*

that are not stated on the data sheet. For the bolt hole distance, choose a distance that falls between the stated minimum and maximum values. *Note:* You may need to erase some of the existing sprocket geometry to perform this step.

3. Use **XLINE** to create horizontal construction lines at the top and bottom quadrants of the bore in the front view of the bearing to locate the shaft in the left-side view. Transfer all other key points from the front view to the left-side view using construction lines or object snap tracking. Complete the bearings in the left-side view, as shown in Fig. 2.2.12. Estimate any dimensions that are not provided by the data sheet or Fig. 2.2.12.

4. The shaft is 31″ long and begins flush with the left bearing. Trim the left side of the construction lines that represent the shaft to the leftmost lines of the left bearing, as shown in Fig. 2.2.12.

5. Offset the leftmost line of the left bearing **31″** to the right to establish the other end of the shaft. Trim away the construction lines to the right of the offset line. Then trim the offset line as necessary to form the right end of the shaft.

6. Break the shaft lines at the edges of both bearings and the sprocket. Move the portions of the shaft lines that are hidden by the bearings and sprocket to the **Hidden** layer.

7. Zoom in so that the front and top views of the detail are clearly visible.

8. Transfer key points and distances on the bearings and shaft from the front view to the top view.

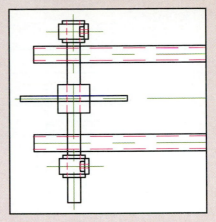

Fig. 2.2.13 Bearings and shaft in the top view.

9. Complete the top view of the bearings and shaft, as shown in Fig. 2.2.13. Estimate any dimensions that are not available.

To drive the #120 roller chain and its sprocket, an approximately 18″ #60 sprocket is needed at the end of the shaft. The shaft is driven by a smaller #60 sprocket, which is attached to a hydraulic motor.

10. Zoom in on the front view of the detail.

11. The engineers have chosen Browning part number 60R17 for the larger #60 sprocket. Refer to the data sheet for #60 roller chain sprockets to find the exact dimensions and add it in the front view. *Note:* The pitch circles for the #60 sprocket and the #120 sprocket are so close that they appear to overlap. For clarity, you may delete the pitch circle for the #120 sprocket because it lies behind the #60 sprocket.

12. Offset the centerline of the shaft by **28 1/2** to the right to locate the centerline of the smaller sprocket.

13. Browning part number 60P13 will be used for the smaller sprocket. Refer to the data sheet for #60 roller chain for dimensions. Draw the front view of this sprocket on the vertical centerline you created in step 12. See Fig. 2.2.14. Align the horizontal centerline with the shaft centerline. Use a bore diameter of Ø1.00. Estimate any dimensions that are not given by the data sheet or Fig. 2.2.14.

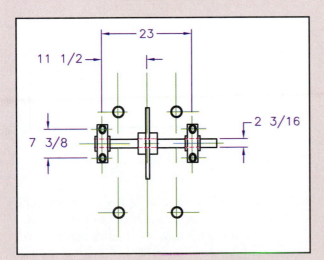

Fig. 2.2.12 Bearings and shaft in the left-side view.

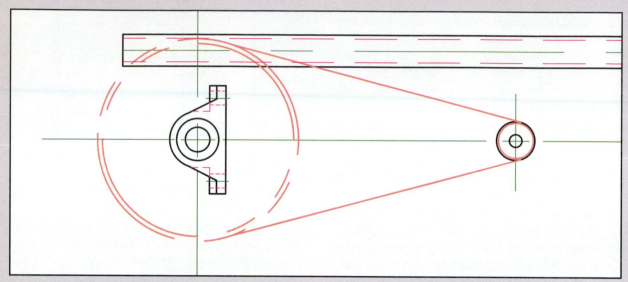

Fig. 2.2.14 *Front view of the #60 sprockets and chain.*

14. Add the roller chain that runs between the two #60 sprockets, as shown in Fig. 2.2.14. Represent the chain using tangent lines that run from the O.D. of the large sprocket to the O.D. of the small sprocket. Place the chain on the **Phantom** layer.

15. Zoom in on the left-side view of the detail. In this view, the smaller #60 sprocket is hidden by the larger #60 sprocket, so the smaller sprocket will not be drawn.

16. Add the larger #60 sprocket in the left-side view. Center the sprocket's teeth **16 7/16"** to the right of the center of the #120 sprocket, as shown in Fig. 2.2.15. Refer to the #60 data sheet for necessary dimensions.

17. Break the shaft lines at the #60 sprocket and move the portion of the shaft lines hidden by the sprocket to the **Hidden** layer.

18. Add a representation of the two drive chains in the left-side view. Use a phantom linetype. Place it .50" outside the sprocket lines in each case, as shown in Fig. 2.2.15.

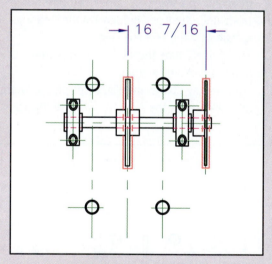

Fig. 2.2.15 *Place the larger #60 sprocket in the left-side view as shown here.*

19. Zoom in on the top view of the detail and add both #60 sprockets, as shown in Fig. 2.2.16. Refer to the #60 data sheet as necessary. Add both drive chains in this view also.

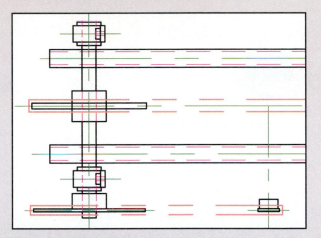

Fig. 2.2.16 *Placement of the #60 sprockets in the top view.*

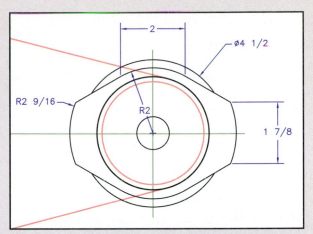

Fig. 2.2.17 *In the front view, the hydraulic drive motor appears as shown behind the smaller #60 sprocket.*

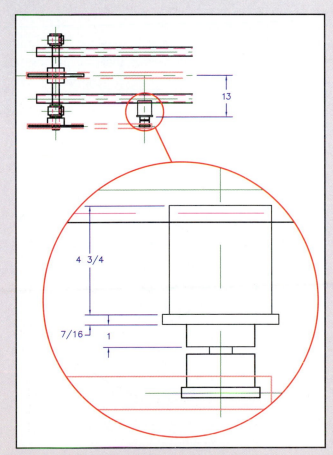

Fig. 2.2.18 *Placement of the hydraulic motor in the top view.*

The conveyor will be powered by a hydraulic drive motor. The motor will be a purchased part, so you only need to draw a basic representation of it for your set of working drawings.

20. Add the hydraulic motor to the front view using the dimensions shown in Fig. 2.2.17. One of the simplest ways to construct the face of the motor is to create circles with the radii shown. Then offset the horizontal and vertical centerlines as shown by the linear dimensions to find points of tangency. Join the points of tangency using straight lines. Trim the circles to the lines to finish the face of the motor. Add the **Ø4 1/2″** body of the motor by creating a circle and then trimming it to the face of the motor.

21. Add the hydraulic motor to the top view of the drawing using the dimensions shown in Fig. 2.2.18. Place the mounting face of the motor **13″** from the centerline of the main drive chain. Take the remaining dimensions from the front view.

Review Questions

1. Which views of an object are included in the normal views?
2. In the tire conveyor, what factor most influences the length of the main drive chain? Why?
3. In AutoCAD, what is an alignment path?
4. What AutoCAD command is used to set up object snap tracking?
5. With what other AutoCAD feature does object snap tracking work?
6. Why are frequent file backups very important when you work on a set of field-developed working drawings?
7. What two methods are commonly used to represent standard parts in a drawing file?
8. Which of the two methods from question 7 is preferred, and why?

Practice Problems

In problems 1 through 4, draw the normal views of the objects shown. Use object snap tracking to align the views correctly. Dimension the drawings.

Portfolio Project

Biomass Conveyor (continued)

Begin working on the geometry for the conveyor you have planned. Due to the nature of this assignment, the set of drawings you have planned will most likely mirror the drawings for Project 2.

1. Determine which of the drawing files you created in Section 2.1 is the most logical file to work on first.
2. Begin the set of working drawings by creating preliminary normal views that you can use as the basis for the required drawings and views.
3. Save your work.

1.

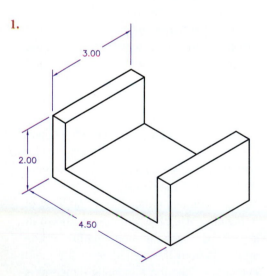

3.00

2.00

4.50

MATL: .5" CARBON STEEL

2.

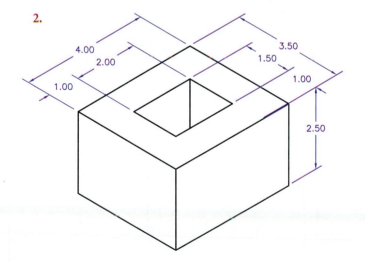

4.00
2.00
1.00
3.50
1.50
1.00
2.50

3.

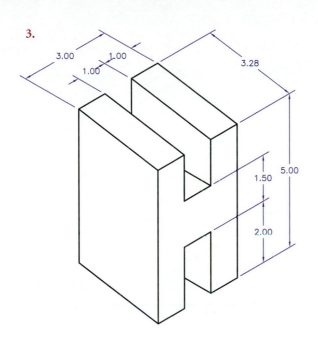

4.

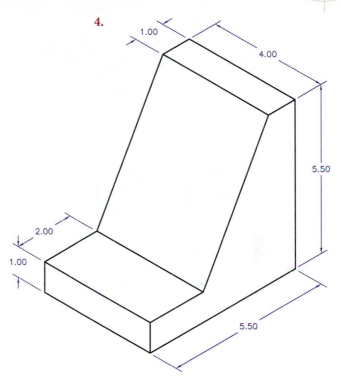

In problems 5 through 10, draw the normal views of the objects shown. Use object snap tracking to align the views. Estimate all dimensions not given.

5.

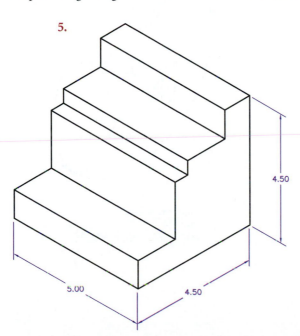

6.

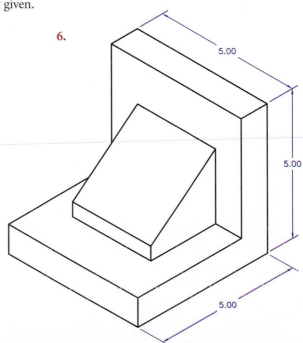

7.

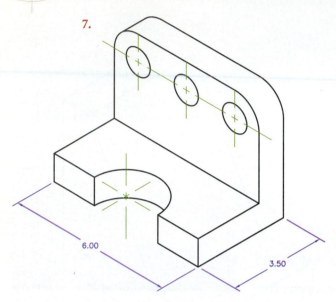

8.

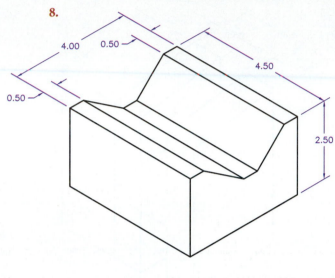

9.

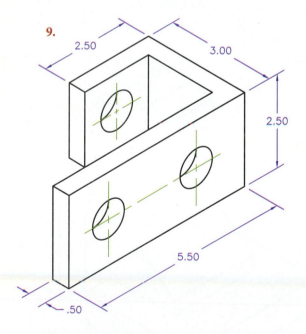

10.

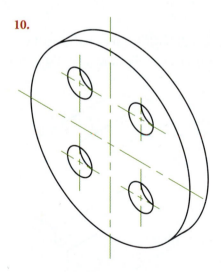

DIAMETER: ⌀4.00
DEPTH: .25
4X ⌀.375 BOLT HOLES

Upper Weldment Drawing

Because the tire conveyor is a long structure, the decision has been made to break it into two weldments for ease of handling. Drawings will be needed for both the upper weldment and the lower weldment. The assembly split line on the top and front views you created in Section 2.2 shows where the upper weldment ends and the lower weldment begins. In this section, you will work on the upper weldment drawing.

The geometry you created in Section 2.2 will be used later to finish developing the reference assembly drawing. In the meantime, you can copy the preliminary views to drawing 0022 and use them as a basis for sheet 1 of the upper weldment drawing. Then you will copy the geometry from drawing 0022 to form the basis for the geometry in drawing 0022-2, which is sheet 2 of the upper weldment drawing.

Section Objectives
- Build on the preliminary views to create drawings 0022 and 0022-2.
- Use object linking and embedding (OLE) to embed a parts list in an AutoCAD file.
- Edit an OLE object in AutoCAD.

Key Terms
- container application
- object linking and embedding (OLE)
- partial parts list
- server application
- visualization

Drawing Analysis

To show all of the information needed to construct the upper weldment, the senior drafter has determined that the following views are necessary:

- top view
- front view
- left-side view (to show the upper end of the conveyor framework)
- detail of the upper end of the upper weldment
- sectional detail A-A taken through the width of the framework to show the cross supports
- section B-B taken longitudinally at the top of the weldment to show the upper end of the lower chain support

These views can be arranged on two D-size sheets. With proper planning, you can place the top, front, and left-side views as well as section B-B on the first sheet. All of the detail views will be placed on the second sheet. This arrangement keeps all of the larger-scale detail views together on the second sheet.

Copy and Prepare the Preliminary Views

In Practice Exercise 2.3A, you will begin work on the upper weldment drawing. Upper weldment drawing sheet 1 is shown in Fig. 2.3.1, and sheet 2 is shown in Fig. 2.3.2. Refer to these drawings as you work through the steps in all of the practice exercises related to the upper weldment drawing.

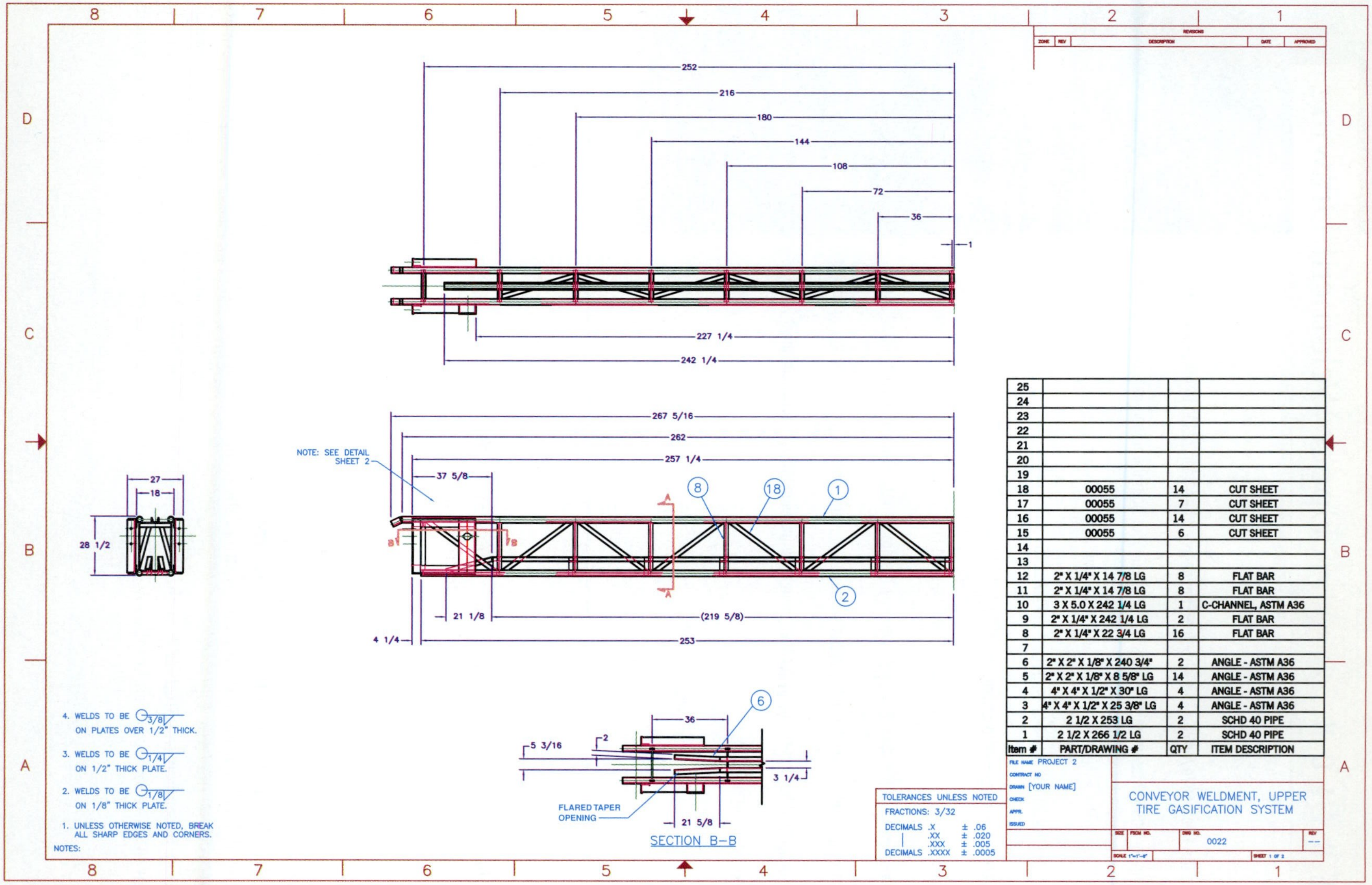

Item #	QTY	PART/DRAWING #	ITEM DESCRIPTION
25			
24			
23			
22			
21			
20			
19			
18	14	00055	CUT SHEET
17	7	00055	CUT SHEET
16	14	00055	CUT SHEET
15	6	00055	CUT SHEET
14			
13			
12	8	2" X 1/4" X 14 7/8 LG	FLAT BAR
11	8	2" X 1/4" X 14 7/8 LG	FLAT BAR
10	1	3 X 5.0 X 242 1/4 LG	C-CHANNEL, ASTM A36
9	2	2 X 1/4" X 242 1/4 LG	FLAT BAR
8	16	2 X 1/4" X 22 3/4 LG	FLAT BAR
7			
6	2	2" X 2" X 1/8" X 240 3/4"	ANGLE - ASTM A36
5	14	2" X 2" X 1/8" X 8 5/8" LG	ANGLE - ASTM A36
4	4	4" X 4 1/2" X 30" LG	ANGLE - ASTM A36
3	4	4" X 4 1/2" X 25 3/8" LG	ANGLE - ASTM A36
2	2	2 1/2 X 253 LG	SCHD 40 PIPE
1	2	2 1/2 X 266 1/2 LG	SCHD 40 PIPE

CONVEYOR WELDMENT, UPPER
TIRE GASIFICATION SYSTEM

DWG NO. 0022

SHEET 1 OF 2

FILE NAME: PROJECT 2
CONTRACT NO.
DRAWN: [YOUR NAME]
CHECK:
APPR.
ISSUED

TOLERANCES: UNLESS NOTED
FRACTIONS: 3/32
DECIMALS: X   ± .06
         .XX   ± .020
         .XXX   ± .005
DECIMALS .XXXX ± .0005

SECTION B—B

FLARED TAPER OPENING

NOTE: SEE DETAIL SHEET 2

NOTES:
1. UNLESS OTHERWISE NOTED, BREAK ALL SHARP EDGES AND CORNERS.
2. WELDS TO BE 1/8 ON 1/8" THICK PLATE.
3. WELDS TO BE 1/4 ON 1/2" THICK PLATE.
4. WELDS TO BE 3/8 ON PLATES OVER 1/2" THICK.

Fig. 2.3.1 Upper weldment drawing sheet 1.

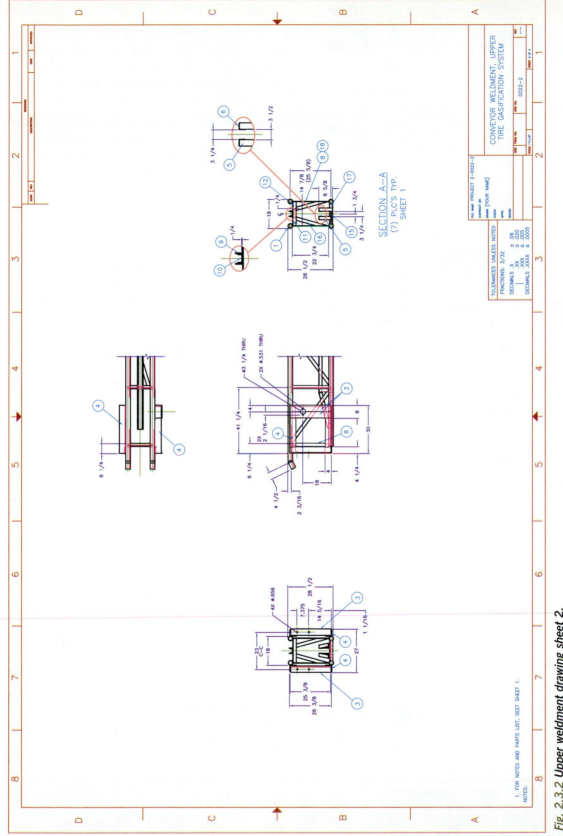

Fig. 2.3.2 Upper weldment drawing sheet 2.

Practice Exercise

2.3A

Turn your attention to the views on upper weldment drawing sheet 1 in Fig. 2.3.1. Notice that the front, top, and left-side views are similar to those you have already created for the reference assembly drawing. You can copy the existing views to create the basis for the upper weldment views.

1. Open both of the following drawing files:
   PROJECT 2-0020.dwg
   PROJECT 2-0022.dwg
   Display PROJECT 2-0020.dwg on the screen and enter **ZOOM All** to see the entire drawing area.

2. Enter **COPYCLIP** (or press **CTRL-C**) and use a crossing window to select all of the geometry you created for the reference assembly.

3. From the **Window** pull-down menu, select the **PROJECT 2-0022.dwg** file to display it on the screen. **ZOOM All**. Enter **PASTECLIP** (or press **CTRL-V**) to paste the geometry into this drawing file. Center the geometry on the drawing area.

4. Using the assembly split lines in the front and top views as cutting edges, trim away all of the geometry to the right of the split lines. Delete any remaining geometry to the right of the split lines. The expected result is shown in Fig. 2.3.3.

The purpose of the weldment drawings is to show the placement of the welds and the structural composition of the framework. For clarity, parts that are not needed to show this information are omitted or shown only by centerlines. Therefore, the next step in creating the upper weldment drawing is to delete the information that is needed only in the reference assembly drawing.

5. Erase the basic geometry of all of the sprockets in the top view. Leave only the centerlines representing the location of the sprockets and roller chains.

6. Erase the drive chain representations in the top view. The drive chains will be represented in this drawing only by centerlines, so you will need to extend the centerline of the larger #60 sprocket so that it runs through both #60 sprockets, as shown in Fig. 2.3.4.

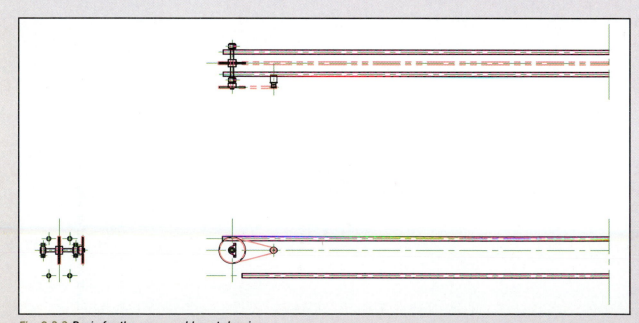

Fig. 2.3.3 Basis for the upper weldment drawing.

7. Erase the basic geometry and centerlines of the roller bearings in the top view. Do not erase the shaft centerlines.

8. Erase the basic geometry of the hydraulic drive motor in the top view. Leave the vertical centerline in place.

9. Again in the top view, erase the shaft on which the sprockets are mounted. Leave the centerline in place. The upper end of the top view should now look like the illustration in Fig. 2.3.4.

10. Zoom in on the front view of the upper weldment drawing. Delete all of the sprockets, but leave the circular phantom line for the #120 sprocket in place. Also leave the vertical and horizontal centerlines in place.

11. Erase the hydraulic motor in the front view, but leave the vertical centerline in place.

12. Erase the roller bearing in the front view, but leave the horizontal centerlines of the mounting holes in place. The upper end of the front view of the upper weldment drawing should now look as shown in Fig. 2.3.5.

13. Zoom in on the left-side view. Erase the sprockets and drive chains, but leave their centerlines in place.

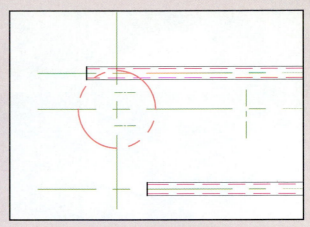

Fig. 2.3.5 **Upper end of the front view of the upper weldment drawing with unnecessary information deleted.**

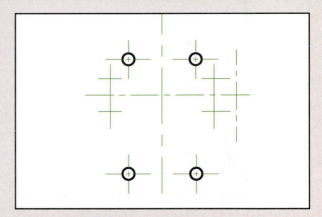

Fig. 2.3.6 **Left-side view of the upper weldment drawing with unnecessary information deleted.**

14. Erase the roller bearings, but leave their centerlines, as well as the centerlines for the bolt holes, in place.

15. Erase the shaft on which the larger sprockets are mounted, but leave the centerline in place. When you finish, the left-side view of the upper weldment drawing should look as shown in Fig. 2.3.6.

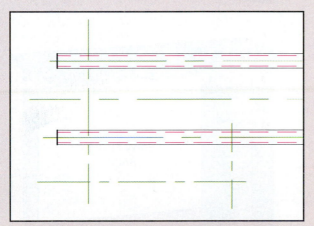

Fig. 2.3.4 **Upper end of the top view of the upper weldment drawing with unnecessary information deleted.**

Fig. 2.3.7 *Upper end of the upper weldment.*

important drafting skill and one you will practice in this project. To help you visualize and understand the spacial relationship of the upper end of the framework to the 2½″ pipes, two views of a 3D model are shown in Figs. 2.3.9 and 2.3.10. Study the models before you begin the next practice exercise.

Geometry for the Upper Weldment

Now that the copy of the normal views has been properly prepared, you can begin work on the weldment drawings. The upper weldment drawing is the simpler of the two, so you can draw it first as a pattern for the more complex lower weldment in the next section.

Upper End: Front View

Since the upper (left) end of the upper weldment is the most complex, we will tackle it first. A photograph of the top end of the finished prototype is shown in Fig. 2.3.7. In this photograph, the larger #60 sprocket has been removed so that you can see the rectangular frame at the top of the weldment more easily. For strength, the rectangular frame is constructed entirely of structural steel angles.

The structural framework that supports the bearings and motor will be constructed entirely of 4″ × 4″ × ½″ structural steel angles and 2″ × ¼″ flat bars. These will be cut to size and welded in the field.

In Fig. 2.3.8, each piece is shown in a different color and labeled to help you see how the pieces fit together. Note that the pieces labeled D and E are not solid. They are two pieces of angle placed side by side to form a hollow opening. Compare this drawing with the photograph in Fig. 2.3.7 as you visualize the pieces to be drawn. **Visualization**, or seeing the finished product in your "mind's eye," is an extremely

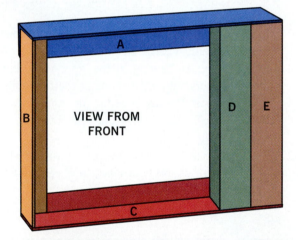

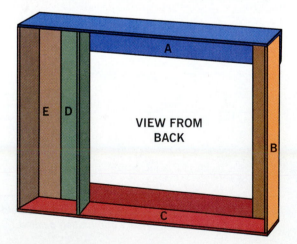

Fig. 2.3.8 *The structural steel angles that make up the rectangular frame shown in Fig. 2.3.7.*

In Practice Exercise 2.3B, you will add the rectangular framework at the top (left) end of the upper weldment to all three views of the upper weldment drawing. It is very important to understand the relationship of the pieces before you draw the framework. Refer to the photograph, models, and Figs. 2.3.1 and 2.3.2 as necessary as you complete the exercise.

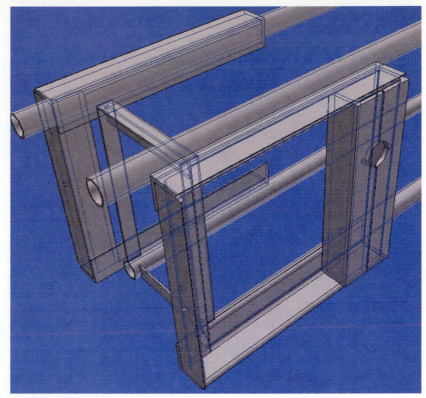

Fig. 2.3.9 A simplified 3D model showing the rectangular frames at the top of the upper weldment from a view similar to that in the photograph in Fig. 2.3.7.

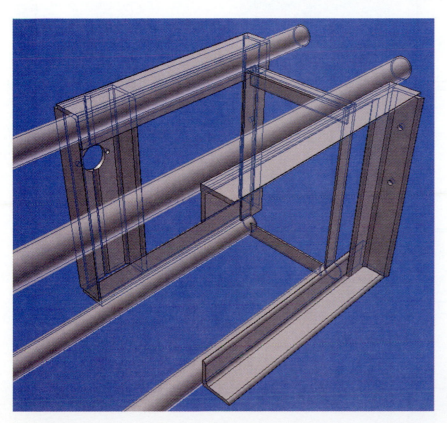

Fig. 2.3.10 A simplified 3D model showing the rectangular frames from the other side. Note the differences in the two frames.

Practice Exercise 2.3B

Begin by adding the rectangular frame in the front view of the upper weldment drawing. Refer to Fig. 2.3.11 as you perform the following steps.

1. Zoom in on the upper (left) part of the front view of the upper weldment drawing.
2. Add the 4" × 4" × 1/2" structural steel angles that form the rectangular frame. Begin by offsetting the vertical centerline of the larger sprocket to the right by **2 1/2"** to determine the location of the left side of the leftmost piece. Offset the sprocket's horizontal centerline up by **8"** to establish the top of the line. (This line coincides with the centerline of the upper pipe.) Then offset the same centerline down by **18"** to establish

the lower end of the line. Trim the vertical line to the two horizontal lines. Move all three of the offset lines to the **Object** layer.

3. Offset the vertical line from step 2 to the right by **30"** to establish the location of the outer edge of the rightmost structural steel angle. Trim the two horizontal lines from step 2 to the two vertical lines to complete the outer lines of the rectangular frame.
4. Offset all four of the lines by **1/2"** to the inside to create the thickness of the structural steel angles. Refer to Fig. 2.3.11 for the direction to offset each line. Trim and extend the lines as necessary to complete the outer framework.
5. Offset the rightmost vertical line by **4"** to the left to create the 4" leg of the structural steel angle. Then offset the new line by **4"** to the left to create the 4" leg of the inner piece.

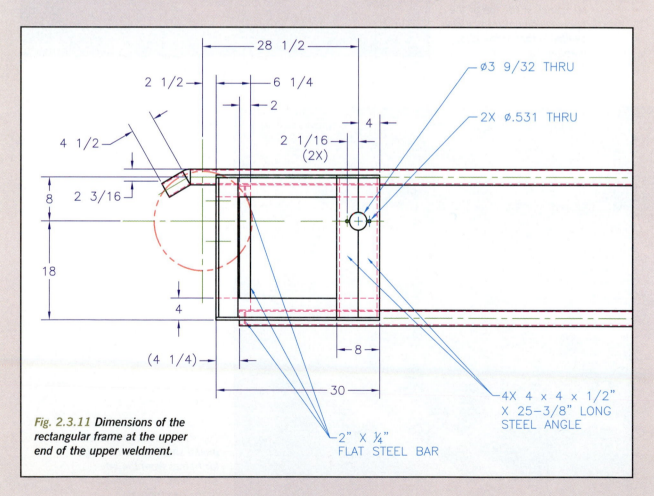

Fig. 2.3.11 Dimensions of the rectangular frame at the upper end of the upper weldment.

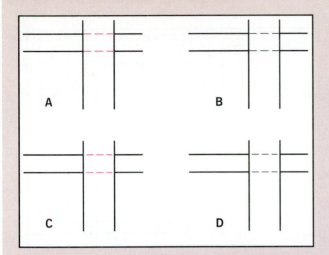

Fig. 2.3.12 Treatment of hidden lines. In parts A and B, the hidden line touches the horizontal continuous lines, making the drawing hard to interpret at a glance. Parts C and D show best-practice examples of the same intersection printed on a color printer and on a black-and-white printer, respectively.

Notice in the photograph and 3D models that the rectangular frames are located outside of the Schedule 40 pipes. This means that part of both the upper and lower pipes are hidden in the front view.

9. Break the pipe lines as necessary and move the portions that are hidden behind the rectangular frame to the **Hidden** layer.

Standard drafting practice dictates that when a hidden line is drawn end-to-end with a (continuous) object line, there should be a small gap between the two. This helps viewers see exactly where the object line ends and the hidden line begins. This may not seem important because lines set on your Hidden layer are a different color from those on the object layer. However, following this rule is especially important when the drawing will be printed on a black-and-white printer, because the color information is lost. See Fig. 2.3.12.

10. With Ortho on, use grips to move the endpoints of the hidden lines that lie end-to-end with continuous lines slightly away from the intersection of the two lines. See Fig. 2.3.13.

6. Offset the last line you created in step 5 to the right by 1/2″ to create the thickness of the structural steel angle. Trim it to the inner surface of the upper and lower pieces as shown in Fig. 2.3.11.

7. Offset the outside lines of the structural angles to create the 4″ legs that show in the interior of the rectangular frame. Refer again to Fig. 2.3.11 for distances. Break the inside lines as necessary and move hidden portions of the angles to the **Hidden** layer.

8. The inner surfaces of the two internal pieces (labeled D and E in Fig. 2.3.8) are hidden in the front view, so move both of the internal lines to the **Hidden** layer.

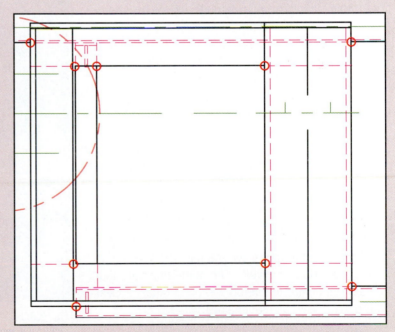

Fig. 2.3.13 Use grips to shorten the hidden lines slightly so that they do not touch the continuous lines at the places identified here with red circles.

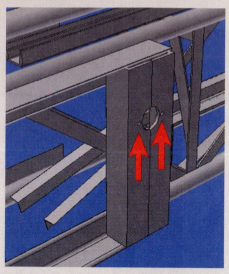

Fig. 2.3.14 *Holes must be created in these two angles for the hydraulic motor and #60 sprocket.*

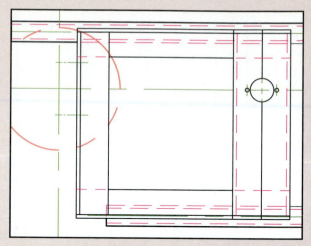

Fig. 2.3.15 *Front view of the rectangular frame.*

Next, you will create the holes for the hydraulic motor in pieces D and E. Refer to Fig. 2.3.14.

11. Create the large hole through which the motor protrudes with a diameter of **3 9/32**.

12. Add the bolt holes with vertical centerlines **2 1/16″** on either side of the centerline of the Ø3 9/32 hole. Their horizontal centerline is the same as that of the Ø3 9/32 hole. The diameter of the bolt holes should be **.531**. Add vertical centerlines to the bolt holes.

13. Trim the vertical line that represents the edges of parts D and E away from the Ø3 9/32 hole. See Fig. 2.3.15.

Notice in Fig. 2.3.16 that the four Schedule 40 pipes are connected at the upper end of the framework by a rectangular brace made of 2″ × 1/4″ flat steel bars. In the front view, the flat (2″) side of one of the steel bars is visible. The stabilizing bars that run at right angles to the top and bottom bars appear in their end view as hidden lines.

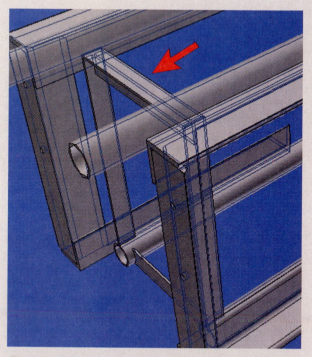

Fig. 2.3.16 *The four Schedule 40 pipes are connected and stabilized by a rectangular frame made of five flat steel bars.*

14. Offset the left side of the rectangular frame **6 1/4″** to the right to create the right edge of the 2″ steel bar. Then offset the new line **2″** to the left to locate the left edge of the bar. Trim both lines to the lower edge of the upper pipe and the upper edge of the lower pipe. See Fig. 2.3.17.

15. Break the lines you created in step 14 at the edges of the angle iron and move the hidden portions to the **Hidden** layer.

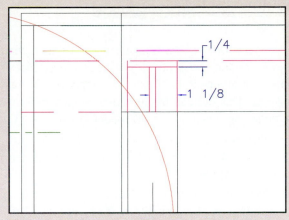

Fig. 2.3.19 **End view of the two flat bars at the top of the rectangular brace.**

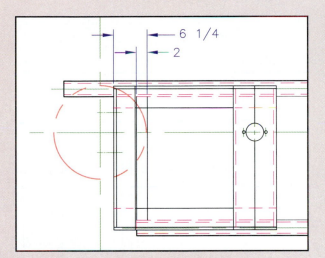

Fig. 2.3.17 **Placement of the flat steel bar.**

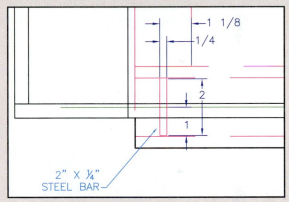

Fig. 2.3.20 **End view of the flat bar at the bottom of the rectangular brace.**

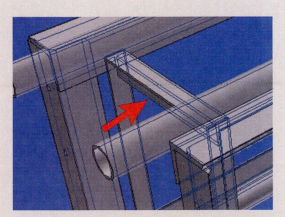

Fig. 2.3.18 **Two steel bars placed at right angles to each other form the top of the supporting structure that connects the Schedule 40 pipes.**

16. Zoom in on the upper end of the flat steel bar. As you can see in Fig. 2.3.18, two steel bars extend from one of the upper Schedule 40 pipes to the other top pipe. These two bars are at right angles, and the ends of both must be shown in the front view. Add the end views of these two bars, as shown in Fig. 2.3.19. Place them on the **Hidden** layer.

17. As you can see in Fig. 2.3.16, the flat steel bar at the bottom of the rectangular brace is positioned vertically on the pipes and is centered on the lower Schedule 40 pipes. Zoom in on the lower end of the rectangular brace in the front view and add the end view of this bottom steel bar. See Fig. 2.3.20.

The final task in the front view is to create the bend in the top two Schedule 40 pipes. These pipes will be cut and bent 4 1/2″ from the upper end of the conveyor. This cut will be filled and welded to form the angle, as shown in the 3D model in Fig. 2.3.21.

18. Offset the left end line of the upper Schedule 40 pipe by **4 1/2″** to the right. Break all five of the lines associated with the top pipes (the two object lines, two hidden lines, and the centerline) at their intersection with the offset line. See Fig. 2.3.22.

The pipe will be bent so that the top left edge of the bent piece is exactly 2 3/16″ below the top of the rest of the pipe. To achieve this without knowing the exact angle, you will need to place a temporary line 2 3/16″ below the top pipe line and then rotate the end of the pipe.

19. Offset the top line of the piece of pipe to be bent down by **2 3/16″**. With Ortho on, extend the left end of this temporary line to the left to form a reference line.

20. Enter the **ROTATE** command and select the left end line of the pipe and the five horizontal lines to be rotated. Do not select the reference line you created in step 19. Press **Enter** and then select the right end of the bottom line as the base point for rotation. Use the cursor to rotate the selected lines counterclockwise until the top left corner of the pipe touches the reference line, as shown in Fig. 2.3.23.

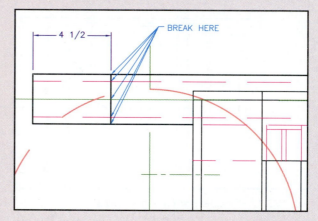

Fig. 2.3.22 *Position of the break in the upper Schedule 40 pipes.*

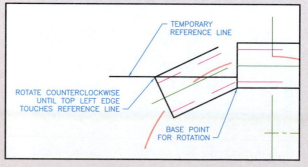

Fig. 2.3.23 *Rotate the end of the pipe as shown here.*

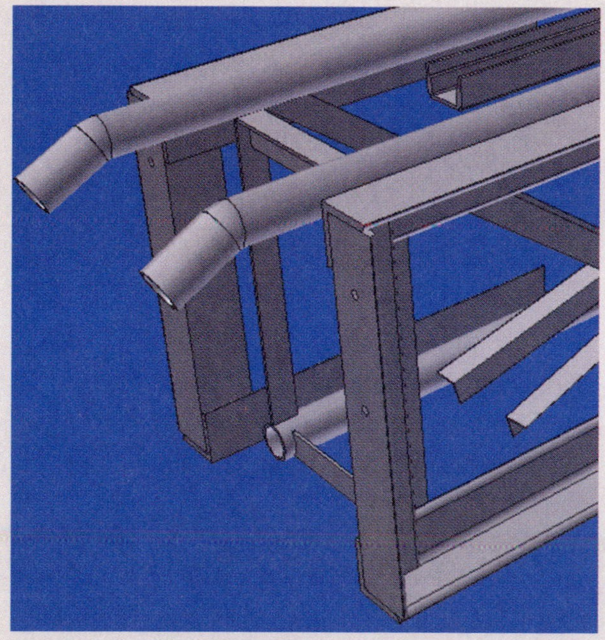

Fig. 2.3.21 *The upper pipes will be broken and filled to create the bends shown in this 3D model.*

21. Delete the temporary reference line.
22. Add the left end of the bent piece of pipe by placing a line on the **Object** layer. Then connect the top pieces of pipe with a line to represent the filled and welded bend. When you finish this step, the upper part of the front view should look like the one in Fig. 2.3.24.

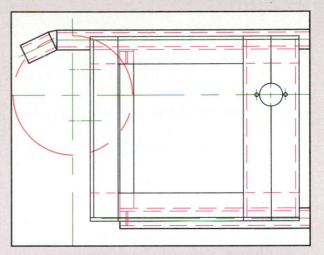

Fig. 2.3.24 *Upper end of the front view of the upper conveyor drawing.*

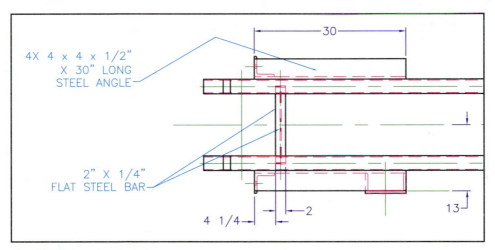

4X 4 x 4 x 1/2"
X 30" LONG
STEEL ANGLE

2" X 1/4"
FLAT STEEL BAR

30

4 1/4

2

13

Fig. 2.3.25 *Upper end of the top view of the upper weldment drawing.*

Upper End: Top and Side Views

Now that the framework at the upper end of the front view has been completed, you can work on the top and side views. They will be easier to create because you are now more familiar with the framework and because you can base the location of lines on those in the front view.

A partially dimensioned version of the top view of the upper weldment drawing is shown in Fig. 2.3.25. Study the illustration before you begin working on the top view.

Visualize the relationship of the pieces. Try to picture how they would look from the top. If necessary, refer again to the 3D models in Figs. 2.3.9 and 2.3.10.

Remember that the top and front views must align exactly. The views may be too far apart to make object snap tracking a realistic option. The easiest way to begin adding the framework to the top view is to zoom in on the front view and create vertical construction lines that will help you place the geometry in the top view. Follow the steps in Practice Exercise 2.3C to add the geometry shown in Fig. 2.3.25.

Practice Exercise 2.3C

1. Zoom in on the front view of the upper weldment drawing. Make sure running object snaps are active and that **Endpoint** is one of the active object snaps.
2. Enter the **XLINE** command and press **V** to select the Vertical option. Pick the points shown in Fig. 2.3.26 to create vertical construction lines.

Notice that step 2 did not direct you to construct lines at all of the pertinent locations—only those required to box in the rectangular frame. Later, you will create additional construction lines to show the depth of the steel angles, etc. The reason these lines were not created now is to avoid confusion when you return to the top view.

3. Offset the main horizontal centerline of the top view down by **13″** to locate the inside front of the hydraulic motor housing. Move the offset line to the **Object** layer. Then offset the new line down by **1/2″** to establish the thickness of the steel angle.
4. Offset the uppermost pipe line up by **4″** to establish the width of the rectangular frame at the back of the structure. The upper end of the top view should now look like Fig. 2.3.27A.

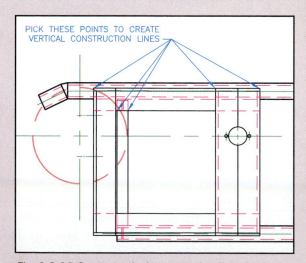

Fig. 2.3.26 Create vertical construction lines in the front view to help place geometry in the top view.

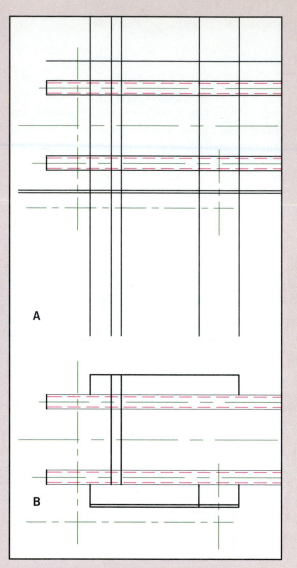

Fig. 2.3.27 Top view before (A) and after (B) the initial trimming operation.

5. Trim the construction lines and the lines you created in steps 3 and 4 to size, as shown in Fig. 2.3.27B.
6. In the top view, the flat steel bar is centered on the two Schedule 40 pipes and is 14 7/8″ long. To establish its ends, you can therefore offset the main horizontal centerline by half the length of the flat bar. Move the offset line to the **Object** layer. Then trim all of the lines to create the edges of the bar.

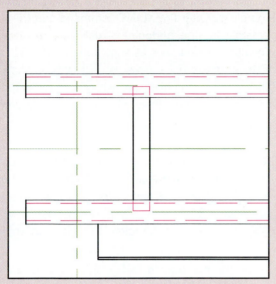

Fig. 2.3.28 *The ends of the flat bar are hidden below the Schedule 40 pipes.*

7. The flat bar is below the Schedule 40 pipe in this view. Break the lines of the flat bar at their intersections with the pipe and move the appropriate parts of the lines to the **Hidden** layer. Be sure to move the ends of the hidden lines slightly away from the intersection so that they can be distinguished from the object lines. See Fig. 2.3.28.

8. Zoom in on the left end of the front view once again and add the rest of the construction lines to locate the framework geometry in the front view. Enter the **XLINE** command and the **V** option. Use the **Endpoint** and **Midpoint** object snaps to place the construction lines as shown in Fig. 2.3.29.

9. Zoom in on the top view. Offset the top Schedule 40 pipe line up by $1/4''$, and offset the bottom Schedule 40 pipe line down by $1/4''$. You now have all of the lines you need to finish the framework in the top view. Use the **TRIM**, **EXTEND**, and **BREAK** commands to finish the framework. Move hidden lines to the **Hidden** layer. When you finish, the left end of the top view should look like Fig. 2.3.30.

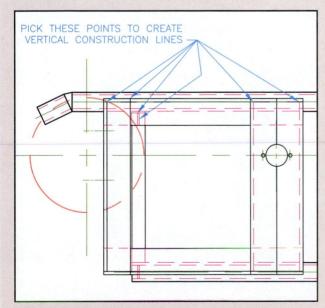

Fig. 2.3.29 *Place the secondary construction lines at these locations.*

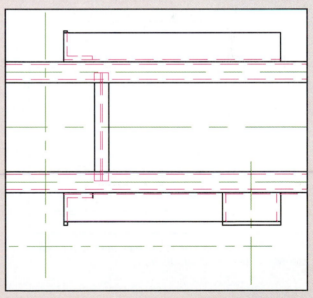

Fig. 2.3.30 *Framework lines trimmed and moved to the correct layers. Technically, the top view of the angle iron should show the curved surfaces. In the interest of time, many drafters choose to leave the edges rectangular, as shown here.*

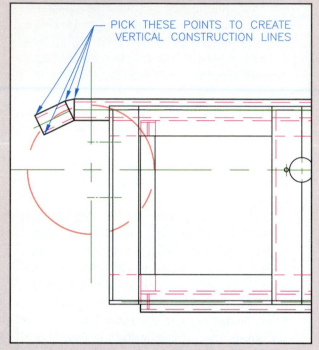

PICK THESE POINTS TO CREATE
VERTICAL CONSTRUCTION LINES

Fig. 2.3.31 *Placement of construction lines for the bend in the top view.*

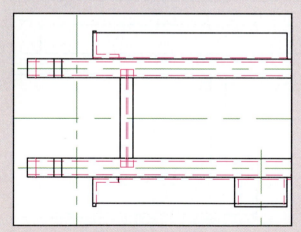

Fig. 2.3.32 *Trim and extend the pipe lines to incorporate the welded bend into the top view.*

Fig. 2.3.33 *End view of the 3D model showing the bolt holes in the rectangular frames. Note that circular holes may appear polygonal, as shown here, due to the resolution at which the model is created.*

10. Zoom in on the left end of the front view. The last two construction lines needed for the top view are those that locate the bend and weld in the top pipes. Enter the **XLINE** command and the **V** option. Use the **Endpoint** object snap to place the construction lines as shown in Fig. 2.3.31.

11. Extend and trim the lines as necessary to show the bend and the filled and welded wedge in the top view. Move the lower pipe line to the **Hidden** layer. See Fig. 2.3.32.

This completes, for now, your work on the top view of the upper weldment drawing. Turn your attention next to the side view. Refer to the 3D model of the end view in Fig. 2.3.33 and to the dimensions and notes in Fig. 2.3.34 as you work on the side view. You should be able to complete this view on your own.

12. Given the information in Fig. 2.3.34, complete the side view of the upper weldment drawing. Use construction lines from the front view as necessary to align the views. When you finish, the side view should look like Fig. 2.3.35.

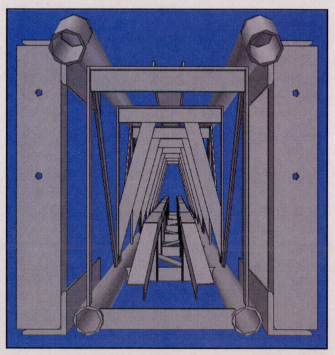

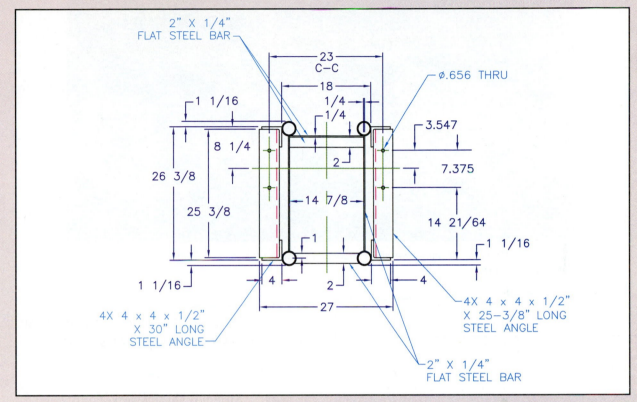

2" X 1/4"
FLAT STEEL BAR

23
C–C

18

1/4
1/4

Ø.656 THRU

1 1/16

3.547

8 1/4

2

7.375

26 3/8

25 3/8

14 7/8

14 21/64

1

1 1/16

1 1/16

4

2

4

27

4X 4 x 4 x 1/2"
X 30" LONG
STEEL ANGLE

4X 4 x 4 x 1/2"
X 25–3/8" LONG
STEEL ANGLE

2" X 1/4"
FLAT STEEL BAR

Fig. 2.3.34 *Left-side view of the upper weldment drawing.*

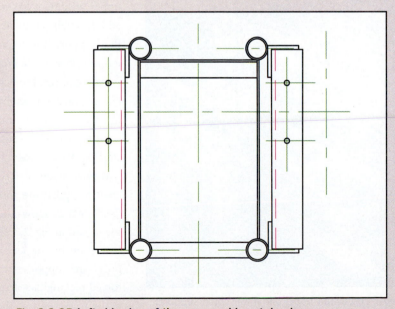

Fig. 2.3.35 *Left-side view of the upper weldment drawing.*

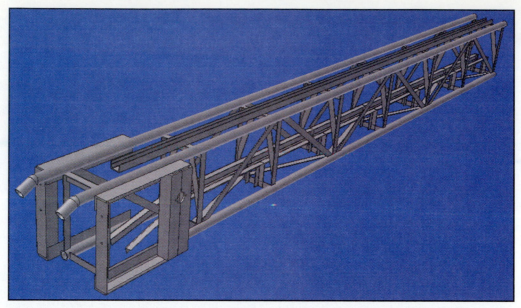

Fig. 2.3.36 **3D model of the entire upper weldment, showing the trussed framework and the chain supports.**

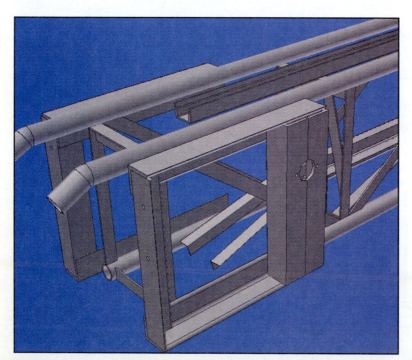

Fig. 2.3.37 **Detail of the upper end of the upper and lower chain supports.**

Structural Framework: Front View

Now that the basic structure of the top of the upper weldment has been established, you can add the chain supports and the steel framework that supports the conveyor. A 3D model of the entire upper weldment is shown in Fig. 2.3.36. As you can see, the Schedule 40 pipes are connected and stabilized by a series of flat steel bars in a trusslike configuration.

The upper chain support is made of channel steel, and the lower chain support consists of two rails made of structural steel angles. The chain supports are shown more clearly in Fig. 2.3.37, which shows the upper end of the weldment, and in Fig. 2.3.38, which shows the lower end. Both the upper and the lower chain supports are supported at 36″ intervals by reinforced rectangular structures similar to the one at the upper end of the weldment, which connects the four Schedule 40 pipes.

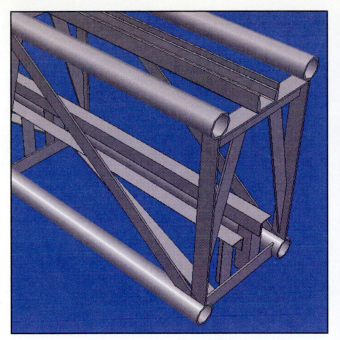

Fig. 2.3.38 *Detail of the lower end of the upper weldment, showing the chain supports.*

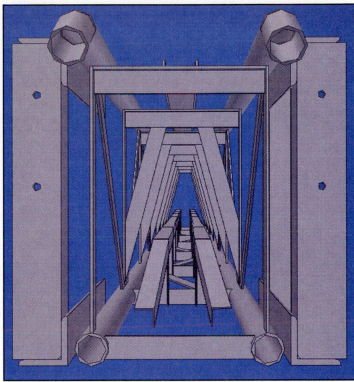

Fig. 2.3.39 *Upper end view of the weldment showing the relationship of the upper and lower chain supports to their supporting structures.*

Notice that for most of its length, the lower chain support is raised 8 5/8″ from the bottom of the support frame by steel angles that are fastened to the frame. At the upper end of the weldment, the lower chain support tapers down toward the bottom of the frame and flares slightly. The upper and lower end views shown in Fig. 2.3.39 and Fig. 2.3.40, respectively, will help you visualize the chain supports in relation to their support structure and the Schedule 40 pipes.

In Practice Exercise 2.3D, you will add the chain supports and construct the supporting framework. Refer to the 3D models shown here as necessary as you complete the exercise.

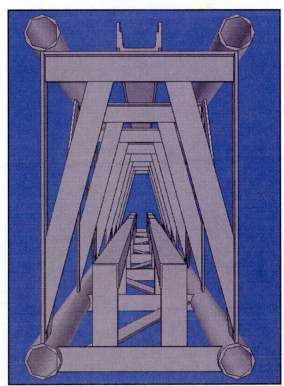

Fig. 2.3.40 *Lower end view of the weldment. Notice that on this end, the lower chain support is not tapered. The apparently solid surface at the front of each lower chain guide rail is actually one leg of its steel angle support post.*

The first step will be to add the vertical supports in the front view of the upper weldment drawing. These supports are made of 2″ × 1/4″ flat steel bars and are placed at 36″ intervals along the entire length of the weldment.

1. The vertical centerline of the first steel bar is located **41 1/4″** to the right of the roller bearing mount surface, as shown in Fig. 2.3.41. You can therefore offset the roller bearing mount surface by this distance to locate the centerline of the steel bar. Then offset the new line to both sides by **1″** to form the steel bar. Move the centerline to the **Center** layer, and trim the bar as shown in Fig. 2.3.41.

2. The vertical bars are stabilized by 2″ × 1/4″ flat steel bars placed diagonally from the bottom of the vertical bars to the upper bars of the framework. Refer again to Fig. 2.3.40. Both the upper bars and the diagonal bars show as hidden lines in the front view. Add the hidden lines on the **Hidden** layer as shown in Fig. 2.3.41. The diagonals are centered on the vertical bars.

The upper chain support is hidden in the front view, but the lower chain support needs to be added. It is supported 8 5/8″ from the bottom of the frame by 2″ × 2″ × 1/8″ steel angles fastened to the left side of the lower steel bars, as shown in Fig. 2.3.42. The horizontal part of the lower chain support begins 37 5/8″ from the roller bearing mount surface, as shown in Fig. 2.3.43.

3. Zoom in on the upper (left) end of the front view. Offset the roller bearing mount surface to the right by **37 5/8″** to locate the left edge of the horizontal portion of the lower chain support.

As you can see in Fig. 2.3.42, the lower edge of the support posts align with the lower edge of the flat steel bar at the bottom of the frame. Recall that the bar is centered on the 2″ Schedule 40 pipe, so the lower edge is 1″ below the pipe's horizontal centerline. You can use this information to position the lower chain support and its support posts accurately.

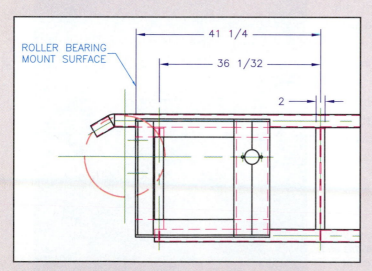

Fig. 2.3.41 *Position of the first vertical bar in the structural framework.*

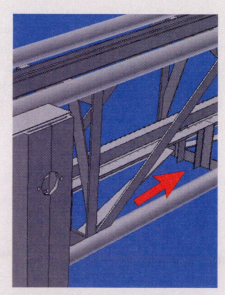

Fig. 2.3.42 *The supports for the lower chain rails are fastened to the left side of the lower 2″ steel bar.*

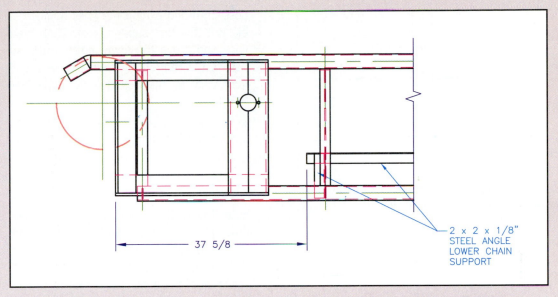

Fig. 2.3.43 *Position the left end of the horizontal portion of the lower chain support as shown here.*

4. Offset the centerline of the lower Schedule 40 pipe down by **1″** to locate the bottom of the support posts. Move the offset line to the **Object** layer.

5. Offset the line you created in the previous step up by **8 5/8″** to establish the top edge of the support posts, which is also the bottom edge of the top leg of the angle steel that makes up the lower chain support.

6. The steel angle is 1/8″ thick and sits on top of the support posts, so offset the line you created in step 5 up by **1/8″** to establish the thickness of the top leg of the steel angle.

7. Each leg of the steel angle is 2″, so offset the top line down by **2″** to create the other leg of the chain support.

8. Trim the vertical left end line and the three horizontal lines of the lower chain support. See Fig. 2.3.44. Because the length of the lines is the same as the centerline from which they were originally offset, there is no need to edit the right end of the chain support.

The same line that defines the left side of the vertical bar in the front view also defines the right side of the steel angle support posts. The top edge is located at the bottom of the upper leg of the chain support, so it too has already been created. Recall that you located the bottom edge of the support posts in step 6. To finish the first support post in the front view, all you need to do is create the left edge of the left leg of the support post and trim the bottom edge.

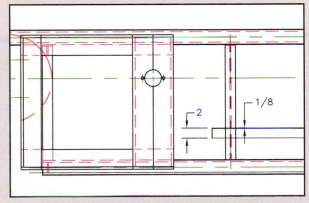

Fig. 2.3.44 *Top end of the lower chain support in the front view.*

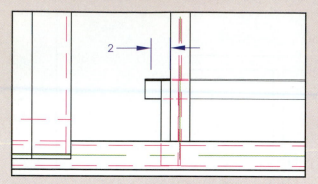

Fig. 2.3.45 *Front view of the steel angle that forms the first support post for the lower chain support.*

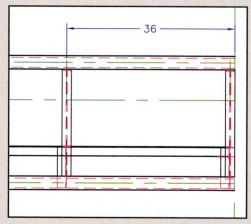

Fig. 2.3.46 *Edit the last support frame so that only half of it appears in the upper weldment.*

9. Offset the left side of the vertical steel bar to the left by **2″** to locate the left edge of the support post. Trim and extend the lines as necessary to block in the first support post. Then break the lines and move the lower edge of the support post to the **Hidden** layer where they are hidden from view by the Schedule 40 pipe. See Fig. 2.3.45.

The support posts and frames occur at regular 36″ intervals through the entire length of the upper weldment, with the exception of the last one. Therefore, the easiest and fastest way to place the remaining support posts and frames is to array them.

10. Enter the **ARRAY** command. In the Array window, set up an array of **1** row and **7** columns. Set the row offset to **0** and the column offset to **36**. Pick the **Select objects** button. In the front view, select all of the lines that represent the vertical, diagonal, and upper bars of the frame, the vertical centerline, and the support post. Be sure to include the hidden lines where the chain support is hidden by the bars of the frame. Pick **OK** to complete the array.

At the extreme right of the upper weldment, the support post and frame need to be moved 1″ to the left to lie completely within the upper weldment. This frame will be bolted to a similar frame on the lower weldment to increase the strength of the joint between the two weldments.

11. Enter the **MOVE** command and select all of the pieces that make up the last support post and frame. Move them **1″** to the left, as shown in Fig. 2.3.46. The front view should now look like the illustration in Fig. 2.3.47.

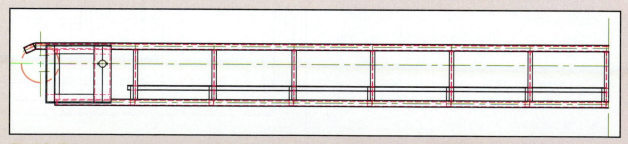

Fig. 2.3.47 *Front view of the upper weldment after the vertical supports have been arrayed.*

At the top end of the conveyor, the lower chain support tapers downward. The taper begins at the left end of the horizontal portion of the chain support and extends 21 5/8″ to the left, as shown in Fig. 2.3.48. Like the rest of the chain support, this portion is crafted from 2″ × 2″ × 1/8″ steel angles.

12. To establish the left end of the tapered portion, offset the left end of the horizontal portion of the chain support to the left by **21 5/8**″. With Ortho on, use grips to bring the lower endpoint of the line down below the bottom of the rectangular frame. Then offset the bottom line of the rectangular frame up by **2 1/4**″. Enter the **LINE** command and use object snaps to draw a line from the upper corner of the left end of the chain support to the intersection of the two lines you just offset. Then erase the two offset lines.

13. Offset the top line of the tapered portion down by **1/8**″. Then offset the same line down by **2**″. At the left end of the tapered portion, add an end line from the top to the bottom of the steel angle. Use the **Endpoint** object snap to place the line accurately.

14. Zoom in on the right end of the tapered portion to edit the bend in the steel. Enter the **FILLET** command and enter **R** to select the Radius option. Enter a radius of **3**. For the first object, select the top line of the tapered portion of the steel. For the second object, select the top line of the horizontal portion. Repeat the fillet for the other two lines that make up the bend. *Important:* If you receive an error message saying that the radius is too large, fillet the bottom line of the taper with the hidden line at the bottom of the horizontal portion. Trim or extend the lines as necessary to clean up the intersection.

15. The chain supports are located behind the rectangular frame and motor housing in this view. Break all three of the tapered lines at the right side of the housing and move the hidden lines to the **Hidden** layer. When you finish, the tapered portion of the lower chain support should look like the illustration in Fig. 2.3.49.

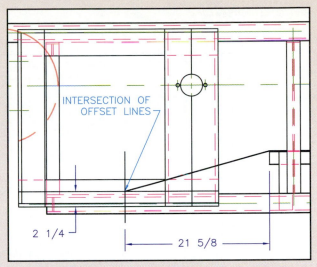

Fig. 2.3.48 *Determining the endpoint of the tapered portion of the lower chain support.*

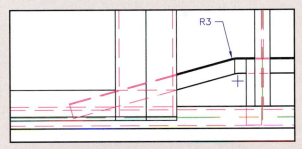

Fig. 2.3.49 *Tapered end of the lower chain support.*

Next, add the vertical supports that run between the upper and lower Schedule 40 pipes. These supports are made of 2″ × 1/4″ flat steel bars and extend between the vertical steel bars of the framework. Refer again to the model in Fig. 2.3.36.

16. Begin at the lower end of the first vertical post to the right of the rectangular frame and motor housing. The steel bar extends upward and to the right at an angle of 53.25°, as shown in Fig. 2.3.50A. Remember that in AutoCAD, 0° is east of the origin on the Cartesian grid, so to achieve the correct angle, you will need to add 90° to the 53.25°. Enter the **LINE** command, select the intersection of the vertical steel bar and the lower Schedule 40 pipe, and use polar coordinates to extend the line: **@40<143.25**.

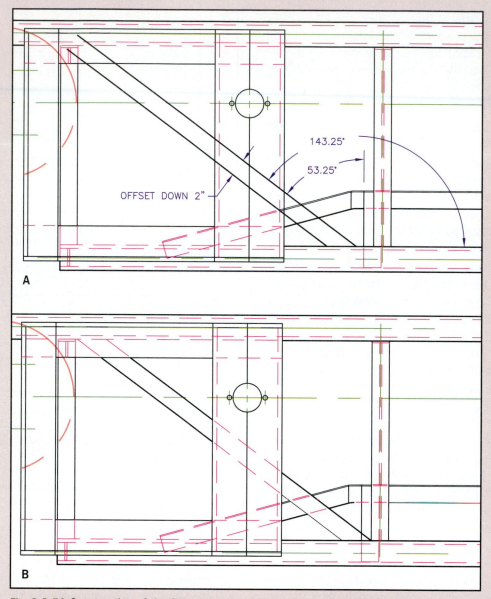

143.25°

53.25°

OFFSET DOWN 2"

A

B

Fig. 2.3.50 Construction of the first cross support.

17. Offset the line you created in Step 16 down by **2″**, as shown in Fig. 2.3.50A.

18. Trim and extend the two lines as necessary so that they meet the lower edge of the upper Schedule 40 pipe and the upper edge of the lower pipe.

19. In the front view, the cross support runs in front of the chain support but behind the rectangular frame. Break the lines and move the hidden portions of the cross support and the lower chain support to the **Hidden** layer. The first cross support should now look like Fig. 2.3.50B.

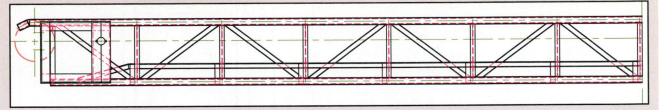

Fig. 2.3.51 Finished cross supports in the front view of the upper weldment drawing.

20. Create the remaining cross supports. Start each line at the intersection of the lower Schedule 40 pipe and a vertical bar. Alternate the direction of the supports. When you finish, the front view should look like the one in Fig. 2.3.51.

Additional cross supports stabilize the bottom of the conveyor by extending from one bottom pipe to the other in a manner similar to that of the cross sup-ports on the sides. They are constructed of 2″ × 1/4″ flat steel bars. In the front view, these bars appear as an end view. They are hidden by the lower Schedule 40 pipe.

21. Offset the centerline of the lower Schedule 40 pipe up and down by **1/8″** to establish the end view of the stabilizers. Move both offset lines to the **Hidden** layer.

Structural Framework: Top and Side Views

The front view of the upper weldment drawing for sheet 1 is now almost complete. We will add sections, dimensions, and notes at a later time. For now, move on to the side view of the upper weldment drawing. The upper chain support and the conveyor chain show clearly in this view.

The upper chain support consists of C 3 × 5 channel steel with 2″ × 1/4″ flat steel bars attached to both vertical sur-faces. See Fig. 2.3.52. The C 3 × 5 designation means that the total distance from the outside face of one flange of the C-channel to the outside face of the other flange is 3″, and each foot of the steel weighs 5 pounds.

Additional supports are also visible in the side view. These, too, consist of 2″ × 1/4″ flat steel bars. They are attached to the frame as shown in Fig. 2.3.52.

In Practice Exercise 2.3E, you will work on the side view of the upper weldment. You will add the upper chain support, the conveyor chain, the diagonal supports, and the lower chain support as they appear from the upper (left) end of the tire conveyor.

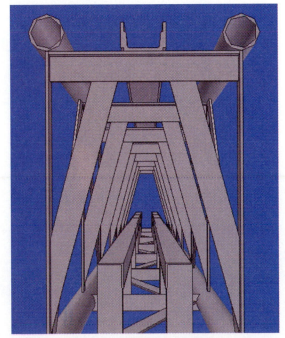

Fig. 2.3.52 End view of the upper weldment showing the upper chain support and the diagonal structural supports.

1. Zoom in on the side view of the upper weldment drawing. Begin by offsetting the vertical center-line to the right by **2 1/16"** to locate the upper end of the left side of the right diagonal support. Move this temporary line to the **Object** layer so that you can see it more clearly.

2. Use the intersection of the temporary offset line with the upper flat steel bar as the first point of the first line. See Fig. 2.3.53. Snap to the inter-section of the right steel bar and the lower right Schedule 40 pipe as the endpoint for the line.

3. Offset the line by **2"** to form the width of the flat steel bar. Trim the offset line to finish the diago-nal support on the right side of the view. Delete the temporary offset line you created in step 1.

4. Mirror the two lines that represent the right diag-onal support to create the left diagonal support. Use the main vertical centerline as the mirror line. The side view should now look like Fig. 2.3.54.

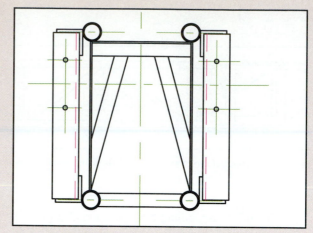

Fig. 2.3.54 *Side view with diagonal supports in place.*

Next you will complete the lower chain support in the side view. Look closely at the model shown in Fig. 2.3.55. At the point where the two rails begin taper-ing down toward the lower Schedule 40 pipes, they also begin to flare slightly. This is difficult to see in the side view, but they must still be drawn. Later, you will create a sectional view to document this more clearly.

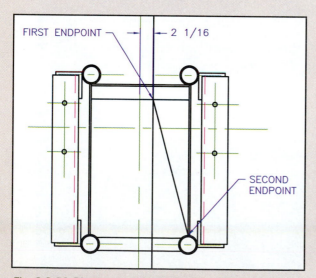

Fig. 2.3.53 *Placement of the first line of the right diagonal support.*

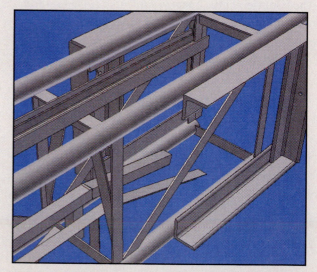

Fig. 2.3.55 *At the upper end of the conveyor, the two rails of the lower chain support taper downward and flare apart.*

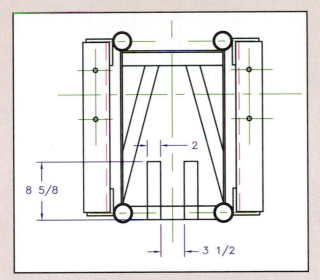

Fig. 2.3.56 *First steps in creating the side view of the support posts.*

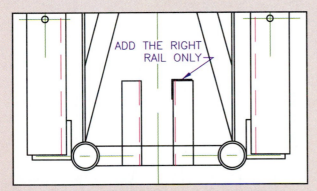

Fig. 2.3.57 *Add the steel angles that make up the rails of the lower chain support.*

5. Begin by representing the lower chain support and its support posts as they appear before the flare begins. The inside edges of the support posts are $3\frac{1}{2}''$ apart and are centered on the vertical centerline. To locate them, offset the vertical centerline to both sides by half of this distance. Move the offset lines to the **Object** layer. Locate the outside edges of the support posts by offsetting each of the lines to the outside by **2**".

The support posts extend upward $8\frac{5}{8}''$ from the bottom of the steel bar that runs between the two lower Schedule 40 pipes. Offset the bottom edge of the steel bar up by **$8\frac{5}{8}''$** to locate the top of the support posts. Trim all of the lines as shown in Fig. 2.3.56.

From this point, work only on the right rail of the lower chain support. It is much easier—and less confusing—to construct the flared portion once and then mirror it to create the other rail.

6. On the right support post, add the right rail of the lower chain support. Recall that the rails consist of $2'' \times 2'' \times \frac{1}{8}''$ steel angles that rest on top of the support posts. Add the rail as shown in Fig. 2.3.57.

At the end of the flared portion, the inside edges of the rails are $5\frac{3}{16}''$ apart, still centered on the main centerline. At this point, the upper edges of the rails are $2\frac{1}{4}''$ above the bottom of the rectangular frame. Using this information, offset lines to block in the end of the steel angle that makes up the right rail.

7. Offset the main centerline and the bottom of the frame as shown in Fig. 2.3.58A to block in the position of the top edge of the angle at the end of the rail. Then trim away all of the lines except the portion of the horizontal line that represents the top edge of the angle, as shown in Fig. 2.3.58B. Then enter the **LINE** command and create lines connecting the top edge of the angle at the top of the support post with the top edge of the angle at the end location you just established. See Fig. 2.3.58C. Trim away support post and lower angle lines as shown in Fig. 2.3.58D.

8. From the lower left corner of the flared portion of the rail, extend a vertical line down by **2**" to establish the side of the steel angle. Connect the lower end of this line with the outside corner of the angle at the top of the support post, as shown in Fig. 2.3.59A. Trim away the interior lines as shown in Fig. 2.3.59B. Then offset the edges of the steel inward by $\frac{1}{8}''$ to form the thickness of the steel angle. Trim the lines as necessary to finish the front of the right rail.

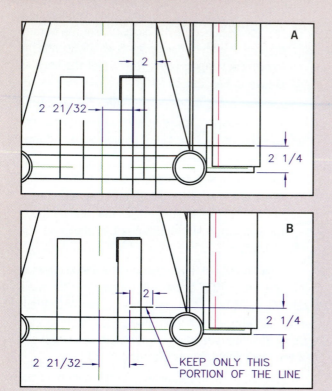

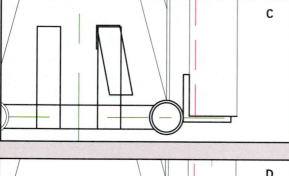

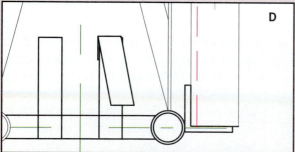

Fig. 2.3.58 *Build the top edge of the flared portion of the right rail by following these steps.*

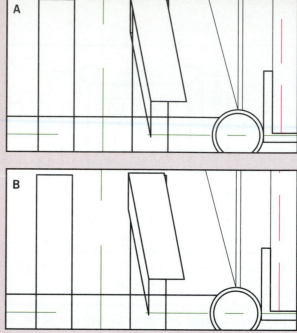

Fig. 2.3.59 *Finish the right rail by connecting the lower edge with the upper edge (A) and trimming the inside lines (B).*

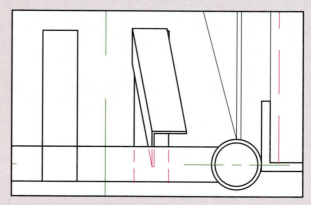

Fig. 2.3.60 *The finished right rail.*

9. Break the lines at the top of the bottom rail and move the part of the post and rail that is behind the lower steel bar to the **Hidden** layer. When you finish, the rail should look like the illustration in Fig. 2.3.60.

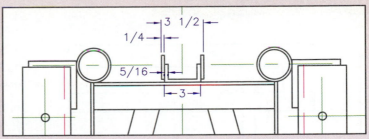

10. The left rail is a mirror image of the right rail, so you can add it easily. First, to keep the area uncluttered, erase the three lines that currently represent the right rail.

11. Enter the **MIRROR** command and select all of the lines that make up the left rail as the items to be mirrored. Select the vertical centerline as the mirror line. (Be sure to use object snaps to snap to the centerline exactly.) Choose not to delete source objects. The left rail should appear as shown in Fig. 2.3.61.

12. Draw in the upper chain support. It is made up of a length of 3 × 5.0 steel C-channel flanked by 2″ × 1/4″ flat steel bars on both sides. Do this step on your own, referring to Fig. 2.3.62 for dimensions as necessary. Look up any dimensions not given. Note that the upper chain support is centered on the main vertical centerline of the side view.

13. Finally, add the hidden lines at the bottom of the tire conveyor in the side view. To do this, zoom in on the front view and make **Hidden** the current layer. Enter the **XLINE** command and enter **H** for Horizontal. Use the **Nearest** object snap to snap to the two horizontal lines indicated in Fig. 2.3.63.

Fig. 2.3.62 Add the upper chain support in the side view.

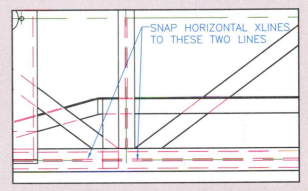

Fig. 2.3.63 Snap horizontal construction lines to these two horizontal lines. Use the Nearest object snap.

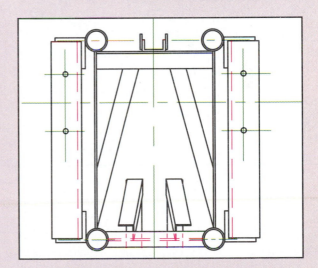

Fig. 2.3.64 The finished side view.

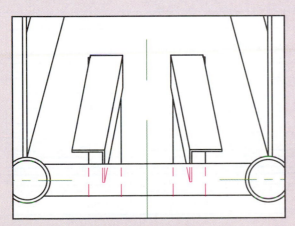

Fig. 2.3.61 The finished left and right rails.

14. Zoom in on the side view. Use the inner edges of the lower Schedule 40 pipes as boundaries to trim the construction lines so that they run only between the pipes. The finished side view should look like the illustration in Fig. 2.3.64.

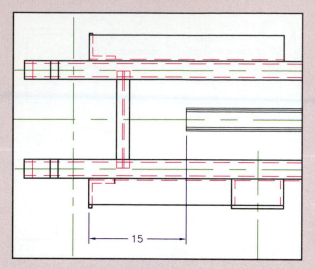

Fig. 2.3.65 *Position of the upper chain support in the top view.*

Next, move to the top view of the upper weldment drawing. In the following steps, you will add the upper chain support and the cross supports to the top view.

15. Consult Fig. 2.3.62 for the dimensions of the C-channel and flat steel bars that make up the upper chain support. As you can see in Fig. 2.3.65, the chain support begins 15″ from the

left end of the rectangular frame. Use this information to construct the upper chain support at this time. Note that the chain support should run to the right all the way to the break line that represents the right end of the upper weldment.

Support structures occur at 36″ intervals along the length of the weldment, as shown in Fig. 2.3.66. The easiest way to create these is to array the first structure, which you have already drawn. Notice, however, that starting with the second support structure, the upper chain support crosses above the structures in the top view, hiding part of the structures from view. Therefore, before you perform the array, it is advisable to copy the first structure to create the second one. Alter the second one to show the hidden lines. Then array the second support structure to create the remaining six.

16. Zoom in on the upper (left) end of the top view. Add a vertical centerline to the first cross support structure, as shown in Fig. 2.3.67.

17. Offset the vertical centerline you created in the previous step **36″** to the right to create the centerline for the second support structure. Then copy the first structure to create the second one. Break the object lines of the second structure and place the part that is hidden by the upper chain support onto the **Hidden** layer, as shown in Fig. 2.3.68.

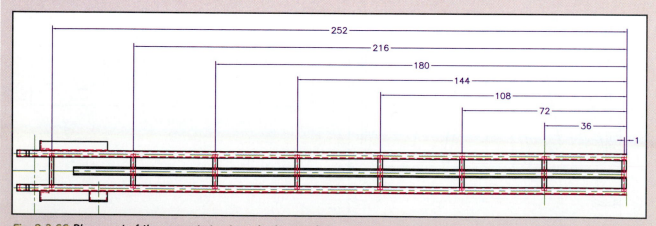

Fig. 2.3.66 *Placement of the support structures in the top view.*

18. Enter the **ARRAY** command and then enter **R** for Rectangular. Set up the array to have **1** row and **7** columns, a row offset of **0**, a column offset of **36**, and an angle of **0**. Pick the **Select objects** button and select all of the lines that make up the second support structure. Preview the array, and if the supports appear like those in Fig. 2.3.66, accept the array.

19. Zoom in on the right side of the top view. Recall that the support structure spans the upper and lower weldments, so only the left half of the last support is shown in the upper weldment drawing. Delete/trim all of the lines to the right of the centerline that forms the right end of the upper weldment.

Like the sides, the bottom of the framework is stabilized by cross supports made of $2'' \times {}^{1}\!/{}_{4}''$ steel bars. These bars run between the two lower Schedule 40 pipes and are attached to the support structures as shown in Figs. 2.3.69 and 2.3.70.

20. Add the bottom cross supports. Refer to the dimensions shown in Fig. 2.3.69. Use a method similar to the one you used to add the side cross supports. Delete the portion of the cross supports that are hidden by the upper chain drive to avoid confusion in the finished drawing. Hidden lines are often removed in this manner when including them adds clutter that makes the drawing hard to read. When you finish, your drawing should look like the one in Fig. 2.3.70.

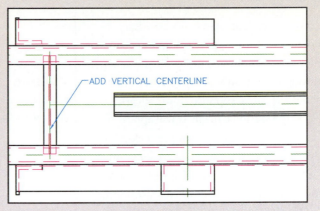

Fig. 2.3.67 *Add a vertical centerline.*

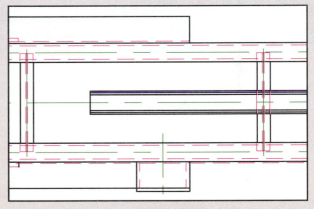

Fig. 2.3.68 *Alter the second support structure to show the hidden lines.*

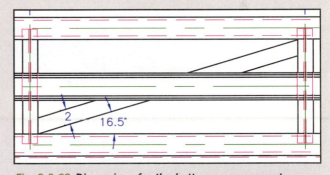

Fig. 2.3.69 *Dimensions for the bottom cross supports.*

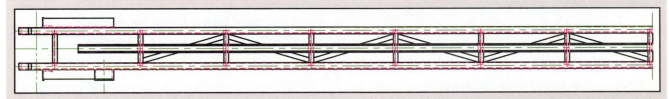

Fig. 2.3.70 *Top view showing the bottom cross supports.*

Section B-B

Two sectional views are necessary to document the upper weldment completely. Section A-A, which is taken through the width of the framework to show the cross supports more clearly, is actually a sectional detail. It will appear on the second sheet of the upper weldment drawing. Section B-B is taken longitudinally at the top of the weldment to show the upper end of the lower chain support.

We will work on Section B-B first, because it appears on the first drawing sheet. After Section B-B is completed, we can finish the first drawing sheet.

As shown in Fig. 2.3.71, Section B-B shares quite a bit of geometry with the top view of the upper weldment drawing. You can therefore copy and alter it in much the same manner as you copied the normal views to create the upper weldment drawing in Practice Exercise 2.3A.

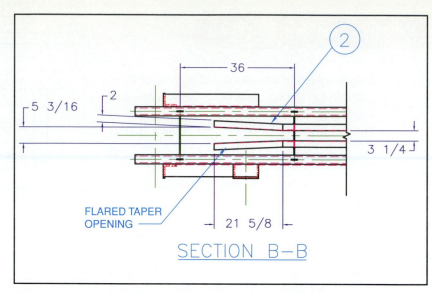

Fig. 2.3.71 *Section B-B.*

In Practice Exercise 2.3F, you will copy the appropriate geometry from the upper weldment drawing and use it to create Section B-B. Refer to Fig. 2.3.71 as necessary as you complete the exercise.

Practice Exercise 2.3F

Before you begin the actual section, you should indicate on the upper weldment drawing where the section will be taken. In this case, Section B-B will be taken as a plan view below the top of the conveyor framework, looking down at the lower chain support.

1. Zoom in on the upper (left) end of the front view of the upper weldment drawing. Place the section line as shown in Fig. 2.3.72.

Now you can create Section B-B. Perform the following steps to copy the relevant geometry from the top view of the upper weldment drawing.

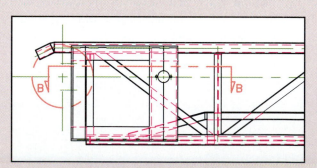

Fig. 2.3.72 *Show Section B-B on the front view as indicated here.*

2. Zoom out to see the entire upper weldment drawing. Position the viewing window so that you have some space below the front view. This is where you will place Section B-B. Refer again to Fig. 2.3.1 at the beginning of this chapter for placement information.

3. Enter the **COPY** command and select the entire top view as the object(s) to be copied. Place the copy so that the left side of the copy is located approximately as shown in Fig. 2.3.1 (page 126). From now on, the copy will be referred to as Section B-B.

4. Zoom in on the top (left) end of Section B-B. With Ortho on, enter the **LINE** command and create a vertical line between the second and third support structures, as shown in Fig. 2.3.73. The exact placement of the line does not matter because this will later become the break line for the section.

5. Trim away all of the long horizontal lines to the right of the line you created in the previous step.

6. Zoom out so that you can see the entire sectional view. Erase all of the remaining parts to the right of the line from step 4. Then zoom back in to the remaining geometry for Section B-B.

7. Erase the upper chain support, since it is not visible in this view. Be careful not to erase the main centerline. Also erase the cross supports to avoid clutter.

8. Add the horizontal part of the lower chain support. This portion of the chain support begins 37 5/8″ from the left end of the roller bearing mount surface. The two rails that make up the lower chain support are 3 1/4″ apart and are centered on the main horizontal centerline. Extend the rails to the vertical break line at the right end of the sectional view, as shown in Fig. 2.3.74.

9. Add the bent sections of the angle steel that make up the lower chain support. Use the dimensions in Fig. 2.3.75.

10. Only the lower Schedule 40 pipes are visible in this view. Trim the pipes back to the hidden line that represents the lower Schedule 40 pipe, and erase the bend lines, as shown in Fig. 2.3.75.

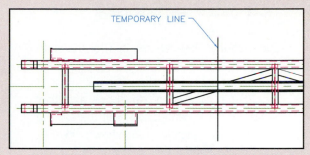

Fig. 2.3.73 *Create a line to which you can trim away the right end of the long horizontal lines.*

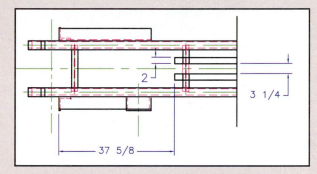

Fig. 2.3.74 *Add the horizontal portions of the lower chain support.*

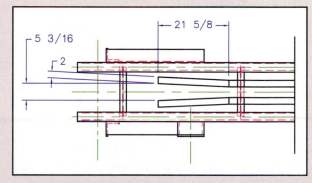

Fig. 2.3.75 *Add the bent portions of the angle steel as shown here.*

11. Zoom in on the break line at the right of the sectional view and add the break symbol to show that this is a break line.

12. Zoom so that you can see both of the vertical support structures. These do not appear in the sectional view. However, the edges of the 2″ flat steel bars do show, and they share the same centerlines. Edit the existing geometry to match that shown in Fig. 2.3.76. Be careful not to erase the top and bottom lines of these structures. You will need them in the next step.

13. Use the top and bottom lines of the original vertical support structures as the basis for the end views of the flat steel bars that are visible in this section. First move the lines to the **Object** layer. Then offset each one by **1/4″** to the outside so that the bars are centered on the horizontal centerlines. Add the short end segments and the hatch to complete all four end views, as shown in Fig. 2.3.77.

14. Focus once again on the lower chain support. Add hidden lines to represent the thickness of the steel angles. Also add the end view of the angles that form the posts that hold the lower chain support in place. See Fig. 2.3.78.

15. Create a new dimension style and name it **Fractions**. Specify fractional units and a precision of **1/16**. Make the arrow size **2 1/2**, extend **2** beyond dim lines, offset text from origin **2**, text size **2 1/2**, offset from dim line **1 1/4**. Suppress leading zeros.

16. Make **Fractions** the current dimension style. Add the dimensions and notes as shown in Fig. 2.3.71 at the beginning of this exercise. Put the dimensions on the **Dimensions** layer and the notes on the **Text** layer. Be sure to include the identification number and its bubble. The diameter of the bubble is **9 1/2″**, and the text size is **4″**.

17. Add the title **SECTION B-B**. Use the **MTEXT** command with the underscore feature. Use a text height of **4″**. When you finish, Section B-B should look like the one in Fig. 2.3.71.

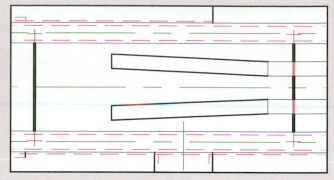

Fig. 2.3.76 Edit the vertical support structures as shown here. Notice that the top and bottom lines from the original structures have not been deleted.

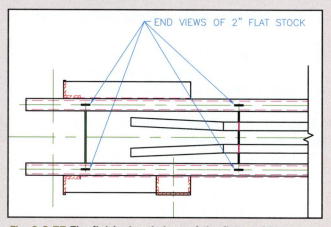

Fig. 2.3.77 The finished end views of the flat steel bars.

Fig. 2.3.78 Add the hidden lines to represent the inside edges of the angles that form the lower chain support and the angles of the support post.

Finish Drawing 0022

You have now completed all of the basic geometry necessary for sheet 1 of the upper weldment drawing. You will need to make some minor adjustments to the final drawing. Because much of the same geometry will be used in the three-view detail on sheet 2, you will copy the geometry to drawing 0022-2 before you make any adjustments.

Adjustments, Dimensions, and Notes

Besides the minor adjustments to geometry, only a few tasks remain to finish drawing 0022. In the following series of exercises, you will add dimensions and notes, identification numbers, and the parts list. You will then complete the title block and tolerance block to finish the drawing.

Practice Exercise 2.3G steps you through the process of making all of the necessary adjustments to the upper weldment drawing. You will also add the dimensions, any notes that are specific to the views, and identification numbers to the three main views of the upper weldment. (Section B-B already has dimensions and identification numbers, so it is ready for use in the layout.) Refer again to the finished drawing in Fig. 2.3.1 as necessary to judge spacing and other characteristics of drawing sheet 1 as you follow the steps in Practice Exercise 2.3G.

Practice Exercise 2.3G

Before you begin the final layout for sheet 1, copy the geometry into the drawing file for sheet 2. This will allow you to reuse the geometry on sheet 2.

1. Open **PROJECT 2-0022-2.dwg**. In PROJECT 2-0022.dwg, **ZOOM All** to see the entire drawing area. Use **COPYCLIP** to copy the three normal views of the upper weldment drawing. Use **PASTECLIP** to place the copy in PROJECT 2-0022-2.dwg.

2. Close PROJECT 2-0022-2.dwg and return your attention to PROJECT 2-0022.dwg. Make **Dimensions** the current layer, and add the overall dimensions shown in Fig. 2.3.79.

3. Zoom in on the upper end of the top view of the upper weldment. Add the centerlines for the bolt holes in the rectangular frames at the upper (left) end of the top view. See Fig. 2.3.80. You may want to decrease their linetype scale to show the centerline linetype.

To avoid cumulative error, long distances in the top and front views will be dimensioned using baseline dimensioning. The baseline will be the right end of the weldment.

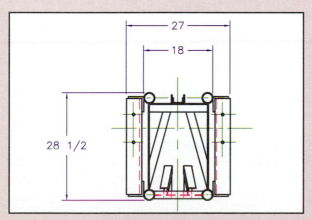

Fig. 2.3.79 *Add the overall dimensions in the side view.*

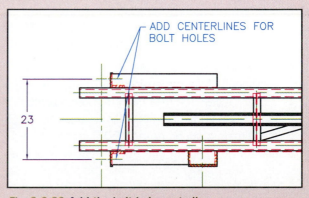

Fig. 2.3.80 *Add the bolt-hole centerlines.*

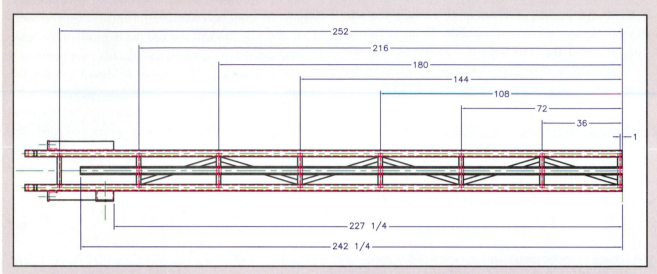

Fig. 2.3.81 **Dimension the top view.**

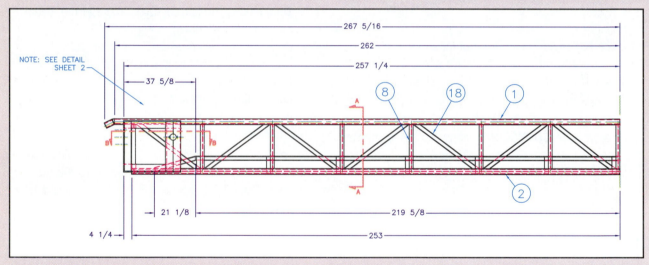

Fig. 2.3.82 **Dimensions, notes, and identification numbers for the front view.**

4. Zoom out to see the entire top view of the upper weldment. Add dimensions as shown in Fig. 2.3.81. *Note:* You may need to zoom repeatedly to pick the endpoints accurately. Consider using the **Dynamic** option of the **ZOOM** command.

5. Zoom in on the left end of the front view of the upper weldment drawing. Erase the phantom circle that represents the #120 sprocket and the associated vertical and horizontal centerlines.

6. On the **Phantom** layer, add the cutting-plane line to show where Section A-A will be taken. See Fig. 2.3.82.

7. Add the overall dimensions in the front view, as shown in Fig. 2.3.82.

8. Make **Text** the current layer. Add the note and the bubbles with their identification numbers. Use the same text and bubble sizes you used in Section B-B. See Fig. 2.3.82.

Partial Parts List

All of the remaining tasks for upper weldment drawing sheet 1 must be accomplished in the layout view. You will alter the existing layout tab for upper weldment sheet 1, which is drawing 0022.

The parts list for both sheets will be placed on sheet 1. As you can see in Fig. 2.3.83, this is a **partial parts list**. It contains only those parts that need to be called out in the upper weldment drawing. The main parts list, which includes parts for the entire tire conveyor, will be placed on the assembly reference drawing later in this project.

Notice that part numbers 7, 13, 14, and 19 through 25 are not specified in the parts list. Parts 7, 13, and 14 were specified in the original plans but were later deleted. Their descriptions were removed from the parts list, but the lines dedicated to these parts were left blank. This practice saves the designer a great deal of time that would otherwise be spent renumbering bubbles and rearranging the parts list.

There are no parts that correspond to lines 19 through 25. These lines were added to the drawing to accommodate any future additions of parts as construction of the tire conveyor continues.

Item #	PART/DRAWING #	QTY	ITEM DESCRIPTION
25			
24			
23			
22			
21			
20			
19			
18	00055	14	CUT SHEET
17	00055	7	CUT SHEET
16	00055	14	CUT SHEET
15	00055	6	CUT SHEET
14			
13			
12	2" X 1/4" X 14 7/8 LG	8	FLAT BAR
11	2" X 1/4" X 14 7/8 LG	8	FLAT BAR
10	3 X 5.0 X 242 1/4 LG	1	C-CHANNEL, ASTM A36
9	2" X 1/4" X 242 1/4 LG	2	FLAT BAR
8	2" X 1/4" X 22 3/4 LG	16	FLAT BAR
7			
6	2" X 2" X 1/8" X 240 3/4"	2	ANGLE - ASTM A36
5	2" X 2" X 1/8" X 8 5/8" LG	14	ANGLE - ASTM A36
4	4" X 4" X 1/2" X 30" LG	4	ANGLE - ASTM A36
3	4" X 4" X 1/2" X 25 3/8" LG	4	ANGLE - ASTM A36
2	2 1/2 X 253 LG	2	SCHD 40 PIPE
1	2 1/2 X 266 1/2 LG	2	SCHD 40 PIPE
Item #	PART/DRAWING #	QTY	ITEM DESCRIPTION

Fig. 2.3.83 *Partial parts list for the upper weldment drawing.*

OLE Objects

In Project 1, you used the ARRAY command to set up the structure of the parts list and DTEXT to add the text. That method has the advantage of requiring only the AutoCAD software. However, if you have access to the Microsoft Excel® spreadsheet software, you can easily create a well-organized parts list that is also easy to maintain. Easy maintenance is an advantage for a field project such as the tire conveyor because there is a good chance that the parts list will need frequent revision as the project progresses.

In this project, you will create the parts list using Excel and insert it as an object into your AutoCAD file. If you do not have access to the Excel software, you can substitute any other spreadsheet program that conforms to the requirements for the Windows **object linking and embedding (OLE)** feature.

OLE requires both a server application and a container application. The **server application**, also called the *source*, is an OLE-compliant program in which you create the object you want to link or embed into another document. The **container application**, also called the *destination application*, is the OLE-compliant program into which you place the linked or embedded object. For our purposes, the server application is Excel, and the container application is AutoCAD.

As the name implies, you can either link an object to the AutoCAD drawing or embed it in the drawing. When you *link* the Excel file to AutoCAD, any changes you make in the Excel file appear in the AutoCAD drawing also. When you *embed* the source file into the AutoCAD file, it becomes a standalone Excel object. Changes in the original Excel file are not reflected in the embedded object in AutoCAD.

In Practice Exercise 2.3H, you will create an appropriate layout for sheet 1 and place the geometry on it. You will then create the partial parts list using the Excel spreadsheet program and embed the parts list to your AutoCAD drawing as an OLE object.

Practice Exercise 2.3H

The first task is to lay out the geometry on a layout sheet. Because you based the drawing file on an ANSI D template, AutoCAD has already provided one layout view named ANSI D Title Block. You will customize this layout for use with upper weldment drawing sheet 1.

1. Pick the **ANSI D - Title Block** tab at the bottom of the drawing area to display the layout and switch to paper space. Select and erase the standard viewport that appears.
2. Make **Viewport** the current layer.
3. From the **View** menu, pick **Viewports** and **1 Viewport** to create a new viewport. Pick a point at the top left corner of the drawing area and another point at the bottom right so that the new viewport covers the entire drawing area.

4. Doubleclick the viewport to display its properties. Pick the space next to **Standard scale** and scroll down the list of scales to choose **1:20**. (This is AutoCAD's nearest standard scale to represent a scale of $1''=1'\text{-}6''$.)
5. Pick the **PAPER** button at the bottom of the drawing area to switch to model space. Enter the **PAN** command and pan to place the geometry as shown in Fig. 2.3.84. If the spacing of the views needs adjusting to fit on the sheet properly, adjust the spacing as necessary, but be sure to maintain the alignment among the top, front, and side views. You may need to try several times to achieve a pleasing look. Then pick the **ANSI D Title Block** tab to return to the layout view.
6. Right-click the **ANSI D - Title Block** tab to present a shortcut menu, and pick **Rename**. Rename the tab to **0022**. Save your work.

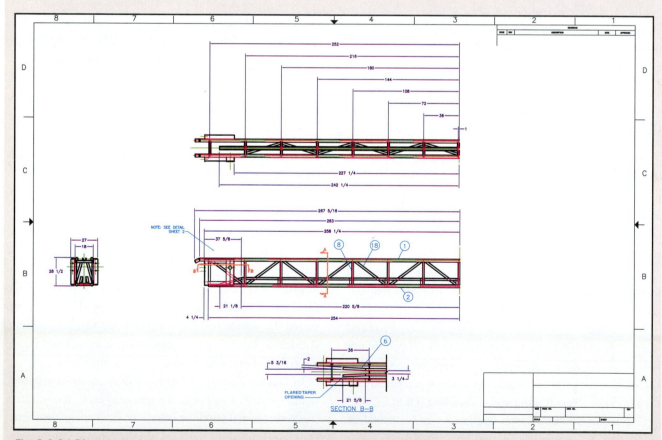

Fig. 2.3.84 Placement of the geometry on sheet 1 of the upper weldment drawing.

7. Open Excel (or a similar OLE-compliant spreadsheet program) and begin a new worksheet. *Note:* If you do not have access to an OLE-compliant spreadsheet, use the procedure outlined in Practice Exercise 1.2R in Project 1.

8. In the first column, create a numbered list from **25** to **1**. In the 26th cell, enter the text **Item #**. Use a font that is similar to the AutoCAD font. *Note:* In Excel, you can create the numbered list automatically by entering the first few numbers and dragging the mouse downward to fill in the numbers.

9. Position the cursor at the top of the column at the line dividing column A and column B until you see a double arrow. Drag the cursor to the left to make the entire column thinner. Make the column just wide enough to accommodate the text in the last line.

10. Proceed to column B. Fill in this column using the data shown in Fig. 2.3.83 on page 161. Be sure to place the text in the correct rows. Since some of the text rows are long, you might want to change the column width to 23 before you start. You can do this either by using the procedure described in the previous step or by highlighting the entire column and selecting **Column** and **Width** from the **Format** menu. Enter **23** in the window that appears and pick **OK**.

11. Follow the same procedure to add the text to columns C and D. Use the data shown in Fig. 2.3.83, and adjust the column widths as necessary. The illustration uses a column width of **8** for column C and **23** for column D.

12. Next, set the alignment for the columns. Column A should already be right-aligned. If not, reset the column alignment by selecting **Cells...** from the **Format** menu. The Format Cells window appears. Pick the **Alignment** tab and set the horizontal alignment to **Right**. See Fig. 2.3.85. Highlight columns B, C, and D and change their format to **Center**. Finally, select cell A26 (the first cell in row 26) and change its alignment to **Center**.

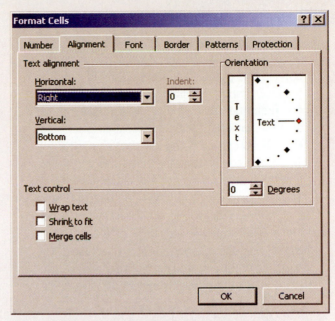

Fig. 2.3.85 *On the Alignment tab of the Format Cells window, set the horizontal alignment to Right.*

13. Highlight all of the cells included in the parts list, from A1 to D26. From the **Format** menu, select **Cells...** to display the Format Cells window. Pick the **Border** tab and pick the **Outline** and **Inside** buttons. This automatically creates the gridlines around the text. Pick **OK** to close the window.

14. Save the Excel file with the name **Partial Parts List.xls** and exit Excel. *Note:* Depending on your current computer settings, you may not be able to see the file extension ".xls."

This completes the construction of the parts list. Now all you have to do is insert it as an object into the AutoCAD drawing file.

15. Return to the AutoCAD file. Make sure the **0022** layout is the current display, and **ZOOM All** to see the entire drawing sheet.

16. From the **Insert** menu, select **OLE Object...** to display the Insert Object window. At the left side of the window, select the **Create from File** radio button. This creates an object from the contents of the file you select. Then pick the **Browse...** button. Navigate to the folder where you stored the **Partial Parts List.xls** file, and doubleclick to select the file.

17. Pick **OK** to make the Insert Object window disappear. The parts list appears on the layout, and the OLE Properties window appears. Leave all of the properties at their default settings, but be sure the **Lock Aspect Ratio** checkbox is selected. Pick **OK** to close the window.

18. Move the cursor over the parts list and notice that the cursor changes to a directional indicator. You can pick with the mouse button and move the parts list anywhere on the drawing sheet. Move it so that the left edge of the parts list aligns with the left edge of the title block. The parts list is currently too big, so it will extend over the border of the drawing sheet.

19. Now size the parts list to fit on the drawing sheet. To do this, position the cursor over the top right corner of the parts list so that a diagonal double arrow appears. Pick with the mouse and drag downward and to the left. Drag until the right edge of the parts list aligns with the right edge of the title block. (You may need to zoom in to do this accurately.) The entire parts list will shrink, but it will maintain its aspect ratio (ratio of length to width).

20. Check the size of the text. It should be about the same size as the text on the dimensions. When it prints on a D-size sheet, it will be approximately $1/8''$ in accordance with ANSI standards. The drawing should now look like the one in Fig. 2.3.86.

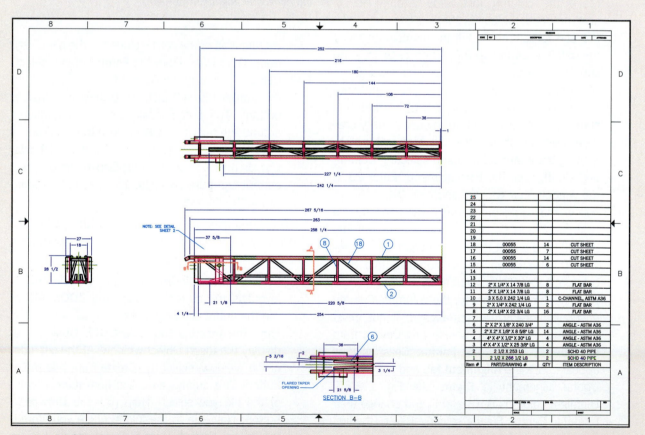

Fig. 2.3.86 Size the parts list to fit above the title block.

Drawing Notes

The drawing notes listed on sheet 1 will apply both to sheet 1 and to sheet 2. A single note on sheet 2 will refer readers to the notes on sheet 1. This practice originally saved drafting time when drafters created drawings manually using drafting instruments. However, the practice persists even in CAD drawings, where it is possible simply to copy the notes from one sheet or file to another.

Welding Notes

Because this is a weldment drawing, you might think that the finished drawing would contain many individual welding symbols like those needed for the hydroelectric turbine nozzle assembly in Project 1. This is not the case, however, because the drafter would need to include so many symbols that the drawing would be difficult to read.

Instead, the drafter realized that all of the welds are similar. They differ only in size, depending on the thickness of the plate being welded. Since only three plate thicknesses are used in the weldment, the most efficient way to describe the welds is to use general drawing notes. Therefore, the drawing notes for the upper weldment drawing consist mostly of welding specifications.

Note Sequence

The notes for this drawing will be numbered from the bottom in a manner similar to the parts list. This is different from the top-down numbering used in Project 1. The reason the notes are numbered from the bottom is that this is a field project. Additional notes are likely to be added to the drawing as construction progresses. It is much easier to add notes to the top of the list than to move the existing notes to make space for additional ones.

In drawings that have several notes, it is sometimes faster and more convenient to create the notes using text editing software and then embed them in the AutoCAD file as OLE objects. In this case, however, the notes contain welding symbols, which are difficult to create in a text document, so it is better to use either DTEXT to create the notes.

Practice Exercise 2.31

The drawing notes for the upper weldment drawing are shown in Fig. 2.3.87. Refer to the discussion of weld symbology that begins on page 91 if you need help understanding the welding symbols.

All of the welds are fillet welds, so you need only build the symbol one time. You can then copy it and merely change the text for the other two symbols.

1. Make **ROMANS** the current text style, and make **TEXT** the current layer.
2. Enter **DTEXT** and create the **NOTES** title and the text for notes 1 and 2. Place them in the lower left corner of the drawing.
3. Create the welding symbol for note 2. You may use a welding symbol library if one is available. If not, create your own symbol. Place the welding symbol on the **TEXT** layer because it is part of the notes.

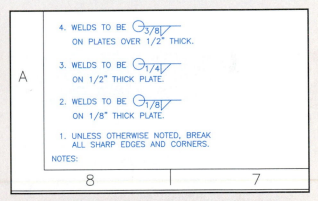

Fig. 2.3.87 *Place these drawing notes in the lower left corner of the UPPER WELD 1 layout.*

4. Copy note 2, including both lines of text as well as the welding symbol. Place the copy above note 2. Edit the text to make this note 3. Be sure to edit the text in the welding symbol as well. Then repeat this procedure to create note 4. Space the notes as shown in Fig. 2.3.87.

Title Block

The title block for this drawing will be similar to that used in Project 1. However, because Project 2 involves a larger number of drawings, drawing numbers become a more important feature of the title block.

Tolerances for Field Projects

It may surprise you to find that tolerancing is a gray area in field welding projects. Recall that the drafter must work very closely with the workers on-site as work progresses on the project. Exact tolerances are difficult to achieve on welds, so even when the drafter specifies a tolerance, the welders may ask the drafter if it is really necessary to spend the time and money to achieve the exact tolerance level. Often, the welders use the tolerance statement as a general guide and simply create the weld as close as they can to that tolerance.

The best way to approach tolerancing in a field project is to add specific tolerances where they are needed. The drafter can then add a loose general tolerance in the tolerance block.

For the tire conveyor, critical tolerancing was not required because the welders were actually the designers. Specific tol-erances may need to be added in the future if the conveyor is constructed by someone else. For now, however, a general tolerance will suffice.

It might be tempting to specify a general tolerance of ±1/16″ for the fractional dimensions on the tire conveyor. However, this is too tight a tolerance for the field welders to hold and still construct the conveyor in a timely and cost-efficient manner. On the other hand, a tolerance of ±1/8″ might be too loose. We will therefore specify a general toler-ance of ±3/32″.

Note that decimal tolerances are also included in the tol-erance block. There are no decimal dimensions on sheet 1, but sheet 2 contains decimal drill dimensions that must be toleranced. The tolerances are shown both on sheet 1 and on sheet 2.

In Practice Exercise 3.2J, you will complete the title block and add the general drawing tolerances. You will also assign drawing numbers to all of the drawings in this set. These numbers will be different from those on the original Emery Energy Company, LLC, drawings, so you will also need to edit the parts list to change the drawing number for parts 15 through 18. This will give you experience in editing a linked OLE object in AutoCAD.

Practice Exercise  2.3J

1. Zoom in on the title block in the lower right cor-ner of the drawing sheet. Make **TEXT** the current layer.
2. Add the information on the left side of the title block, as shown in Fig. 2.3.88. *Hint:* Use AutoCAD's **COPYCLIP** command to copy the text from the title block in Project 1. Insert it into the title block in Project 2 using **PASTE-CLIP**. Then just edit the text to enter the information for Project 2.
3. Add the name of the drawing. Use **MTEXT** for this so that you can control the text alignment more easily. Use a text alignment of **Top Center** and a text size of **3/16**.
4. Add the drawing number (**0022**) for upper weld-ment drawing sheet 1.

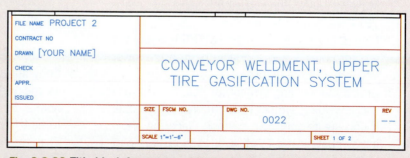

Fig. 2.3.88 *Title block for upper weldment drawing sheet 1.*

5. Add the sheet number and scale information. The sheet number is **SHEET 1 OF 2**, and the scale is **1″=1′-6″**.

6. Zoom to the blank area to the left of the title block. Use the **LINE** command and either **DTEXT** or **MTEXT** to create the tolerance block. Use the general drawing tolerances shown in Fig. 2.3.89 with a text size of **1/16**.

When you created the parts list, you were directed to use Emery's original drawing number (00055) to refer to parts 15 through 18. Changing the number to the series used in this project will give you experience in editing an embedded object in AutoCAD. Follow these steps.

7. Doubleclick the parts list on the 0022 layout. Excel opens automatically and displays the parts list for editing, as shown in Fig. 2.3.90. Note

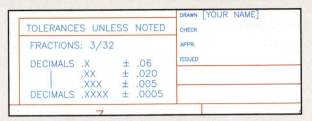

Fig. 2.3.89 *Tolerances for the upper weldment drawing.*

that this is not your original partial parts list.xls file. It is a standalone Excel object because we embedded the file into AutoCAD. Change all occurrences of 00055 to **0025**. Save the Excel file and close it. The change you made is reflected in the AutoCAD file.

8. Save your work and close the file.

This completes drawing 0022, sheet 1 of the upper weldment drawing. When you finish, your drawing should look similar to the one in Fig. 2.3.1 on page 126.

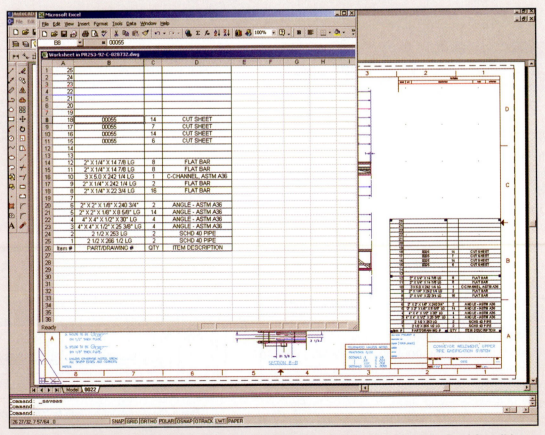

Fig. 2.3.90 *Doubleclicking the parts list in the AutoCAD file automatically opens it in Excel for editing.*

Finish Drawing 0022-2

Sheet 2 contains details of the top end of the upper weldment. You have already drawn most of the geometry, so your work on sheet 2 will go quickly.

Dimensions and Identification Numbers

Before you work further on the existing views, you must modify the front and top views in drawing 0022-2. Notice in Fig. 2.3.2 that only the upper ends of these views are included on sheet 2. This was done so that the upper end of the tire conveyor can be shown at a larger scale for dimensioning and identification purposes.

Drawings are generally dimensioned before the identification bubbles are added because the dimensions must be placed according to ASME Y14. Identification bubbles can be placed anywhere as long as their reference is clear.

Section A-A

The side view that you completed in previous practice exercises includes details specific to the top end of the upper weldment. A sectional view is therefore needed to show a typical cross-section of the weldment. In addition to the sectional view itself, Section A-A includes details to show the upper and lower chain supports more clearly. Because Section A-A is itself a detail section, the drafter can also use it to augment the identification information on the reference assembly drawing.

In Practice Exercise 2.3K, you will modify the views you copied to drawing 0022-2 at the beginning of Practice Exercise 2.3G. Then you will add the dimensions and identification numbers to complete these views. Next, you will copy the existing side view and modify it to form Section A-A. Finally, you will add the drawing note and title block information to finish sheet 2 of the upper weldment drawing.

Practice Exercise

1. Open the **PROJECT 2-0022-2.dwg** file.

   Only the upper portions of the front and top views are needed for sheet 2. Therefore, the first task is to break the front and top views and delete the geometry to the right of the break.

2. Zoom in on the front and top views and make **Dimensions** the current layer.

3. Enter the **XLINE** command and **V** for the Vertical option. Place the construction line halfway between the support structures, as shown in Fig. 2.3.91.

4. Trim/erase all of the geometry to the right of the construction line in the front and top views.

5. Break the construction line to create break lines for the top and front views. Add the break symbol to finish the break lines, as shown in Fig. 2.3.92.

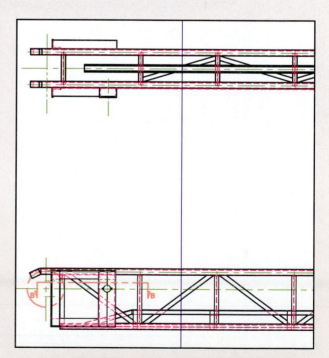

Fig. 2.3.91 *Place the construction line halfway between the second and third support structures.*

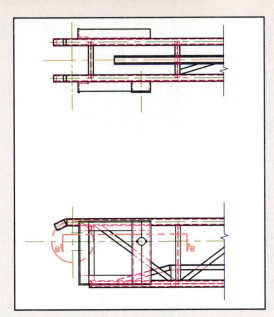

Fig. 2.3.92 *Finished break lines for front and top views.*

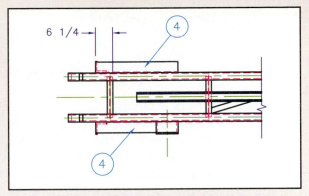

Fig. 2.3.93 *The finished top view for sheet 2.*

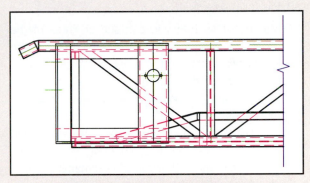

Fig. 2.3.94 *The front view after step 13.*

6. Zoom in on the top view. Erase the vertical centerline that represents the shaft for the larger sprockets.

7. Open **DesignCenter** and navigate to the folder in which you store your AutoCAD drawings. Select **PROJECT 2-0022.dwg** and display its dimension styles. Drag the **Fractions** dimensioning style from DesignCenter into the current drawing (PROJECT 2-0022-2.dwg). Close DesignCenter.

8. Create a new dimension style based on the **Fractions** style. Name the new style **Detail**. Change the text size and arrow size to **2**, and change the offset from dim line, extend beyond dim lines, and offset from origin to **1**.

9. Make **Detail** the current dimension style. Add the dimension on the **Dimensions** layer, as shown in Fig. 2.3.93.

10. Add the identification numbers in the top view, as shown in Fig. 2.3.93. Place the identification numbers on the **Text** layer. Give the text a height of **2 1/2** and the bubbles a diameter of **7**. This completes the top view.

11. Zoom in on the front view.

12. Erase the vertical sprocket centerline, the phantom circle that represents the large sprocket, and the cutting-plane line for Section B-B.

13. Trim the horizontal centerline to the holes for the motor in the rectangular framework. The front view should now look like Fig. 2.3.94.

14. Add the fractional dimensions as shown in Fig. 2.3.95.

15. Create a new dimension style based on the existing Detail style. Name the new style **Detail Decimal**. Leave all of the settings the same, except change the primary units from fractional to **Decimal** with a precision of three decimal places. Make **Detail Decimal** the current dimension style and add the decimal dimensions.

16. Add the identification numbers as shown in Fig. 2.3.95. Use the same text and bubble sizes you used for the top view.

17. Zoom in on the side view.

18. Erase all of the horizontal centerlines except those that run through the bolt holes. Leave the vertical centerlines in place.

19. Add the dimensions as shown in Fig. 2.3.96. Use the **Detail** dimension style for fractional dimensions, and use the **Detail Decimal** dimension style for decimal dimensions.

The next step is to create Section A-A and its associated details. Much of the geometry is the same as that in the side view, so you can use the side view as a basis for the section.

20. Zoom out so that you can see all three of the existing views for sheet 2, as well as some of the blank drawing area to the right of the front view.

21. Make **Object** the current layer. Then freeze the **Dimensions** and **Text** layers to remove them temporarily from the drawing. With Ortho on, enter the **COPY** command and select the entire side view. Place the copy to the right of the front view. Ortho will ensure that the view remains in alignment with the other views.

22. Thaw the **Dimensions** and **Text** layers and zoom in on the copy that will become Section A-A.

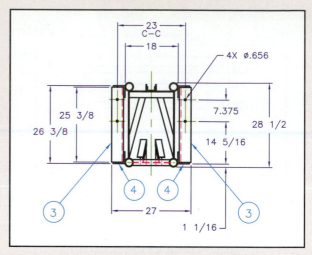

Fig. 2.3.96 *The finished side view for sheet 2.*

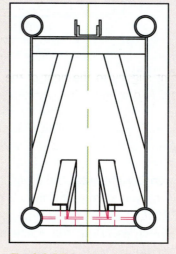

Fig. 2.3.97 *Section A-A without the rectangular framework.*

23. Erase all of the pieces of the rectangular framework at the top of the weldment, as well as the associated steel angles. When you finish erasing, Section A-A should look like Fig. 2.3.97.

24. The rails that make up the lower chain support need to be edited to show a typical cross section. Zoom in on the the lower chain support.

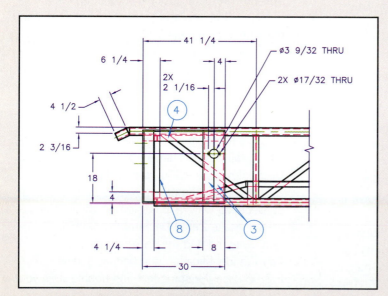

Fig. 2.3.95 *The finished front view for sheet 2.*

25. Edit both rails and their support posts to appear as shown in Fig. 2.3.98. Refer to Figs. 2.3.56 and 2.3.57 on page 151 for the dimensions and placement of the rails and support posts.

Recall that two flat bars run across the bottom of the weldment at right angles to each other. Sectional detail A-A will show only part of the flat bar that appears in edge view. This bar will run in front of the support posts. The other flat bar will appear behind the support posts. Remain at the same zoom magnification to edit this part of the section.

26. Select the two hidden lines that represent the edge view of the horizontal flat bar at the bottom of the weldment and change them to the **Object** layer.

27. Trim away the part of the *left* support post where it is hidden by the edge view of the flat bar.

28. Break the lines that make up the flat bar at a point slightly to the right of the vertical centerline. Add a break line between the two lines to show that this is a broken section. See Fig. 2.3.99.

29. Trim away the parts of the other flat bar that are hidden by the support posts, as shown in Fig. 2.3.99.

This completes the alterations necessary to create the geometry for Section A-A. Before you create the associated details, it is a good idea to add the dimensions and identification numbers to the main view. Then you will know better how to place the details.

30. On the **Center** layer, add horizontal and vertical centerlines for the end views of the pipes. You can do this easily by editing the Detail dimension style. On the **Lines and Arrows** tab, set the center mark type to **Line** and its size to **2**. Then, with Detail as the current dimension style, just pick the **Center Mark** button on the Dimension toolbar and pick each circle to add the centerlines. Also mark the main vertical centerline with the centerline symbol, as shown in Fig. 2.3.100.

31. Add the dimensions for Section A-A, as shown in Fig. 2.3.100. Use the **Detail** dimension style.

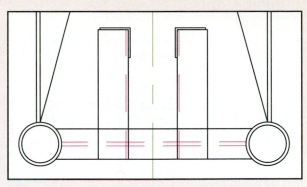

*Fig. 2.3.98 **Edit the rails that make up the lower chain support.***

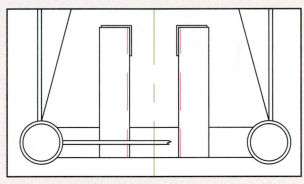

*Fig. 2.3.99 **Edit the two flat bars that run across the bottom of the weldment.***

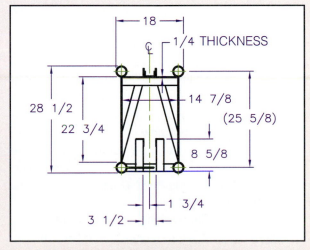

*Fig. 2.3.100 **Add centerlines and dimensions.***

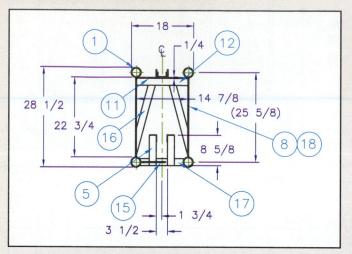

Fig. 2.3.101 *Identification numbers for Section A-A.*

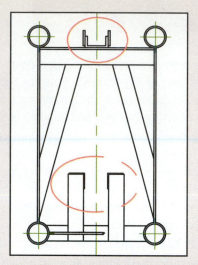

Fig. 2.3.102 *Place the phantom circles to identify the geometry to be shown in detail.*

32. Add the identification numbers, as shown in Fig. 2.3.101.

Now that the main section is finished, the details will be easy to create. You can just copy the appropriate parts of the main section and enlarge them.

33. Zoom out so that Section A-A occupies the lower half of the drawing screen, and make **Phantom** the current layer.

34. Temporarily freeze layers **Dimensions** and **Text**.

35. Use the **ELLIPSE** command to place two ellipses on Section A-A, as shown in Fig. 2.3.102. These ellipses define the areas to be included in the details.

36. Enter the **COPY** command and use a crossing window to select the geometry contained in the upper ellipse, including the ellipse. Note that some of the geometry will extend outside the ellipse. Move the copy to the empty drawing area above the sectional view.

37. Trim the horizontal lines to the edge of the ellipse. Use grips to shorten the vertical centerline, but allow it to extend a short distance above and below the ellipse.

38. Thaw layers **Dimensions** and **Text**. If the dimensions or identification bubbles interfere with the detail geometry, move the detail geometry clear of the main section.

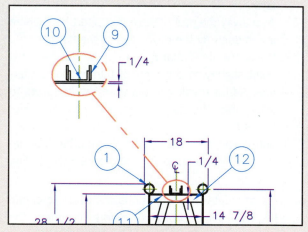

Fig. 2.3.103 *Place the phantom circles to identify the geometry to be shown in detail.*

39. Enter the **SCALE** command, select the detail geometry, and enter a scale of **2**. Move the detail again if necessary to clear the main section.

40. Add the identification bubbles and dimension as shown in Fig. 2.3.103 to finish the detail of the upper chain support. On the **Phantom** layer, add a line to connect the ellipse in the main sectional view to the ellipse in the detail. *Note:* Because you have increased the actual scale of the geometry, the default text of the dimension will be 1/2. You will need to override the text in the Properties window to make it 1/4.

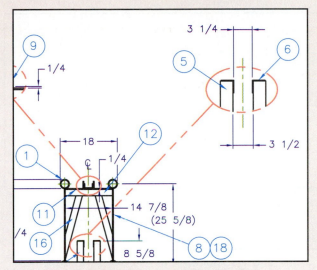

*Fig. 2.3.104 **Add the detail of the lower chain rail. Remember to override the default dimensions.***

45. Pick the **PAPER** button near the bottom of the drawing area to enter model space. Then enter the **PAN** command and use the cursor to position the geometry in its approximate location on the layout sheet. You do not need to place the geometry precisely at this time. Pick the **MODEL** button to return to paper space.

46. Open the **PROJECT 2-0022.dwg** file to display the 0022 layout sheet. Zoom in on the lower left corner of the drawing sheet. Enter **COPYCLIP** and select all of the title block and tolerance block information you created for sheet 1 as the items to copy. Return to the **PROJECT 2-0022-2.dwg** file and enter **PASTECLIP** to paste the information into the appropriate place on the 0022-2 layout. Close PROJECT 2-0022.dwg.

47. Change the sheet number to **SHEET 2 OF 2**, and change the name of the drawing file to **PROJECT 2-0022-2.dwg**. Change the drawing name to **0022-2**, and change the scale to **1″ = 1/8″**. See Fig. 2.3.106.

41. Freeze layers **Dimensions** and **Text**. Then repeat the previous steps to create the detail of the lower chain support. Add the identification numbers and dimensions as shown in Fig. 2.3.104.

42. To complete the section, add the text below the main sectional view to identify Section A-A. Use the **MTEXT** command and use **Top Center** for the justification. Use a text height of **4** for the first line and a text height of **3** for the other two lines. Underscore the first line. When you finish, Section A-A should look like Fig. 2.3.105.

This finishes the geometry for sheet 2 of the upper weldment drawing. Now all you need to do is create a layout for this sheet, place the geometry on the layout, and complete the title block and drawing notes.

43. Use the existing **ANSI D - Title Block** layout. If the Viewport layer is frozen, thaw it now. Erase any existing viewports, and create a new viewport that encompasses the entire drawing area.

44. Right-click the **ANSI D - Title Block** tab. Rename the layout to **0022-2**.

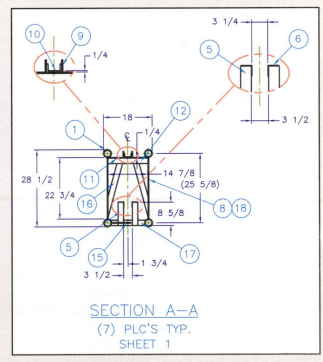

*Fig. 2.3.105 **Section A-A.***

Fig. 2.3.106 *Completed title block for drawing PROJECT 2-0022-2.dwg.*

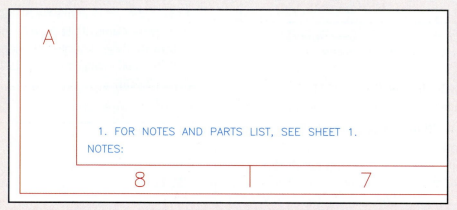

Fig. 2.3.107 *Note text for sheet 2 of the upper weldment drawing.*

48. Zoom in on the lower left corner of the drawing sheet. Using **DTEXT**, create the **NOTES** head at the bottom of the drawing area. Above that, enter the single note for this sheet, as shown in Fig. 3.2.107.

49. **ZOOM All** to see the entire layout. Now there is room to show the views on this detail sheet at a larger scale, as planned. To change the scale, you must first thaw layer **Viewport**. Then doubleclick the edge of the viewport to display its properties. Click **Standard scale** and choose **1:16** from the menu.

This completes your work on drawing 0022-2. You have now finished two of the eight drawings required for the tire conveyor documentation. You have also generated much of the geometry required for the remaining drawings, so your work in this project is about half done.

To check your work so far, compare your drawings to those in Figs. 2.3.1 and 2.3.2, which show the finished upper weldment drawings. Your drawings should look very similar to these.

50. Save and close all of your drawing files.

Review Questions

1. What two commands allow you to copy objects in an AutoCAD drawing to the Windows Clipboard and then paste those objects into other drawing files?
2. Explain why the two commands described in question 1 are useful when a set of working drawings encompasses more than one drawing file.
3. Why is visualization an important skill for drafters?
4. What is the purpose of Section B-B in drawing 0022?
5. Why is a partial parts list needed for the upper weldment drawing?
6. What does the acronym OLE stand for?
7. In OLE, what is the difference between a container application and a server application?
8. What is the major advantage of using a spreadsheet to create a parts list for a field project and then embedding the parts list in the drawing file as an OLE object?
9. Briefly explain how to edit a parts list that has been created in Excel and inserted into the AutoCAD file as an OLE object.
10. Why are the welding specifications listed in the general drawing notes for the upper weldment drawing instead of being specified at each weld location?
11. Why are the notes on drawings for a field project numbered from the bottom up?
12. Discuss the ramifications of choosing a general tolerance for welding drawings in a field project.

Practice Problems

1. To practice visualization, observe the shaded cutting planes on the object in Fig. 2.3.108. Draw a plan view of the object showing the correct location of the cutting-plane lines. Do not dimension. On the plan view, label the cutting-plane line for the section in Fig. 2.3.108A as Section A-A. Label the cutting-plane line for the section in Fig. 2.3.108B as Section B-B. Then create both sectional views.

Portfolio Project

Biomass Conveyor (continued)

Continue your work on the set of working drawings for the biomass conveyor you have designed.
1. Decide which drawing to work on next and open the appropriate drawing file. Remember that you do not have to finish one drawing before starting work on a related drawing.
2. Copy the relevant geometry from the preliminary views to the next drawing you intend to work on.
3. Try to complete at least one drawing at this time.
4. Save your work.

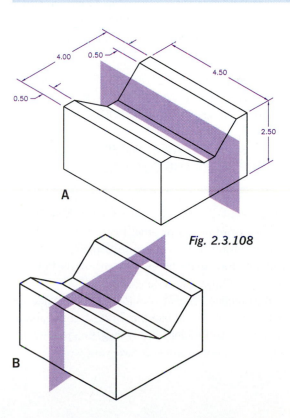

Fig. 2.3.108

2. Open PROJECT 2-0022-2.dwg and use the SAVEAS command to save it with a new name of Practice Problem 2.3.2.dwg. Use the existing geometry to create Section C-C at the top of the upper weldment, as shown in Fig. 2.3.109. Save your work.

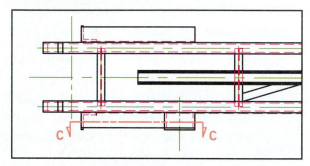

Fig. 2.3.109 *Section C-C for Practice Problem 2.*

3. Open PROJECT 2-0022.dwg and use the SAVEAS command to save it with a new name of Practice Problem 2.3.3.dwg. Pick the 0022 tab to switch to the layout view. Delete the notes from the lower left corner of the layout. Then open an OLE-compliant word processor such as Microsoft Word® and create the following notes:

> 3. GENERAL CONTRACTOR TO FIELD-CHECK AND VERIFY ALL WELDS ON-SITE.
> 2. STRUCTURAL STEEL CONFORMS TO ASTM A36.
> 1. UNLESS OTHERWISE NOTED, BREAK ALL SHARP EDGES AND CORNERS.
> NOTES:

Save the text file as practice note 2.3.3.doc. *Note:* If you are using a word processor other than Microsoft Word, the file extension will be different. Using OLE, embed the NOTES object in the lower left corner of the Practice Problem 2.3.3.dwg file. Save your work.

4. Open the Practice Problem 2.3.2.dwg you created for Practice Problem 2 and use the SAVEAS command to save it as Practice Problem 2.3.4.dwg. Edit the NOTES object to delete note number 3. Save your work.

5. Open PROJECT 2-0022.dwg and use SAVEAS to save it with a new name of Practice Problem 2.3.5.dwg. Edit the parts list to add five additional blank lines at the top of the list, as shown in Fig. 2.3.110. Save your work.

30			
29			
28			
27			
26			
25			
24			
23			
22			
21			
20			
19			
18	0025	14	CUT SHEET
17	0025	7	CUT SHEET
16	0025	14	CUT SHEET
15	0025	6	CUT SHEET
14			
13			
12	2" X 1/4" X 14 7/8 LG	8	FLAT BAR
11	2" X 1/4" X 14 7/8 LG	8	FLAT BAR
10	3 X 5.0 X 242 1/4 LG	1	C-CHANNEL, ASTM A36
9	2" X 1/4" X 242 1/4 LG	2	FLAT BAR
8	2" X 1/4" X 22 3/4 LG	16	FLAT BAR
7			
6	2" X 2" X 1/8" X 240 3/4"	2	ANGLE - ASTM A36
5	2" X 2" X 1/8" X 8 5/8" LG	14	ANGLE - ASTM A36
4	4" X 4" X 1/2" X 30" LG	4	ANGLE - ASTM A36
3	4" X 4" X 1/2" X 25 3/8" LG	4	ANGLE - ASTM A36
2	2 1/2 X 253 LG	2	SCHD 40 PIPE
1	2 1/2 X 266 1/2 LG	2	SCHD 40 PIPE
Item #	PART/DRAWING #	QTY	ITEM DESCRIPTION

Fig. 2.3.110 *Modified parts list for Practice Problem 5.*

6. Open PROJECT 2-0022.dwg and save it with a new name of Practice Problem 2.3.6.dwg. Then open PROJECT 2-0022-2.dwg and use COPYCLIP to copy the entire drawing to the Windows Clipboard. Insert the copy into Practice Problem 2.3.6.dwg. Experiment with using multiple layouts with OLE objects. Describe at least one potential disadvantage of using an OLE parts list with multiple layouts intended for two or more related drawing sheets. Save your work.

By design, the lower weldment contains the same basic structural framework as the upper weldment. In fact, you may consider the lower weldment to be a mirror image of the upper weldment, with the exception of the lower end, which contains several modifications. The views needed for the lower weldment mirror therefore those for the upper weldment and should be arranged similarly on two drawing sheets. The similar placement helps reduce confusion when others read the drawings later.

Because drawings 0022 (upper weldment) and 0021 (lower weldment) are so similar, you can copy the geometry from 0022 and use it as a basis for drawing 0021. Later in this section, you will use drawing 0021 as a basis for the details in drawing 0021-2.

Section Objectives
- Apply the STRETCH command to alter the overall dimensions of the weldment geometry.
- Explore the associativity of dimensions in AutoCAD.
- Create drawings 0021 and 0021-2 based on the geometry from drawings 0022 and 0022-2.

Key Terms
- associative dimensions
- dimension variables
- expanded metal mesh

Drawing Analysis

A 3D model of the entire lower weldment is shown in Fig. 2.4.1. As you can see, the structural framework is almost identical to that of the upper weldment, except that it is a mirror image. You can use this fact to your advantage by:

- copying the geometry from drawing 0022 (upper weldment sheet 1) into drawing 0021.
- mirroring the geometry to create the basic framework for the lower weldment.

However, the details at the lower (right) end of the lower weldment are quite different from those at the upper (left) end of the upper weldment. Most of your work on drawings 0021 and 0021-2 will therefore be focused on the geometry at the lower end of the weldment.

Practice Exercise 2.4A focuses on the steps required to adapt a copy of the structural framework for the upper weldment into usable geometry for the lower framework. Sheets 1

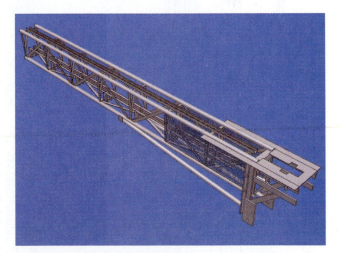

Fig. 2.4.1 **The lower weldment.**

and 2 of the lower weldment drawings are shown in Figs. 2.4.2 and 2.4.3 on the following pages. Refer to them as necessary as you complete the practice exercises.

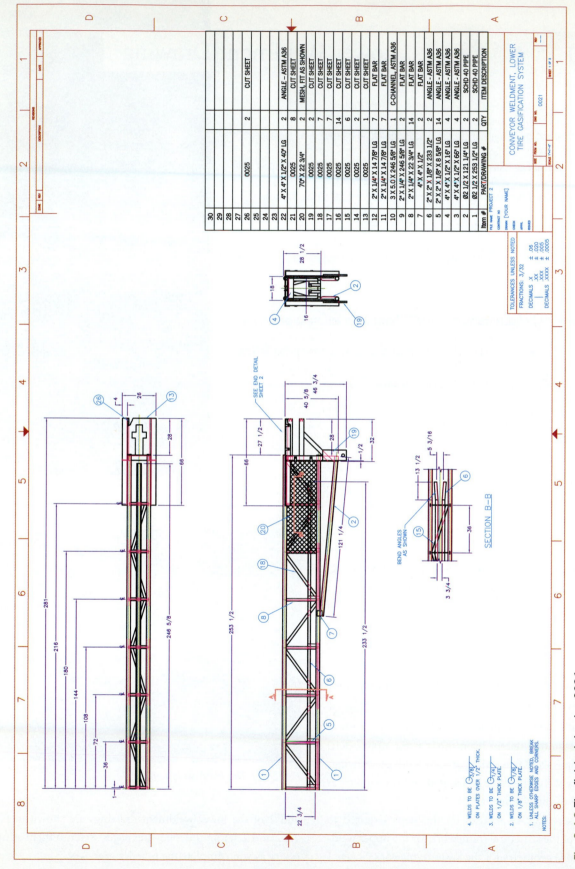

Item #	PART/DRAWING #	QTY	ITEM DESCRIPTION
30			
29			
28			
27	0025	2	CUT SHEET
26			
25			
24			
23			
22	4" X 4" X 1/2" X 40" LG	2	ANGLE - ASTM A36
21	0025	8	CUT SHEET
20	70" X 22 3/4"	2	MESH, FIT AS SHOWN
19	0025	2	CUT SHEET
18	0025	7	CUT SHEET
17	0025	7	CUT SHEET
16	0025	14	CUT SHEET
15	0025	6	CUT SHEET
14	0025	2	CUT SHEET
13	0025	1	CUT SHEET
12	2 X 1/4" X 14 7/8" LG	7	FLAT BAR
11	2 X 1/4" X 14 7/8" LG	7	FLAT BAR
10	3 X 5.0 X 246 5/8" LG	1	C-CHANNEL, ASTM A36
9	2" X 1/4" X 246 5/8" LG	2	FLAT BAR
8	2" X 1/4" X 22 3/4" LG	14	FLAT BAR
7	4" X 4" X 1/2"	2	FLAT BAR
6	2" X 2" X 1/8" X 233 1/2"	2	ANGLE - ASTM A36
5	2" X 2" X 1/8" X 8 5/8" LG	14	ANGLE - ASTM A36
4	4" X 4" X 1/2" X 16" LG	4	ANGLE - ASTM A36
3	4" X 4" X 1/2" X 66" LG	4	ANGLE - ASTM A36
2	Ø2 1/2 X 121 1/4" LG	2	SCHD 40 PIPE
1	Ø2 1/2 X 253 1/2" LG	2	SCHD 40 PIPE

CONVEYOR WELDMENT, LOWER
TIRE GASIFICATION SYSTEM

DWG. NO. 0021

SECTION B–B

BEND ANGLES
AS SHOWN

SEE END DETAIL
SHEET 2

NOTES:
1. UNLESS OTHERWISE NOTED, BREAK ALL SHARP EDGES AND CORNERS.
2. WELDS TO BE 1/8" ON 1/8" THICK PLATE.
3. WELDS TO BE 1/4" ON 1/2" THICK PLATE.
4. WELDS TO BE 3/8" ON PLATES OVER 1/2" THICK.

TOLERANCES UNLESS NOTED
FRACTIONS: 3/32
DECIMALS .X   ± .06
          .XX   ± .020
          .XXX ± .005
          .XXXX ± .0005

Fig. 2.4.2 *The finished drawing 0021.*

Fig. 2.4.3 The finished drawing 0021-2.

Practice Exercise 2.4A

There are actually two ways to go about copying the geometry from drawing 0022 to drawing 0021. You can use the COPYCLIP and PASTECLIP commands to make an actual copy, or you can simply use the SAVEAS command to save PROJECT 2-0022.dwg with a new name of PROJECT 2-0021.dwg. This new file would replace the drawing file you set up in Section 1.

We will use the second method because it offers several advantages. For example, the dimension styles you set up in Section 2.3 will be included automatically, so you won't have to import them.

1. Open the **PROJECT 2-0022.dwg** file.
2. Enter the **SAVEAS** command and save the file with a new name of **PROJECT 2-0021.dwg**.
3. Pick the **Model** tab to enter model space and **ZOOM All** to see the entire drawing.
4. The first task is to remove the dimensions, text, and cutting-plane lines from the drawing. To do this quickly, open the Layer Properties Manager by clicking the **Layers** button. Make **Dimensions** the current layer. Then freeze all of the layers *except* Dimensions, Phantom, and Text and pick **OK**. With only these layers visible in the drawing area, enter the **ERASE** command and enter **All** to erase everything on these three layers. Then pick the **Layers** button again and thaw all of the layers to display the remaining geometry.
5. With Ortho on, enter the **LINE** command and create a vertical line just to the right of the existing views. This will be the mirror line. Enter the **MIRROR** command. In response to the Select objects prompt, enter **All**. Snap to the ends of the vertical line at the prompt for the mirror line. Then enter **Y** to delete the source objects. Erase the mirror line.
6. Enter the **GRID** command and specify a grid spacing of **50**. Then **ZOOM All**. The grid appears only within the drawing limits, so you can use it to determine whether the mirrored geometry lies completely within the drawing limits. If it does not, enter **MOVE All** to place the geometry within the drawing limits. The drawing should now look like Fig. 2.4.4.

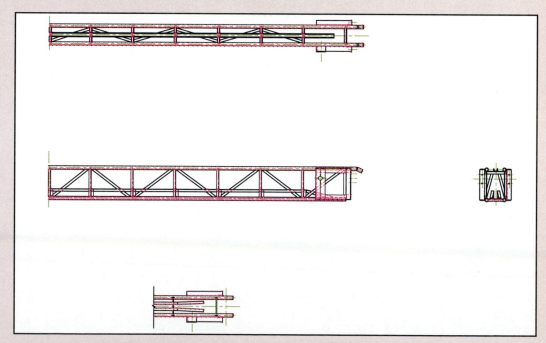

Fig. 2.4.4 The mirrored geometry that forms the basis for drawing 0021.

STRETCH Command

The overall lengths of the pieces in the lower weldment vary slightly from those in the upper weldment. For example, the Schedule 40 pipes at the bottom of the weldment are 1/2″ longer in the lower weldment.

The easiest way to change the overall length of the lower weldment is to use AutoCAD's STRETCH command. To use STRETCH, you select objects using a crossing window that includes only that part of the object that needs to be changed. This process is easier to demonstrate than to describe, but the STRETCH command can be a very valuable tool once you understand how to use it.

For demonstration purposes, we will use the template shown in Fig. 2.4.5A. This template can be modified in many ways using STRETCH. For example, suppose the guide at the upper left of the template needed to be deeper. You could enter the STRETCH command, create a crossing window as shown in Fig. 2.4.5B to select the parts to be stretched, and enter a base point. (Recall that a crossing window is one in which you pick a right corner before choosing the opposite left corner.) Then you could stretch the lines by moving the cursor or, for more precise control, specify the amount of movement using polar coordinates.

For this example, the chosen base point was the lower left corner of guide 1. A polar coordinate of @10<270 was entered to make the guide 10″ deeper than its original depth. Study the illustrations in Fig. 2.4.5C through F to see other ways in which the template can be changed using only the STRETCH command.

Associative Dimensions

By default, dimensions in AutoCAD are **associative dimensions**. In other words, when you change the size of the geometry to which a dimension refers, the dimension updates to reflect the new size. Fig. 2.4.6 shows what happens to example dimensions when the stretch described in Fig. 2.4.5D is performed.

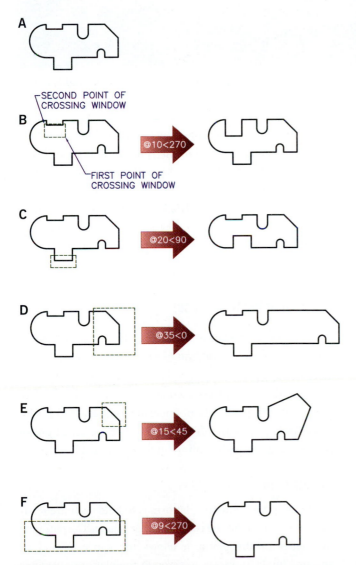

Fig. 2.4.5 *Use of the STRETCH command.*

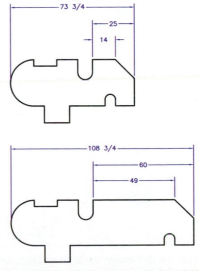

Fig. 2.4.6 *Associativity of dimensions in AutoCAD.*

We did not mirror the dimensions from drawing 0022 because the text would have been mirrored also. However, using associative dimensions can save a drafter a great deal of time in field projects in which the dimensions may change several times as the project progresses. Also, for simple, nonmirrored copies, you can copy both the part and its dimensions and then adjust the geometry and dimensions simultaneously using this feature. You should therefore be familiar with the concept and use it whenever appropriate.

Dimensions that have been exploded lose their associativity. This is why it is considered poor practice to explode dimensions in order to change their appearance. Instead, drafters always should use AutoCAD's **dimension variables** to change the appearance of dimensions. The functions of the most commonly used dimension variables are included in the Properties window, so dimensions are easy to modify without exploding. You can also change a variable by entering its name directly at the keyboard and then entering the appropriate value.

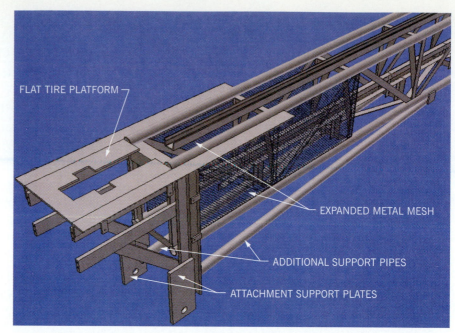

Fig. 2.4.8 Purpose of the additional pieces at the lower end of the lower weldment.

When you set up a dimension style, the dimension variables are changed according to your specifications. These variable settings apply only to that dimension style. To see the dimension variables and their settings for a given dimension style, enter the Dimension Style Manager. Choose the dimension style in which you want to view the variable settings and pick the Compare button. The variables and their settings are displayed as shown in Fig. 2.4.7.

Function of the New Pieces

Most of the changes to be made at the lower end of the conveyor involve the addition of new pieces. A basic knowledge of the function of each additional piece to be drawn will help you draw the pieces accurately and avoid inadvertent drawing errors due to lack of understanding. Refer to the 3D model in Fig. 2.4.8 and the photograph of the actual prototype in Fig. 2.4.9 as you read about the function of each piece.

Tire Platform

The purpose of the tire platform is to hold each tire ready for the next hook on the conveyor to engage it. The cutout at the center of the platform allows the conveyor and its attached hooks to move up through the platform, carrying the tire with it as the chain moves up the conveyor.

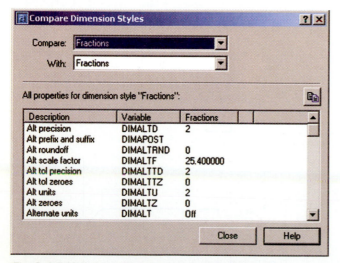

Fig. 2.4.7 Listing the dimension variable settings for a specific dimension style.

Fig. 2.4.9 **Lower end of the tire conveyor.**

Support Pipes

Additional support pipes have been added at the lower end to increase the strength of the support system. When the conveyor is in place, most of the weight of the system rests on this area.

Attachment Support Plates

The flat plates at the bottom of the weldment provide support for the weldment when it is raised to its working position. A shaft is passed through the holes. Then the flat plates, or mounting feet, are attached on the shaft. These feet allow the conveyor to pivot on the shaft as it is tilted up into its place. In the final installation, the plates will be fixed to the foundation to lock the conveyor in place. The shaft and the mounting feet will be shown on the reference assembly drawing as items 33 and 3, respectively.

Expanded Metal Mesh

When designing any product, it is important to keep the safety of the end users in mind. The designers of the tire conveyor realized that when the conveyor is in use, people need to stand near the lower end of the conveyor to operate it. The potential for accidents is relatively high at this location.

To help ensure the safety of these workers, an **expanded metal mesh** has been added to both sides of the lower end. Expanded metal is made from a thin, solid sheet of metal. Slits are created at regular intervals in the metal, and then the metal is stretched to create a pattern of diamond-shaped holes. Mesh made of expanded metal is extremely strong, so it makes a good safety fence for this application. Workers can see through it to monitor the conveyor's performance, but it protects against accidents.

Drawing 0021

Now that you have analyzed the drawing task and the item to be drawn, you can begin the actual work of modifying the geometry to create drawing 0021. You will need to add all of the new pieces discussed in the previous section.

Front View

In Practice Exercise 2.4B, you will add the new components and modify the existing geometry in the front view. Refer to Figs. 2.4.2, 2.4.8, and 2.4.9 as necessary as you complete the exercise.

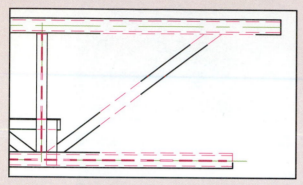

Fig. 2.4.10 *Appearance of the lower end of the lower weldment after step 2.*

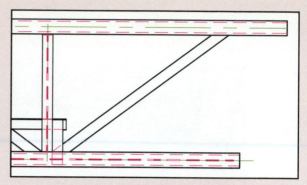

Fig. 2.4.11 *Lower end of the lower weldment with object lines repaired.*

The geometry in the PROJECT 2-0021.dwg file still contains parts that are specific to the upper weldment. The first task will be to delete the unnecessary geometry.

1. Open **PROJECT 2-0021.dwg** and zoom in on the lower (right) end of the front view of the lower weldment.

2. Erase all of the lines that make up the rectangular framework, the motor housing, and the bent portion of the upper Schedule 40 pipes. Also erase the portion of the lower chain support that bends downward, the centerlines for the bolt-holes, and the last vertical flat-bar cross-support structure. When you finish, the lower end of the front view should look like Fig. 2.4.10.

Notice that some parts of the lines that are now visible are still in the Hidden linetype. Even though portions of these lines may be hidden in the final drawing, it is good practice to update them now so that all object lines that are not hidden from view are solid (unbroken) lines on the Object layer. This prevents the possibility of accidentally leaving an object line hidden.

3. Mend the lines of the upper and lower Schedule 40 pipes and the diagonal flat bar so that each one consists of a single line object on the **Object** layer. Be careful to keep the endpoints of the lines at their correct locations. See Fig. 2.4.11.

Next, turn your attention to the Schedule 40 pipes. In the upper weldment, the pipes at the top of the structural framework are longer than those at the bottom of the framework. This is not so in the lower weldment. The upper and lower pipes are all exactly 253 1/2″ long. You can modify the lengths in one of two ways:

- using the STRETCH command
- offsetting the left end of the pipes by 253 1/2″ and then trimming or extending the horizontal lines to the correct length

4. Use the method of your choice to make all of the Schedule 40 pipes exactly **253 1/2″** long. If you use the **STRETCH** command, you will need to calculate the difference in the lengths of the pipes in the upper and lower weldments. Then use polar coordinates with those numbers to stretch all of the lines to their new size. When you finish, your drawing should look like the one in Fig. 2.4.12.

5. Edit the lower chain support. In the lower weldment, the lower chain support still flares in the top view, but it does not bend down in the front view. Use **STRETCH** to make its new overall length **233 1/2″**.

6. The lower chain support now runs behind the diagonal flat bar. Change the portion of the lower chain support that runs behind the flat bar to the **Hidden** layer.

The structures at the lower end of the lower weldment include a platform for mounting the tires onto the conveyor and a mounting support system. Both of

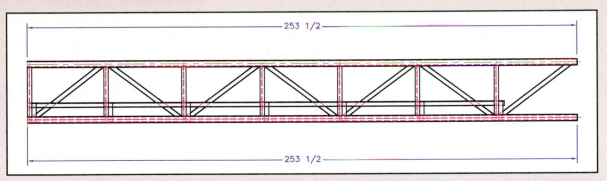

Fig. 2.4.12 *Edit the upper and lower Schedule 40 pipes to the proper length.*

these structures are supported by two $4'' \times 4'' \times 1/2''$ steel angles fastened at right angles to each other, as shown in Fig. 2.4.13. The next task is to add these angles and their associated mounting tabs in the front view of the drawing.

7. The left end of the horizontal **$4'' \times 4'' \times 1/2''$** steel angle support aligns with the rightmost vertical support post. The top edge of the angle is located at the centerline of the upper Schedule 40 pipe, and the angle is **66″** long, extending **27 1/2″** to the right of the right end of the pipes. Use this information to create the horizontal steel angle support.

8. Change the object lines that fall behind the angle to the **Hidden** layer. The lower end of the front view should now look like Fig. 2.4.14.

9. The vertical **$4'' \times 4'' \times 1/2''$** angle steel piece is **40 1/8″** long, and its right edge is located **28″** from the right edge of the horizontal angle. Its top edge butts up against the underside of the horizontal angle. Use this information to create the vertical steel angle.

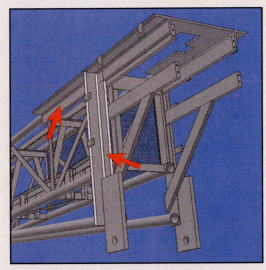

Fig. 2.4.13 *The structures at the lower end are supported on each side by two pieces of angle steel at right angles.*

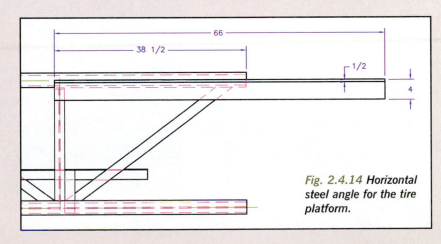

Fig. 2.4.14 *Horizontal steel angle for the tire platform.*

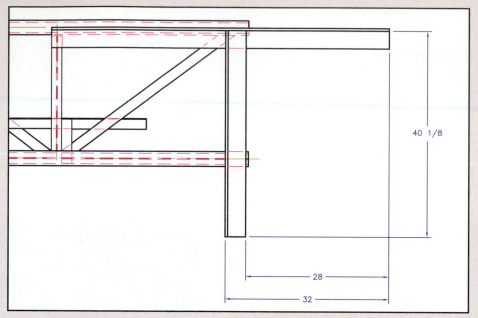

Fig. 2.4.15 *Vertical steel angle support for the tire platform.*

10. Change the object lines of the lower Schedule 40 pipe to hidden lines. For clarity, trim away the existing hidden lines in this portion of the Schedule 40 pipe. Do not trim away the center-line, however. See Fig. 2.4.15.

The additional support pipes on the lower weldment are fastened at one end to the vertical support you created in steps 9 and 10. At the other end, they are fastened to the existing lower Schedule 40 pipe by means of a $4'' \times 4'' \times 1/2''$ plate made from a piece of flat bar, as shown in Fig. 2.4.16.

11. The right edge of the $4'' \times 4'' \times 1/2''$ plate is located **113**″ from the left edge of the vertical support angle. The top of the plate is located **11/16**″ below the top object line of the Schedule 40 pipe. Use this information to create the plate. See Figs. 2.4.17 and 2.4.18.

Now you can create the extra $2\,1/2''$ Schedule 40 pipe supports. One way to position the pipe correctly without taking the time to calculate the angle is described in the following steps.

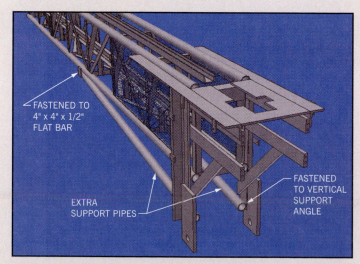

FASTENED TO 4" x 4" x 1/2" FLAT BAR

EXTRA SUPPORT PIPES

FASTENED TO VERTICAL SUPPORT ANGLE

Fig. 2.4.16 *Connections for extra support pipes.*

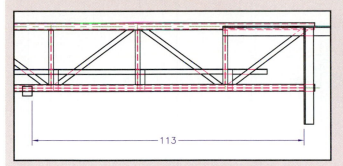

Fig. 2.4.17 *Location of the plate at the left end of the extra support pipes.*

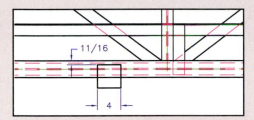

Fig. 2.4.18 *Vertical position of the plate.*

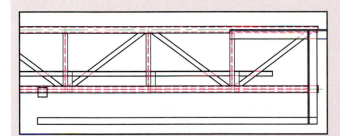

Fig. 2.4.19 *Object lines for the extra support pipe.*

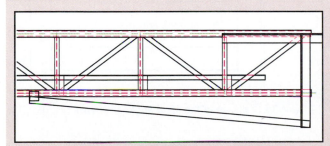

Fig. 2.4.20 *Rotate the lines into place.*

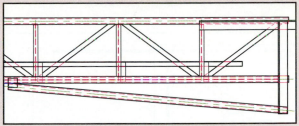

Fig. 2.4.21 *Finished extra support pipe in the front view.*

12. Enter the **PLINE** command and snap to the lower right corner of the vertical support angle for the first point in the polyline. Then enter the following polar coordinates to finish the object lines for the pipe:

    @121-1/4<180
    @2-7/8<90
    @121-1/4<0
    Close

    This creates an orthographic version of the pipe's object lines, as shown in Fig. 2.4.19.

13. If Ortho is on, turn it off before proceeding. Enter the **ROTATE** command and pick the upper right corner of the new Schedule 40 pipe as the base point for rotation. Use the cursor to rotate the pipe so that the upper left corner touches the intersection of the 4″ × 4″ × 1/2″ plate and the existing lower Schedule 40 pipe. When you finish this step, the new pipe should be positioned as shown in Fig. 2.4.20.

14. Enter the **EXPLODE** command and explode the polyline to convert it to four lines.

15. Offset the object lines as required to add the hidden and centerlines to complete the Schedule 40 pipe. Be sure to change the lines to their correct layers. The finished pipe is shown in Fig. 2.4.21.

The flat plates that make up the mounting feet are attached to the vertical angle support. In the set of working drawings, the dimensions of the mounting feet are detailed on the cut sheet, which you have not yet completed. They are to be 8″ × 18″, cut from 1″ flat stock. Follow these steps to place the mounting feet accurately.

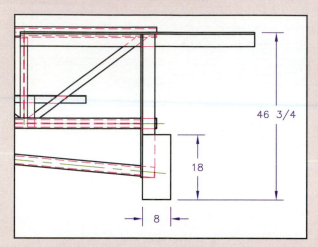

Fig. 2.4.22 *Placement of the flat-plate mounting feet in the front view.*

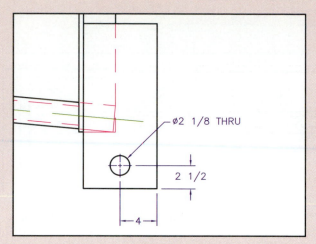

Fig. 2.4.23 *Placement of the shaft hole in the mounting feet.*

16. Offset the top of the horizontal steel angle support down by **46 3/4″** to locate the bottom of the flat plate. The left side of the plate should fit flat against the inside corner of the steel angle. Use this information to draw the flat plate. Refer to Fig. 2.4.22.

17. The plate hides part of the vertical steel angle and the extra support pipe from view. Change the object lines to the **Hidden** layer, and delete the existing hidden lines that represent the inside surface of the pipe.

18. Add the hole through which the shaft will be run. Place the hole as shown in Fig. 2.4.23.

19. Two additional supporting flat bars are needed for stability on each side of the lower end. The left edge of each bar is aligned with the left edge of the vertical steel angle support. The bars are **3 1/2″** wide, **1″** thick, and **32″** long. Create the bars, placing them as shown in Fig. 2.4.24.

20. Notice that the top bar produces some confusion because it lies partly behind the top angle. It is difficult to tell which lines belong to which part. To fix this problem, break away the edge of the angle to reveal the flat bar underneath, as shown in Fig. 2.4.25. Use either the **PLINE** command or the **SKETCH** command with **SKPOLY** set to **1** to produce the break line.

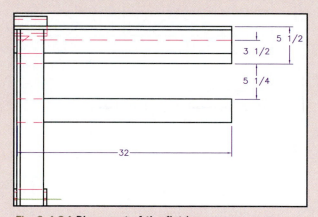

Fig. 2.4.24 *Placement of the flat bars.*

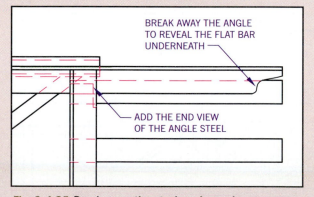

Fig. 2.4.25 *Break away the steel angle as shown.*

21. Add the end view of the 40 × 40 × 1/2″ steel angle, as shown in Fig. 2.4.25.

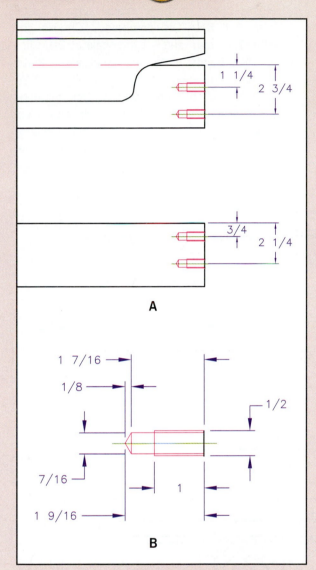

A

B

Fig. 2.4.26 (A) Placement of the two tapped holes in the flat bars. (B) Dimensions for the tapped holes.

22. To finish the two flat bars, add the tapped holes as shown in Fig. 2.4.26A. Use the dimensions shown in Fig. 2.4.26B.

23. A diagonal support runs from the lower flat bar to the metal foot. It is cut from a flat bar and is **3"** wide, **1"** thick, and **32 1/8"** long. It is beveled **3"** at each end to fit. Create the bar and position it as shown in Fig. 2.4.27. Change hidden portions of the lines to the **Hidden** layer.

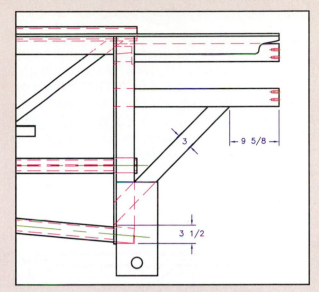

Fig. 2.4.27 Add the diagonal flat-bar support.

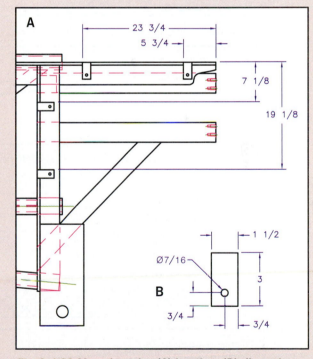

Fig. 2.4.28 Mounting tabs: (A) location; (B) dimensions.

24. Add the four mounting tabs as shown in Fig. 2.4.28. The tabs are cut from **3/8"** flat bar. Place the tabs as shown in Fig. 2.4.28A. Use the dimensions shown in Fig. 2.4.28B.

Finally, you can add the expanded metal mesh. The most efficient way to create the mesh is to use the BHATCH command and the ANSI37 hatch pattern that is provided with AutoCAD. Note that, although you are using a hatch pattern, this hatch belongs on the Object layer. It is being used not as a hatch but as a part of the structure.

25. Zoom out to a magnification at which you can see the rightmost three vertical support posts. Enter the **BHATCH** command. Choose the **ANSI37** pattern, and set the scale at **20**. Click the **Pick Points** button to select areas to be hatched. The heavy red line in Fig. 2.4.29 shows the hatch boundary result; you will need to pick several points within that area to select all of it. Preview the hatch and fix any problems you see. Then pick OK to create the hatch. Your drawing should look like the one in Fig. 2.4.30.

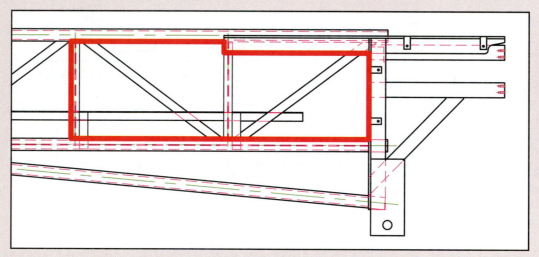

Fig. 2.4.29 *The selection area for the BHATCH operation.*

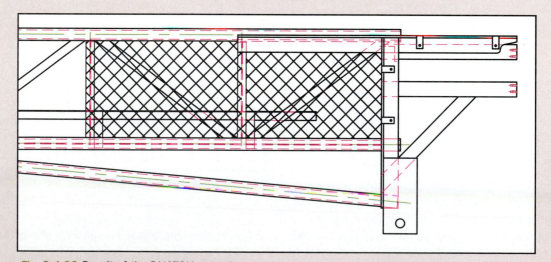

Fig. 2.4.30 *Result of the BHATCH operation.*

Top View

Now that the alterations to the front view are complete, you can use them as a basis to add the geometry in the top view. Again, your changes will be limited to the right (lower) end of the conveyor.

To help visualize the components, refer to Fig. 2.4.31, which contains a 3D model showing the top of the right end of the conveyor. A closeup of the right end of the top view is shown in Fig. 2.4.32. Refer to these illustrations as you work on the top view in Practice Exercise 2.4C.

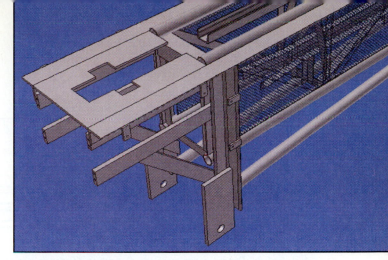

*Fig. 2.4.31 **The top of the lower end of the lower weldment, as seen from above.***

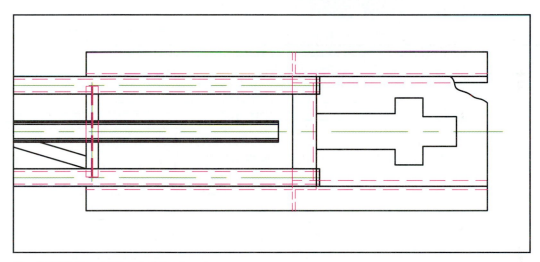

*Fig. 2.4.32 **The right end of the top view.***

Practice Exercise 2.4C

As in the front view, you must erase the unneeded geometry in the top view before you can begin to add the correct geometry. Zoom in on the right end of the top view, and perform the following steps.

1. Erase the rectangular framework, the motor housing, and all of the lines that represent the bearings and shafts. Also delete the rightmost cross support (which appears vertical in the top view). When you finish, the right end of the top view should look like Fig. 2.4.33.

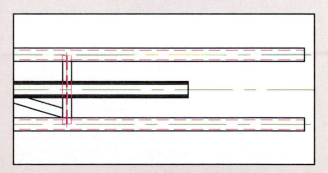

*Fig. 2.4.33 **Top view prepared for new geometry.***

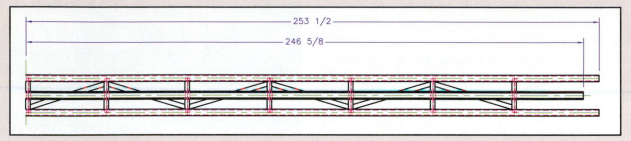

Fig. 2.4.34 *Overall lengths of the Schedule 40 pipes and the upper chain support in the top view.*

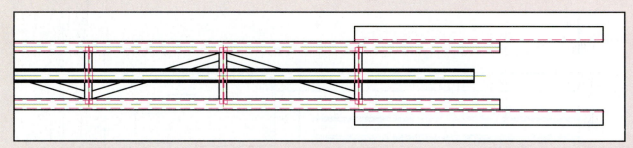

Fig. 2.4.35 *Placement of the horizontal steel angle supports in the top view.*

2. Recall that you spent time on the front view adjusting the lengths of the Schedule 40 pipes. You can save time in the top view by transferring the distances. Use the **XLINE** command to place vertical construction lines at key points or use **Object Snap Tracking** to transfer the endpoints from the front view to the top view.

3. Adjust the right end of the top chain support. Leave the left end where it is, but change its length to **246 5/8″**. You may do this using the **OFFSET** and **TRIM** commands or the **STRETCH** command. Fig. 2.4.34 shows the expected dimensions after you adjust the lengths of the horizontal lines.

4. Transfer the end lines of the horizontal steel angle supports from the front view to the top view. Complete the **4″ × 4″ × 1/2″** angle steel pieces in the top view, as shown in Fig. 2.4.35.

5. Add the vertical steel angle supports as shown in Fig. 2.4.36. Transfer all of the necessary location information from the front view.

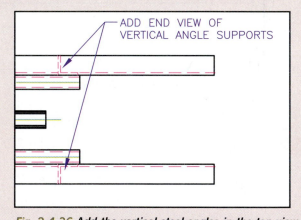

Fig. 2.4.36 *Add the vertical steel angles in the top view.*

6. Add the **32″**-long, **1″** horizontal flat bars. Transfer their location from the front view, as shown in Fig. 2.4.37.

7. The left side of the tire platform rests on a piece of **4″ × 4″ × 1/2″** angle steel that runs between the horizontal flat bars from step 6. Add this angle using the dimensions shown in Fig. 2.4.38.

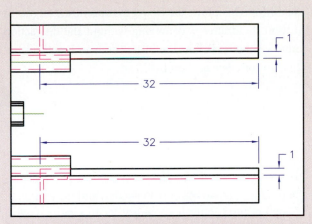

Fig. 2.4.37 *Add the flat-bar supports in the top view.*

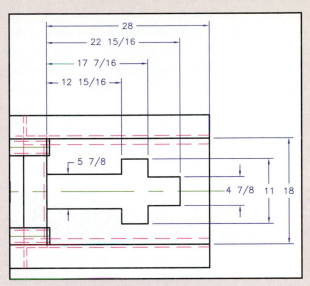

Fig. 2.4.39 *Dimensions for the tire platform.*

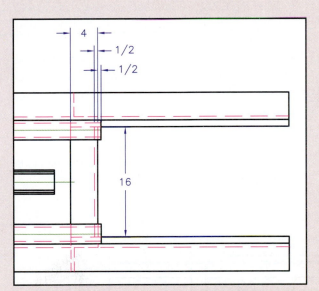

Fig. 2.4.38 *The flat-bar supports run the length of the tire platform.*

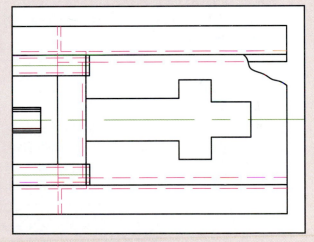

Fig. 2.4.40 *Break away a part of the tire platform to show the flat bar underneath.*

8. Add the **28″ × 18″** tire platform, including the cut-out for the drive chain. Take the dimensions from Fig. 2.4.39. *Note:* Extend the main horizontal centerline before you begin to provide a reference for offset operations. Because the tire platform is above the two flat bars, they are now hidden, so change them to the **Hidden** layer.

9. To reduce confusion caused by all of the aligning edges at the right side of the tire platform, break away a piece of the upper right corner of the platform, as shown in Fig. 2.4.40, to allow part of the 1″ flat bar to show through. Change that part of the flat bar to the **Object** layer. This completes the alterations in the top view.

Section B-B

The main purpose of Section B-B is to document the slight flare of the lower chain support at the right (lower) end of the lower weldment. This feature is hidden in the top view, and the geometry is oblique to the side view, so a section is needed to clarify it.

Unlike Section B-B in the upper weldment drawing, this section has break lines at both ends, indicating that the section contains only a narrow, focused portion of the lower weldment. This allows the viewer to concentrate on the information the drafter is conveying without the distraction of other geometry. Follow the instructions in Practice Exercise 2.4D to complete Section B-B.

Practice Exercise 2.4D

Before you begin drawing Section B-B, you must add the cutting-plane line to show where the section will be taken. In this case, the cutting-plane line appears in the front view.

1. Zoom in on the front view of the lower weldment drawing. Position the cutting-plane line as shown in Fig. 2.4.41. Remember to place the line on the **Phantom** layer.

Section B-B for the lower weldment drawing is similar to Section B-B for the upper weldment drawing. However, the focus is slightly different, and the lengths of some of the features are not the same. Therefore, even though the Section B-B you copied from the upper weldment drawing looks similar to the section you want to achieve, the most efficient method is to erase the copied section entirely and use a copy of the top-view geometry for the lower weldment drawing. The procedure outlined in the following steps is therefore similar to that you used to create Section B-B for the upper weldment drawing.

2. Enter the **ERASE** command and use a window to select all of the Section B-B that you copied from the upper weldment drawing.

3. **ZOOM All** to see the entire drawing area. Use the **COPY** command to make a copy of the entire top view. Be sure Ortho is on to keep the views aligned for now. Place the copy below the front view in the space freed up by deleting the old Section B-B.

4. Zoom in on the copy you made in the previous step. From now on, we will refer to this geometry as Section B-B.

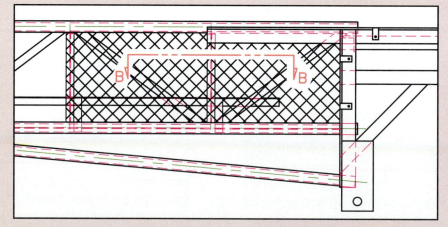

Fig. 2.4.41 *Location of the cutting-plane line for Section B-B.*

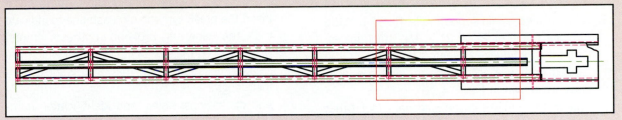

Fig. 2.4.42 Create a cutting box to define the geometry that will be used in the section.

5. Only the geometry enclosed in the heavy red box in Fig. 2.4.42 is needed for Section B-B. Enter the **RECTANGLE** command and create a box, as shown in Fig. 2.4.42. The exact location of the box doesn't matter, but be sure to place it so that a vertical support is included. Also, make sure that the box extends far enough to the right that the right end of the lower chain support is included. Since the views are still aligned, you can determine this by referring to the front view, in which the lower chain support is visible.

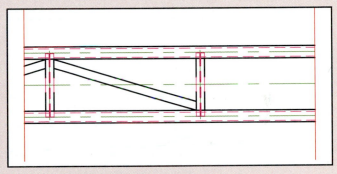

Fig. 2.4.43 Section B-B after step 9.

6. Zoom in on the box and its enclosed geometry. Enter the **TRIM** command and select the box from the previous step as the cutting edge. Trim away all of the geometry outside of the box. Zoom out to erase all of the geometry that did not touch the box. Then zoom back in on the box and its contents.

7. Enter the **EXPLODE** command and select the box to convert its sides to individual lines. Erase the two horizontal (top and bottom) lines, but leave the vertical lines in place. These will later become the break lines.

8. Erase the lines that represent the horizontal steel angle support and the top chain support.

Notice that the lines of the diagonal support are now broken. Although it is not absolutely necessary, it is a good idea to mend these lines now so that each line consists of a single line object. This is because, in the final view, the lines will need to be broken in different places. Mending the lines now helps avoid confusion and extra work later.

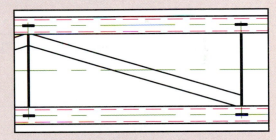

Fig. 2.4.44 The finished vertical supports.

9. Mend the lines of the diagonal support. Each line should be a single object on the **Object** layer. Your drawing should now look like Fig. 2.4.43.

10. Edit the flat-bar vertical support structures. The horizontal end views are of 1/4"-thick flat bar. They are centered on the horizontal centerlines for the Schedule 40 pipes. Mend the lines of the diagonal support. When you finish, the vertical supports should look like Fig. 2.4.44.

11. Add the lower chain support rails. Recall that the rails are made of **2″ × 2″ × 1/4″** steel angles. Use the front view to locate the ends of the flanged portion. The dimensions and specifications are shown in Fig. 2.4.45.

12. Change the portions of the diagonal support that are hidden by the chain support to the **Hidden** layer.

13. Change the vertical lines at the left and right ends of the section to the **Dimensions** layer. Insert the break symbol into each line to indicate that the view is broken at these points. Depending on the size of the box from which you created these lines, you may also want to trim their length.

14. Add dimensions, notes, and identification numbers as shown in Fig. 2.4.46. Use **MTEXT** to add the underscored label **SECTION B-B**.

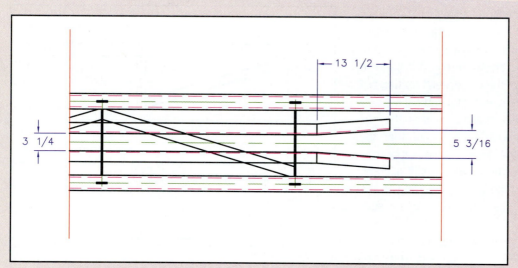

Fig. 2.4.45 *Dimensions for adding the lower chain support.*

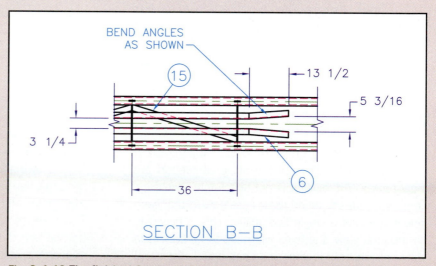

Fig. 2.4.46 *The finished Section B-B for the lower weldment drawing.*

Side View

You can and should base the side view on the geometry from the side view in the upper weldment drawing. However, the side view requires more alterations than the top view. A 3D model showing a side view is shown in Fig. 2.4.47. The view was captured slightly below eye level to give you a perspective of how the lower weldment looks from its lower end.

In Practice Exercise 2.4E, you will alter the geometry in the side view to match the requirements for the lower weldment. Refer to Fig. 2.4.47 as necessary to visualize the various components as you complete this exercise.

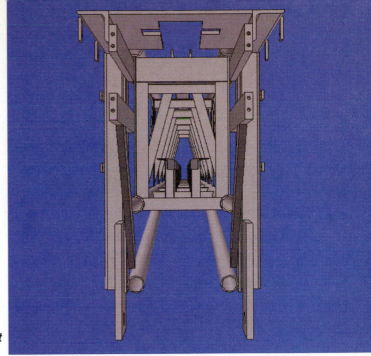

Fig. 2.4.47 *View from the right end of the lower weldment.*

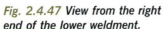

Practice Exercise 2.4E

The side view of the lower weldment looks quite a bit different than the side view of the upper weldment because of the additional support pipes and steel angles at the lower end. However, the basic framework of four Schedule 40 pipes and vertical support structures is the same. The first task is to remove the unneeded geometry from the existing view.

1. Zoom in on the side view.
2. Erase the rectangular framework from both sides, but leave the end view of the steel angles at the top of the rectangular framework in place. (Erase the lower steel angles, though.) Also erase the main horizontal centerline and the flanged portion of the lower chain support rails. Leave the top line and the partial vertical lines of each rail in place for easier location of the revised lower chain support later in this practice exercise. Mend the vertical lines to be single objects on the **Object** layer.

3. Add the vertical **4″ × 4″ × ¹/2″** angle steel supports and the plate that makes up the metal feet, as shown in Fig. 2.4.48. Use construction lines or object snap tracking from the front view to locate the ends of the pieces.

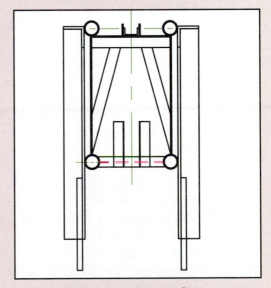

Fig. 2.4.48 *Side view after step 3.*

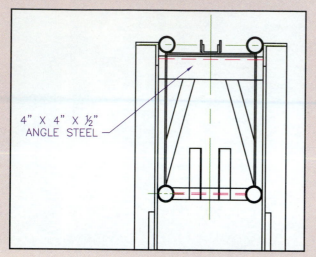

Fig. 2.4.49 *Adjust the angle steel at the top of the cross-support framework.*

4" X 4" X ½" ANGLE STEEL

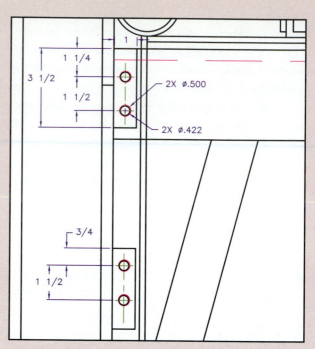

2X ⌀.500

2X ⌀.422

Fig. 2.4.50 *Dimensions for the end view of the flat bars on the left side of the view. Then use the same dimensions for the flat bars on the right side.*

4. Erase the horizontal line that represents the lower edge of the smaller angle at the top of the framework and replace it with a line **4**″ from the top line to represent the **4**″ × **4**″ × **1/2**″ angle steel. Trim the diagonal supports away from the interior of the angle steel. See Fig. 2.4.49.

5. Add the end view of the tire platform. Recall that it is cut from **1/2**″ stock and is flush with the top horizontal support angles.

6. Draw the end view of the horizontal **1**″ flat bars. Obtain their location from the front view. Add the tapped holes in all four flat bars. Use the dimensions shown in Fig. 2.4.50 for the two bars on the left side in this view. The right-side bars are exactly the same. You can save time by drawing the left-side bars and then mirroring them about the main vertical centerline to create the right-side bars.

7. Add the edge view of the mounting tabs. Recall that the tabs are cut from **3/8**″ flat bar. Use construction lines or object snap tracking to place the tabs correctly. Fig. 2.4.51 shows the tabs on the left side of the view. Create them first, and then mirror them about the main vertical centerline to create the tabs on the right side.

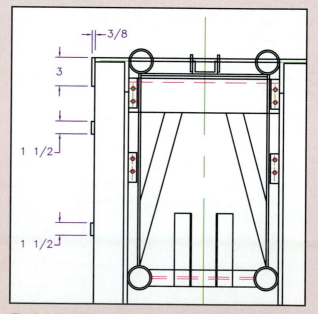

Fig. 2.4.51 *Dimensions of the mounting tabs. Locate them accurately by referring to the front view.*

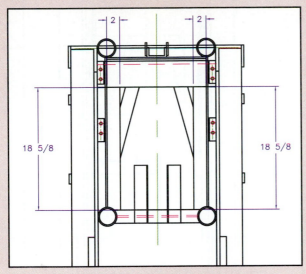

Fig. 2.4.52 Add the vertical steel angle supports.

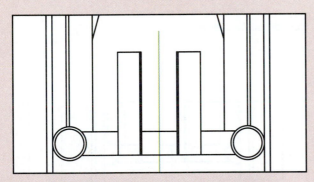

Fig. 2.4.53 The lower flat-bar support after step 9.

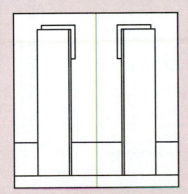

Fig. 2.4.54 Straight part of the rails that form the lower chain support.

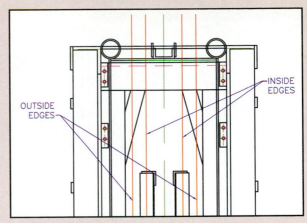

Fig. 2.4.55 Offset the main centerline to find the edges of the flanged portion of the rails.

8. Add the **2″ × 2″ × 1/4″** vertical angle supports, as shown in Fig. 2.4.52. Trim the diagonal supports to meet the steel angle.

9. Now focus on the flat bar that extends between the two lower Schedule 40 pipes. Erase the hidden lines that represented the edge view of another flat bar in the upper weldment drawing. Edit the lines of the flat bar away from the posts for the lower chain support. See Fig. 2.4.53.

The next task is to add the rails of the lower chain support. Because the ends are flared, it is difficult to draw accurately. One reason for creating Section B-B before creating the side view was to gain a better idea of what the flared ends look like.

10. Begin by adding the **2″ × 2″ × 1/4″** angles that actually form the rails, as shown in Fig. 2.4.54.

11. From Section B-B, obtain the distance between the inside surfaces of the rails at the end of the flared portion. Offset the main vertical centerline by half of this distance to find the inside edges of the rails in the current view. Then offset those lines by 2″ to find the outside edges. See Fig. 2.4.55. *Hint:* In offsetting operations, you may find it useful to place the offset lines temporarily on a layer that uses a contrasting color to make them easier to see. In Fig. 2.4.55, the offset lines have been placed on the red Hatch layer. Just be sure to change them to the correct layer when you finish the drawing task.

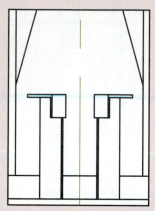

Fig. 2.4.56 *The finished lower chain support.*

12. Connect and edit the lines as necessary to complete the flared ends. The finished lower chain support is shown in Fig. 2.4.56.

13. Add the diagonal flat-bar supports, which appear to be vertical in this view. Create the left flat-bar support according to the dimensions in Fig. 2.4.57. Then mirror it around the main vertical centerline to create the right support.

14. Trim away the parts of the lower Schedule 40 pipes that lie behind the flat-bar supports.

Now you can add the diagonal Schedule 40 pipe supports. In the side view, they appear as oblique cylinders, as shown in Fig. 2.4.58. In the following procedure, you will create the left pipe support and mirror it to create the right support.

15. Create the circles that represent the lower end of the left diagonal Schedule 40 pipe. Use the dimensions shown in Fig. 2.4.58.

16. Copy the circle that represents the outside of the pipe. Use the **Quadrant** object snap to choose the top quadrant of the circle as the base point. Choose the bottom quadrant of the existing lower Schedule 40 pipe as the second point of displacement, as shown in Fig. 2.4.59.

17. Enter the **LINE** command and place a line from the right quadrant of the upper circle to the right quadrant of the lower circle to create the length of the pipe.

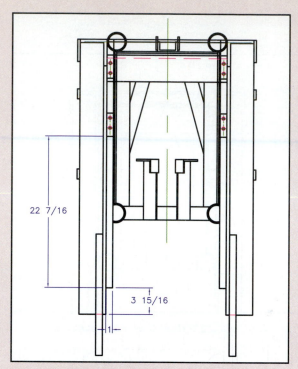

Fig. 2.4.57 *Placement of the diagonal flat-bar support in the side view.*

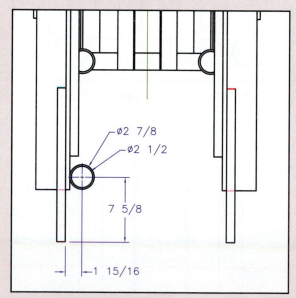

Fig. 2.4.58 *Add the circles that represent the lower end of the diagonal support pipe.*

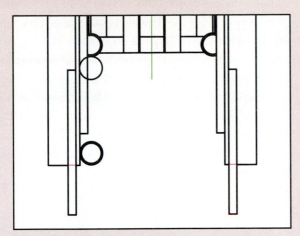

Fig. 2.4.59 **Position of the circle that defines the top end of the diagonal support pipe.**

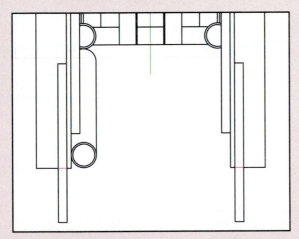

Fig. 2.4.60 **Trim the upper circle to complete the outside of the diagonal support pipe.**

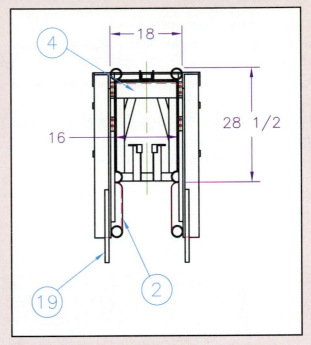

Fig. 2.4.61 **The finished side view.**

18. Trim away the parts of the upper circle that are hidden by the diagonal flat-bar support and the rest of the pipe. See Fig. 2.4.60.
19. Make the line and the arc that compose the body of the pipe into a single polyline. This will make it easier to offset the outside line to create the hidden inner surface of the pipe. To make the line and arc into a polyline, enter the **PEDIT** command and select the line you created in step 17, which represents the long side of the diagonal pipe. Because this is not a polyline, AutoCAD asks if you want to make it into one. Enter **Y** for Yes. Then enter **J** to activate the Join option and pick the curved portion at the top of the pipe (the remains of the circle you trimmed in step 18). End the PEDIT command.

20. Offset the polyline to the inside by **3/16"** to create the pipe wall thickness. Change the offset line to the **Hidden** layer. Trim the bottom of the line to the outside wall of the lower end to finish the left diagonal support pipe.

21. Mirror the entire left diagonal support pipe about the main vertical centerline to create the right support pipe.

This completes the geometry for the side view of the lower weldment. To finish the view, you need only add the dimensions and identification numbers.

22. Add the dimensions and identification numbers as shown in Fig. 2.4.61. Place dimensions on the **Dimensions** layer and identification numbers on the **Text** layer.

Finish Drawing 0021

The geometry for drawing 0021, sheet 1 of the lower weldment drawing, is now complete. To finish drawing 0021, you need to do the following:

- Add dimensions, notes, and specifications to the front and top views.
- Add the cutting-plane line for Section A-A. (The section itself appears on the second sheet.)
- Adapt the existing 0022 layout for use with drawing 0021.
- Generate a parts list.
- Add the general drawing notes.

Before you perform these tasks, however, you should make a copy of drawing 0021. Remember that the geometry from this drawing can and should be used as a basis for drawing 0021-2. By copying the drawing before you add the remaining dimensions, parts list, and other items, you can save the time needed to remove them for the second drawing sheet.

In Practice Exercise 2.4F, you will perform all of the tasks listed here to finish drawing 0021. The completed drawing is shown at the beginning of this section in Fig. 2.4.2 (page 178). Refer to it as necessary as you complete this exercise. When you finish the exercise, your drawing should look very similar to the one in Fig. 2.4.2.

Practice Exercise  2.4F

1. Open **PROJECT 2-0021.dwg**. From the **File** menu, select **Save As...** and save the drawing as a new file named **PROJECT 2-0021-2.dwg**. This will become the drawing file for sheet 2 of drawing 0021. Close the file.

2. Reopen the **PROJECT 2-0021.dwg** file (sheet 1 of drawing 0021). Zoom in on the front view of the lower weldment drawing. Add the cutting-plane line for Section A-A. Place it as shown in Fig. 2.4.62. Place the cutting-plane line on the **Phantom** layer.

3. Add the identification numbers and bubbles, as shown in Fig. 2.4.63. Place these items on the **Text** layer.

4. On the **Dimensions** layer, add the dimensions to the front view, as shown in Fig. 2.4.64.

5. On the **Text** layer, add the note referring readers to sheet 2 for the end detail, as shown in Fig. 2.4.64.

6. Zoom in on the top view of the lower weldment drawing.

7. Add the dimensions and identification numbers on the appropriate layers, as shown in Fig. 2.4.65. Use baseline dimensioning for the long horizontal distances.

This completes your work in model space. Next you will work on the layout. Because you copied this drawing file from drawing 0022, a layout named 0022 already exists. In fact, much of the work you did on the 0022 layout can be reused for drawing 0021. For example, the general drawing notes are

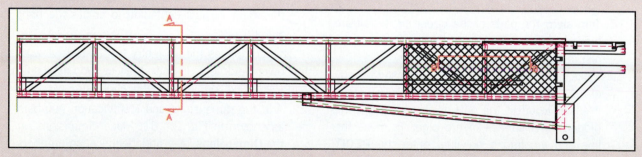

Fig. 2.4.62 Location of the cutting-plane line for Section A-A.

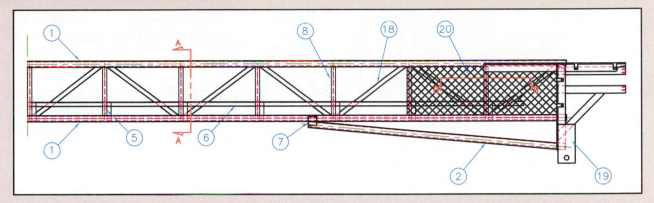

Fig. 2.4.63 *Location of the identification numbers on the front view.*

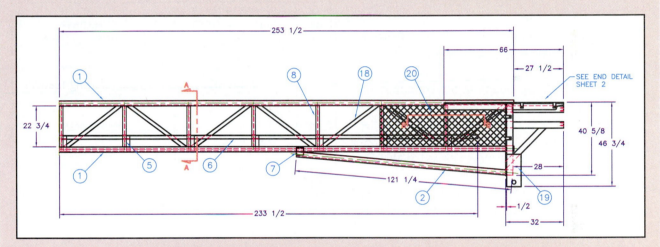

Fig. 2.4.64 *Add the dimensions in the front view.*

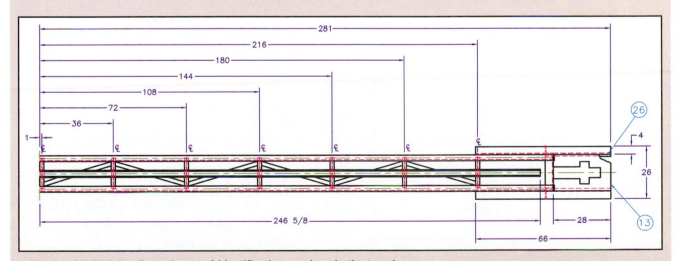

Fig. 2.4.65 *Add the dimensions and identification numbers in the top view.*

similar. You will need to edit them, but that is better than starting from scratch. You can reuse the parts list in a similar manner.

8. Pick the **0022** layout tab at the bottom of the drawing area to display the existing layout. Notice that your current drawing comes in automatically, although it is not yet placed correctly on the layout. Since you will be using the same scale for this drawing that you used for drawing 0021, the drawing is already at the correct scale. All you have to do is move it into position. The general drawing notes and OLE parts list from drawing 0022 are also present.

9. Right-click the tab name and select **Rename** from the shortcut menu. Rename the tab **0021**.

10. Pick the **PAPER** button at the bottom of the screen to enter model space. Enter the **PAN** command and position the drawing on the sheet as shown in Fig. 2.4.2 on page 178. When you finish, be sure to press the **MODEL** button to return to paper space.

11. Zoom in on the general drawing notes in the lower left corner of the layout. Edit the notes to match those in Fig. 2.4.66.

Next, you can revise the parts list for use with this drawing. This will be fairly easy to do because the basic format is already set up.

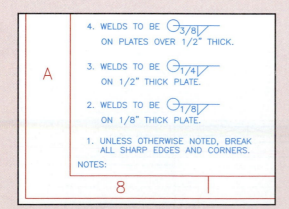

Fig. 2.4.66 *Notes for drawing 0021.*

Item #	PART/DRAWING #	QTY	ITEM DESCRIPTION
30			
29			
28			
27			
26	0025	2	CUT SHEET
25			
24			
23			
22	4" X 4" X 1/2" X 40" LG	2	ANGLE – ASTM A36
21	0025	8	CUT SHEET
20	70" X 22 3/4"	2	MESH, FIT AS SHOWN
19	0025	2	CUT SHEET
18	0025	7	CUT SHEET
17	0025	7	CUT SHEET
16	0025	14	CUT SHEET
15	0025	6	CUT SHEET
14	0025	2	CUT SHEET
13	0025	1	CUT SHEET
12	2" X 1/4" X 14 7/8" LG	7	FLAT BAR
11	2" X 1/4" X 14 7/8" LG	7	FLAT BAR
10	3 X 5.0 X 246 5/8" LG	1	C-CHANNEL, ASTM A36
9	2" X 1/4" X 246 5/8" LG	2	FLAT BAR
8	2" X 1/4" X 22 3/4" LG	14	FLAT BAR
7	4" X 4" X 1/2"	2	FLAT BAR
6	2" X 2" X 1/8" X 233 1/2"	2	ANGLE - ASTM A36
5	2" X 2" X 1/8" X 8 5/8" LG	14	ANGLE - ASTM A36
4	4" X 4" X 1/2" X 16" LG	4	ANGLE - ASTM A36
3	4" X 4" X 1/2" X 66" LG	4	ANGLE - ASTM A36
2	Ø2 1/2 X 121 1/4" LG	2	SCHD 40 PIPE
1	Ø2 1/2 X 253 1/2" LG	2	SCHD 40 PIPE
Item #	PART/DRAWING #	QTY	ITEM DESCRIPTION

Fig. 2.4.67 *Parts list for drawing 0021.*

12. Zoom out to see the entire layout.

13. Doubleclick the parts list to open it in the spreadsheet program that created it.

14. Notice in Fig. 2.4.67 that the parts list for this drawing has 30 lines for parts. To add the five extra lines at the top of the list, press CTRL+A to select the entire spreadsheet. Press CTRL+X to cut the selection. Place the cursor in the first cell on the sixth row, and press CTRL+V to reinsert the entire parts list. This leaves four blank lines at the top of the spreadsheet.

15. Revise the existing text to match the parts list shown in Fig. 2.4.67. Some of the entries are exactly the same as they were for drawing 0022, but others vary only slightly. Check your work carefully when you finish.

TOLERANCES UNLESS NOTED		FILE NAME PROJECT 2		CONVEYOR WELDMENT, LOWER TIRE GASIFICATION SYSTEM
FRACTIONS: 3/32		CONTRACT NO		
		DRAWN [YOUR NAME]		
DECIMALS .X ± .06		CHECK		
.XX ± .020		APPR.		
.XXX ± .005		ISSUED		
DECIMALS .XXXX ± .0005				

SIZE   FSCM NO.   DWG NO. 0021   REV --
SCALE 1"=1'-0"   SHEET 1 OF 2

Fig. 2.4.68 **Title block and tolerance block for drawing 0021.**

16. Save the spreadsheet. From the **File** menu, select **Close & Return to PROJECT 2-0021.dwg**.

17. Because the OLE object is now longer by five text lines, you will need to reposition it in the layout. Position the cursor anywhere on the OLE object, and move it up so that the bottom of the parts list touches the top of the title block.

18. Zoom in on the title block area and change the information as shown in Fig. 2.4.68. Again, most of the work has already been done. You simply need to change the name of the drawing to reflect that this is the lower weldment and change the drawing number to **0021**. Everything else remains the same.

19. Zoom out, save the drawing, and close the drawing file.

This completes your work on sheet 1 of drawing 0021, the lower weldment drawing. Your drawing should look like the one in Fig. 2.4.2 on page 178.

Drawing 0021-2

To finish the lower weldment drawing, you need only adapt the geometry from sheet 1 and add the dimensions, notes, and details to complete sheet 2. This process is very similar to the one described in Practice Exercise 2.3K in Section 3, in which you adapted drawing 0021 to create drawing 0021-2. Like sheet 2 of drawing 0022, sheet 2 of drawing 0021 consists of a series of detail views and a detail section, Section A-A.

In Practice Exercise 2.4G, you will work on the PROJECT 2-0021-2.dwg file that you created at the beginning of Practice Exercise 2.4F. When you complete the exercise, you will have completed the entire lower weldment drawing. The 3D model in Fig. 2.4.69 will help you visualize the parts as you create the details. Also refer to the finished version of sheet 2 in Fig. 2.4.3 (page 179).

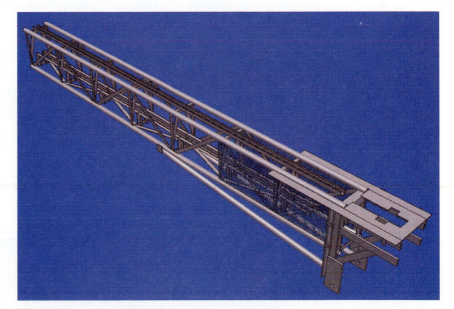

Fig. 2.4.69 **Model of the completed lower weldment.**

Practice Exercise 2.4G

1. Open the **PROJECT 2-0021-2.dwg** file that you created at the beginning of Practice Exercise 2.4F. Zoom in on the front view.
2. With Ortho on, create a vertical line as shown in Fig. 2.4.70 to define the break for the detail view. Place the line on the **Dimensions** layer, and add the break symbol.
3. Use **TRIM** and **ERASE** to remove all of the front view to the left of the break line.
4. Open **DesignCenter** and navigate to your finished **PROJECT 2-0022-2.dwg** file. With this file selected, doubleclick the **Dimstyles** icon in the window to the right to see the dimension styles that were created in that drawing. Drag the styles named **Detail** and **Detail Decimal** into the current drawing. This both saves you the time it would take to set up these styles and ensures that the dimension sizes and attributes are exactly the same in the two drawings. Close DesignCenter.
5. Add the dimensions and identification numbers, as shown in Fig. 2.4.71. Be sure to place them on the correct layers. Use the **Detail** dimension style for fractional dimensions and the **Detail Decimal** style for decimals. For the toleranced dimension, create the dimension; then doubleclick it to display its properties. In the Tolerances portion of the Properties window, set up the tolerance as shown in Fig. 2.4.71.
6. Zoom in on the top view. With Ortho on, create a vertical line as shown in Fig. 2.4.72 to define the break for the detail view. Place the line on the **Dimensions** layer, and add the break symbol. *Note:* Fig. 2.4.72 shows the top view after the trim and erase procedures in step 8.
7. Use **TRIM** and **ERASE** to remove all of the top view to the left of the break line.
8. Add the dimensions and identification numbers as shown in Fig. 2.4.72.
9. Zoom out to see the entire drawing. Enter the **ERASE** command and erase Section B-B, which should not appear on sheet 2.

The only geometry still missing from sheet 2 is Section A-A. Before you spend the time to develop this section, you should think about the set of working drawings as a whole. Recall that you developed a similar Section A-A for drawing 0022-2. Compare the two sections closely and study the information they portray. Drawing 0022-2 is shown in Fig. 2.3.2 on page 127, and drawing 0021-2 is shown in Fig. 2.4.3 on page 179.

Notice that the two sections are actually identical. This makes sense if you think about where the sections are taken. They are actually mirror images; both look toward the joint of the upper and lower weldments. Therefore, you can simply copy the entire section rather than redoing the geometry.

10. Open **PROJECT 2-0022-2.dwg** and switch to model space. Use a window to select all of Section A-A. Enter the **COPYCLIP** command (**CTRL+C**) to copy the selection to the clipboard.

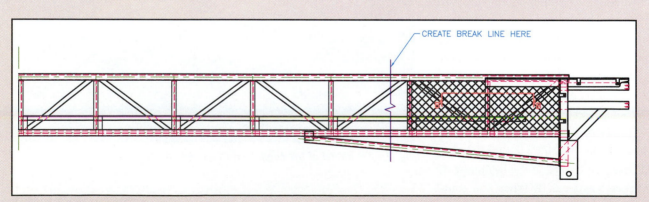

CREATE BREAK LINE HERE

Fig. 2.4.70 Location of the break line in the front view.

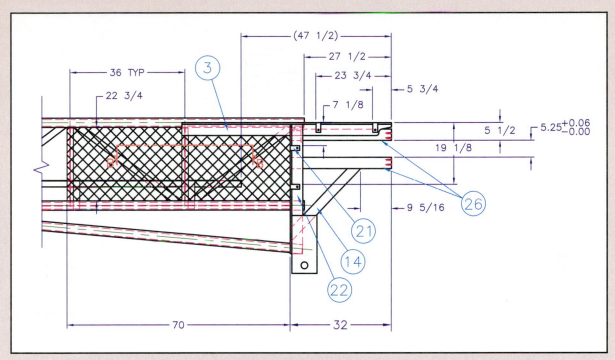

Fig. 2.4.71 *Dimensions and identification numbers for the front view.*

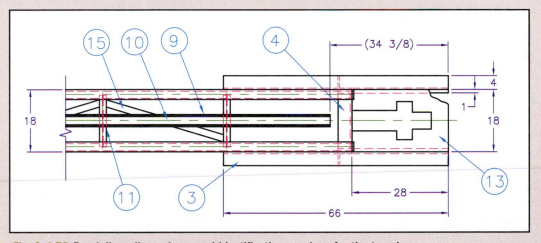

Fig. 2.4.72 *Break line, dimensions, and identification numbers for the top view.*

11. Close PROJECT 2-0022-2.dwg without saving.

12. Paste the copy into the drawing 0021-2 using **PASTECLIP** (**CTRL+V**). Place it to the left of the front view and aligned exactly with it. *Note:* To do this accurately, you may want to create temporary construction lines as guides before you place the copy into the drawing. When you finish, your drawing should look like Fig. 2.4.73.

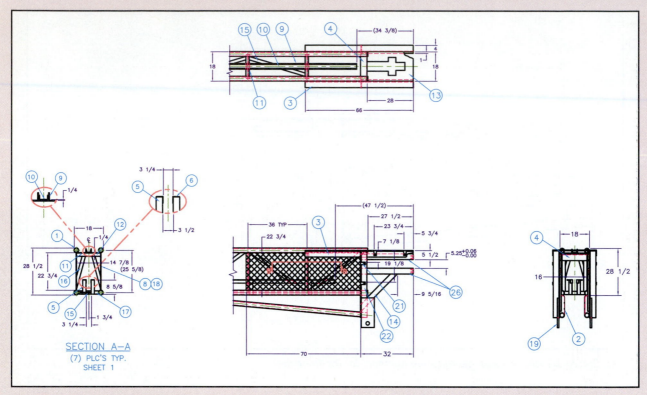

*Fig. 2.4.73 **The finished geometry for sheet 2.***

13. Pick the **0022** tab to enter the layout and paper space. Right-click the tab, select **Rename**, and rename it **0021-2**.

14. Doubleclick one of the lines of the viewport to see its properties. Pick a point in the space next to **Standard scale** and change the scale to **1:16**.

15. Pick the **PAPER** button at the bottom of the screen to enter model space. Enter the **PAN** command and position the drawing in the center of the sheet. Pick the **MODEL** button to return to paper space.

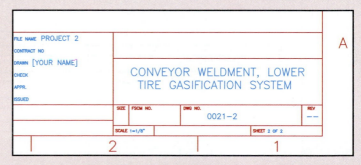

*Fig. 2.4.74 **Title block for drawing 0021-2.***

16. Zoom in on the notes in the lower left corner of the sheet. Change them to a single note: **1. FOR NOTES AND PARTS LIST, SEE SHEET 1.** *Note:* As an alternative, you can use **COPYCLIP** to copy the note from drawing 0022-2.

17. Zoom in on the lower right corner of the layout.

18. Change the information in the title block to reflect the current drawing, as shown in Fig. 2.4.74.

19. Save the drawing file.

Review Questions

1. What fact about the geometry of the upper and lower weldments can you use to speed up the drawing process for the lower weldment, assuming the upper weldment has already been drawn?
2. For the STRETCH command to work properly, what object selection method must be used? Why?
3. Explain how you can maintain precise control over the distance an object is stretched and the angle at which it is stretched.
4. What is an associative dimension?
5. Why do drafters avoid exploding dimensions?
6. Suppose you need to suppress the left (first) extension line of a particular dimension. How can you do this using the Properties window? Does this procedure require exploding the dimension?
7. What are dimension variables?
8. What is the purpose of the tire platform at the lower end of the lower weldment?
9. Why does the lower weldment require more support pieces than the upper weldment?
10. What design feature in the lower weldment was added to promote worker safety as they load tires onto the tire conveyor? Why was this feature considered necessary?

Practice Problems

1. Create a new drawing and name it to identify this project, section, and problem number. Draw Object A using the dimensions shown in Fig. 2.4.75. Include the dimensions. Then copy the entire object to a new location and edit it to look like Object B. Use only the STRETCH command to edit the object.

Portfolio Project

Biomass Conveyor (continued)

Continue your work on the set of working drawings for the biomass conveyor you have designed.

1. Decide which drawing to work on next and open the appropriate drawing file. Remember that you do not have to finish one drawing before starting work on a related drawing.
2. Decide on the most efficient way to begin the drawing. Can it be based on a drawing you have already completed?
3. Try to complete at least one additional drawing at this time.
4. Save your work.

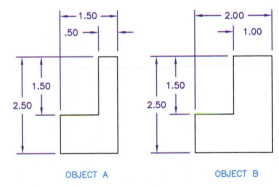

Fig. 2.4.75

2. Create a new drawing and name it to identify this project, section, and problem number. Draw Object C using the dimensions shown in Fig. 2.4.76. Then copy the entire object to a new location and edit it to look like Object D. Use only the STRETCH command to edit the object. Dimension both objects.

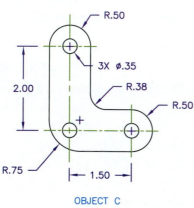

OBJECT C

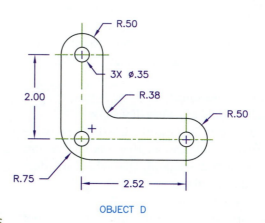

OBJECT D

Fig. 2.4.76

3. Create a new drawing and name it to identify this project, section, and problem number. Draw Object E using the dimensions shown in Fig. 2.4.77. Then copy the entire object to a new location and edit it to look like Object F. Use only the STRETCH command to edit the object. *Note:* This problem requires three STRETCH operations.

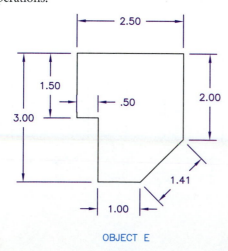

OBJECT E

OBJECT F

Fig. 2.4.77

4. Create a new drawing and name it to identify this project, section, and problem number. Use DesignCenter to import the Fractions dimension style from PROJECT 2-0021-2.dwg to the new file. Make Fractions the current dimension style. Edit the style to make the text size .25. Then create and dimension each of the simple objects in Fig. 2.4.78. Change dimension variables to make your dimensions look like those shown in the illustration. Use the Properties window as appropriate. *Note:* You do not need to create the light blue leaders and leader text. Their purpose is to help you identify the aspects of each dimension that need to be changed.

Fig. 2.4.78

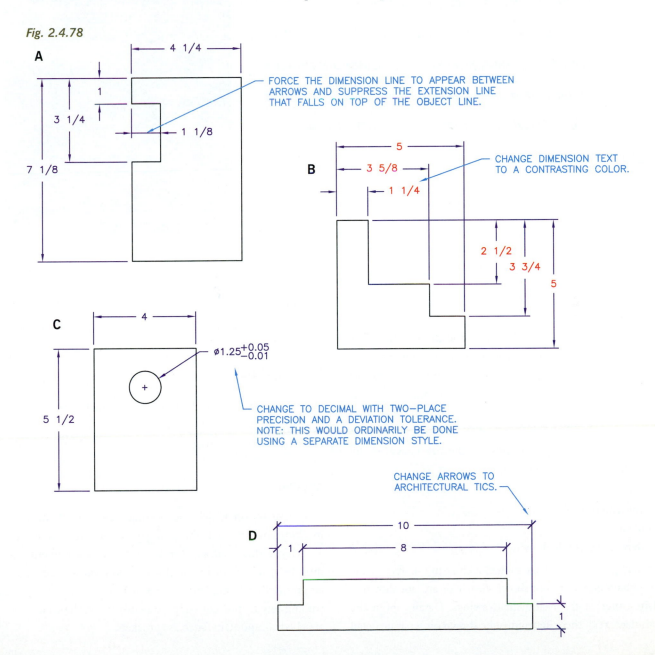

The upper and lower weldment drawings contain the majority of the information needed to construct the tire conveyor for the tire gasification system. However, other drawings are needed to document the tire conveyor completely. For example, details are needed to show the shaft and the drive chain. Also, the sizes of individual pieces of flat bar and angle steel must be specified.

A reference assembly drawing should almost always be included in a complete set of working drawings. Because this is a field assembly project, we have left the reference assembly for last. By the time you create this drawing, you will have the finished weldment drawings to confirm specific sizes and locations. In fact, you can reuse the geometry from the upper and lower weldment drawings to complete the reference assembly quickly and easily.

Section Objectives
- Create appropriate detail drawings to support the reference assembly and the two weldment drawings.
- Create a cut sheet.
- Use existing geometry from several other drawings to complete the reference assembly drawing.

Key Terms
- chain pitch
- chamfer
- cut sheet
- detail identification symbol
- key
- keyseat
- takeup unit

Drawing Analysis

At this point in the field project, the drafter should be fairly certain that the geometry of the two weldments that make up the framework of the tire conveyor will not change further. It is up to the drafter at this point to take stock of the drawings already completed and to plan for the completion of the remaining drawings.

The set of working drawings for the tire conveyor still needs the following drawings:

- shaft detail
- drive chain detail
- cut sheet
- reference assembly drawing

Small pieces and subassemblies such as the shafts and the drive chain either are not shown clearly or are not shown at all in either of the weldment drawings. Details about the manufacture of these items must be shown on separate detail drawings. In the tire conveyor drawings, the shaft details will be drawing 0023. The drive chain detail will be drawing 0024.

Shaft Details

Both of the shafts in the tire conveyor are the same diameter, and both require the same size keyseat. However, other details differ, so both shafts must be shown in the shaft detail.

Keyseat

The shafts on which the bearings are located must be machined accurately if the conveyor is to operate smoothly. To achieve this accuracy, the keyseats and their tolerances must be specified for the machinist. A **keyseat** is a rectangular groove that is machined into a shaft. It mates with a protruding **key** to prevent slippage between the shaft and any attached components, such as sprockets.

Detail Identification

The finished shaft detail drawing is shown in Fig. 2.5.1. Notice the **detail identification symbol** that appears below each shaft. The symbol provides the part number of the part being detailed, the number of the primary drawing on which it is shown, and other pertinent details. Because this project contains a significant number of welds, weight is an important consideration. Therefore, the weight of the shaft is provided on the line that extends to the right of the bubble.

A similar system is used in many sets of architectural drawings, except that the name of the detail is normally given above the line, and the drawing scale is placed below the line. In the shaft detail, the name is not necessary because the shafts are the only parts on this drawing sheet; their name is placed in the title block. The scale is unnecessary because it, too, is the same for both shafts and is listed in the title block.

In Practice Exercise 2.5A, you will create the shaft detail drawing. Refer to Fig. 2.5.1 as you follow the procedure.

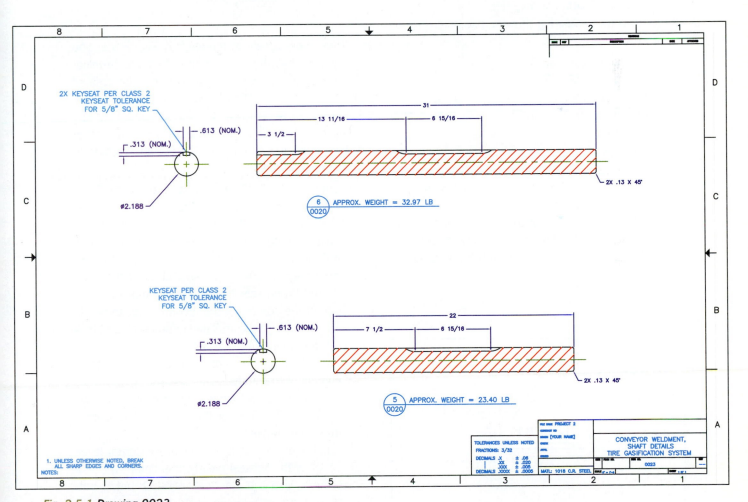

Fig. 2.5.1 **Drawing 0023.**

For drawing 0023 (the shaft detail), we will create an entirely new drawing file. We will then use AutoCAD's DesignCenter to import the layers, dimension styles, and text styles from one of your other drawing files.

1. Create a new drawing file based on AutoCAD's standard **ANSI D - Named Plot Styles.dwt** template. Name the file **PROJECT 2-0023.dwg**. Using **DesignCenter**, import the layers, dimension styles, and text styles from your completed **PROJECT 2-0021-2.dwg** file.

2. Set up the remaining drawing characteristics as follows:

   Limits: **0,0** and **75,50**

   Text style: **ROMANS**

   Units: **Fractional**, with a precision of **1/16**

   Optional: Adjust the grid and snap if you intend to use them.

3. A closeup of the end view is shown in Fig. 2.5.2. Create the side view, dimension it, and add the note as shown. Use appropriate dimension styles for the dimensions.

4. Begin the front view of the shorter shaft by creating a **22″ × 2.188″** rectangle. Use object snap tracking to align it correctly with the end view.

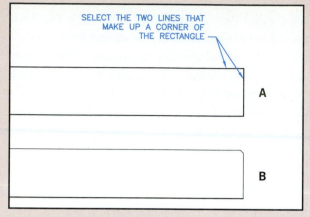

Fig. 2.5.3 *The chamfering process.*

If you look closely at the shaft in Fig. 2.5.1, you will notice that the ends of the shaft are beveled. These bevels, known as **chamfers**, can be created easily using AutoCAD's CHAMFER command. AutoCAD provides several ways to set up a chamfer. You know from Fig. 2.5.1 that these are .13″ chamfers at a 45° angle, so you can use the Distance option to set up the chamfers.

5. Enter the **CHAMFER** command and enter **D** for the Distance option. For the first chamfer distance, enter **.13**. Notice that the second chamfer distance defaults to .13 also. You could override it if the distances were to be unequal, but in this case they are the same, so you can accept the default. Then select the two lines that make up one of the corners of the rectangle, as shown in Fig. 2.5.3A. The completed chamfer should look like the one in Fig. 2.5.3B.

6. Repeat the chamfer operation for the other three corners of the shaft.

7. Create the keyseat using the dimensions shown in Fig. 2.5.4. Transfer the depth of the keyseat from the end view.

8. Add the horizontal centerline on the **Center** layer.

9. To show the keyseat clearly, this detail will be a section. Use the standard **ANSI31** hatch pattern at an appropriate scale to hatch the cut area. Place the hatch on the **Hatch** layer. Do not include the keyseat area in the hatch.

10. Dimension the drawing, as shown in Fig. 2.5.5.

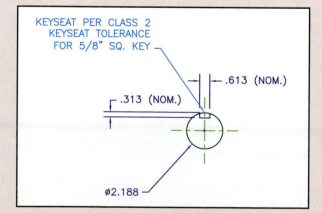

Fig. 2.5.2 *End view of the shorter shaft.*

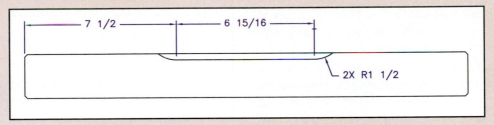

Fig. 2.5.4 *Dimensions for the keyseat. Take the depth from the end view.*

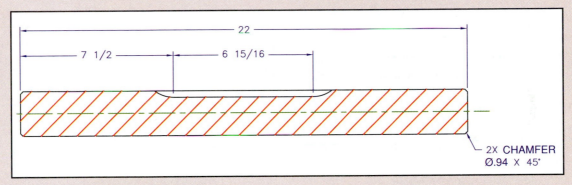

Fig. 2.5.5 *Dimensions for the front view.*

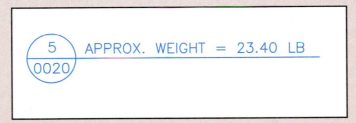

Fig. 2.5.6 *Detail identification number.*

Notice that the chamfer dimensions specify two chamfers, rather than four. This is correct; because this shaft is cylindrical, the bevels you drew at the top and bottom of the front view are actually part of the same chamfer.

11. Finally, add the detail identification symbol, as shown in Fig. 2.5.6. Use a **Ø2** circle and a text size of **.25**. Place the part number (**5**) in the top half of the circle. Place the number of the drawing that refers to this part (**0020**) in the bottom half. Refer to Fig. 2.5.1 for correct placement of the symbol.

The shorter of the two shafts is now complete. The longer shaft will be easy to create, because you can base it on the shorter shaft.

12. Zoom out to see the entire drawing area.

13. Copy both views of the shorter shaft. Place the copy above the original. (The two shafts do not have to align.) From this point, the copy will be referred to as the longer shaft.

14. Zoom in on the end view of the longer shaft. It is identical to the end view of the shorter shaft, except that the longer shaft contains two keyseats, so you must modify the keyseat note.

Doubleclick the text to edit it. Place the cursor at the beginning of the first line and add **2X** before the word KEYSEAT. See Fig. 2.5.7.

15. Now adjust the size of the longer shaft. Enter the **STRETCH** command and use a crossing window to select the right end of the shaft. Be careful not to include the keyseat in the selection, but do include the 22″ horizontal dimension and the chamfer dimension. Notice in Fig. 2.5.1 that the longer shaft is 31″, or 9″ longer than the shorter shaft. Use polar coordinates to stretch the shaft to the right by 9″. Notice that the horizontal dimension and the hatch update automatically.

16. The keyseat on the longer shaft is 6 3/16″ farther from the left side of the shaft than the keyseat on the shorter shaft. To reposition the keyseat, enter the **MOVE** command. Select the lines that make up the keyseat as well as the keyseat dimension. Use a polar coordinate of **@6-3/16<0** to position the keyseat correctly. Again, notice that the hatch boundary and the horizontal dimension update automatically.

17. Add the second keyseat on the left end of the shaft, as shown in Fig. 2.5.8. It is the same depth as the existing keyseat.

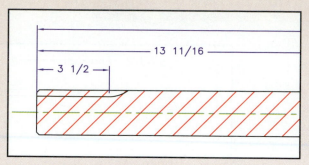

Fig. 2.5.8 *Add the second keyseat.*

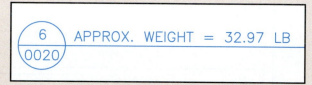

Fig. 2.5.9 *Edit the text to refer to part 6. Change the weight as shown here.*

Notice that the hatch boundary does not update to exclude the new hatched area. There are ways to edit a hatch boundary, but in this case, the simplest method is to delete the hatch and create a new one. You can use the Inherit Properties feature to set up the new hatch quickly.

18. Zoom out so that you can see the front view of both shafts and make **Hatch** the current layer.

19. Erase the hatch on the longer shaft.

20. Enter the **BHATCH** command. Pick the **Inherit Properties** button and pick the hatch on the shorter shaft. When AutoCAD prompts you to select an internal point, pick a point inside the longer shaft. You will need to pick two points—one above the centerline, and one below it. Press **Enter** to return to the Boundary Hatch window. Preview the hatch, and if it is correct, pick OK to place the hatch on the drawing.

21. Zoom in on the detail identification symbol and edit the text as shown in Fig. 2.5.9.

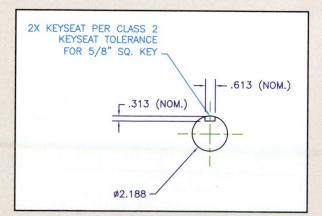

Fig. 2.5.7 *End view of the longer shaft.*

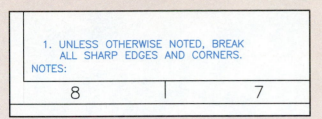

NOTES:

1. UNLESS OTHERWISE NOTED, BREAK
   ALL SHARP EDGES AND CORNERS.

| 8 | 7 |

Fig. 2.5.10 **Note for drawing 0023.**

22. Use **STRETCH** to lengthen the extension lines of the three existing horizontal dimensions to make room for the 3 1/2″ keyseat dimension. Then add the dimension. Refer to Fig. 2.5.1 for placement.

23. Zoom out to see the entire drawing and study its effect. You may want to adjust the position of the longer shaft to center it over the shorter shaft. You may also choose to move the detail identification symbol for the longer shaft to center it better under the front view.

24. Pick the **ANSI D Title Block** tab at the bottom of the drawing area to move into paper space. Rename the tab **0023**. Make **Viewport** the current layer. Delete any existing viewports and create a new one that covers the entire drawing area of the sheet. Doubleclick the viewport to display its properties, and change the scale to **1:2**. If necessary, pick the **PAPER** button to switch to model space and use **PAN** to center the drawing in the drawing area. End the PAN command and pick the **MODEL** button to switch back into paper space.

25. Zoom in on the lower left corner of the drawing area and place the note as shown in Fig. 2.5.10.

26. Zoom in on the lower right corner of the drawing area and complete the title block. Add the tolerance block immediately to the left of the title block. See Fig. 2.5.11. *Note:* Remember that you can copy most of this information from one of the drawings you have already finished. If you use this method, be sure to change the information to match that in Fig. 2.5.11.

27. Save the drawing file and close it.

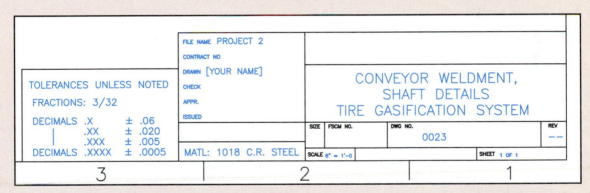

Fig. 2.5.11 **Title and tolerance blocks for drawing 0023.**

Drive Chain Detail

The drive chain that forms the heart of the conveyor is made up of a standard roller chain with the addition of vertical hooks at regular intervals, as shown in Fig. 2.5.12. The purpose of the hooks is to engage the tires and hold them in place during transport up the conveyor to the top of the gasification unit. Although the engineers computed a working angle for the conveyor at which the tires would stay on, they added the hooks as an additional safety feature to protect people working near the conveyor while it operates.

The roller chain is a standard purchased part, so you will create a basic representation according to the manufacturer's specification sheet. If some of the nonessential dimensions are missing from the sheet, it is okay to estimate, but remember that the drawing should be as accurate as possible. The hook is not a standard part and must be shown in detail.

In Practice Exercise 2.5B, you will create the drive chain detail. The finished sheet is shown in Fig. 2.5.13. Refer to this illustration as necessary as you complete the exercise.

Fig. 2.5.12 The drive chain has engagement hooks at regular intervals to keep the tires in place on the conveyor.

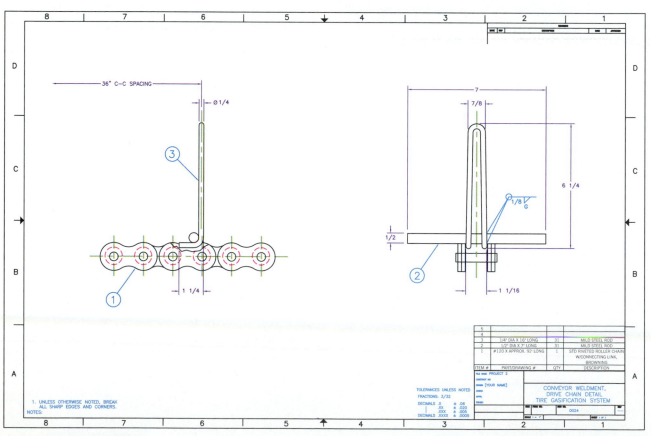

Fig. 2.5.13 Drawing 0024.

For the drive chain, the engineers have specified a #120 single-strand riveted roller chain with connecting links that is manufactured by Browning. The specification sheet for this part is included with the other Browning information in Appendix C. You will need to refer to the specification sheet to complete this practice exercise.

1. Open **PROJECT 2-0023.dwg** and save it with a new name of **PROJECT 2-0024.dwg**.

2. In PROJECT 2-0024.dwg, pick the **PAPER** button if necessary to switch into model space. Enter **ERASE All** to remove all of the objects from the drawing.

In this drawing, we will be creating the side view first. This view will show three complete links in the chain, as well as an engagement hook.

3. Begin by drawing a single link. Refer to the specification sheet to find the **chain pitch**. This is the distance between the centers of the two circular ends of each link in the chain. Place the vertical and horizontal centerlines on the drawing. Build a basic representation of a link around the centerlines, referring to the dimensions on the specification sheet as necessary and estimating any dimensions not given. The finished link should look like the one in Fig. 2.5.14.

4. With Ortho on, make two copies of the link. Place the copies so that the distance between the centers remains constant, as shown in Fig. 2.5.15. As an option, you may prefer to use a rectangular array to create the copies.

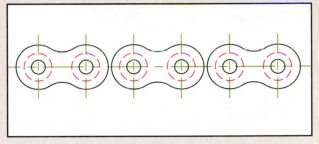

Fig. 2.5.15 *Copy the link twice so that all of the centers are equidistant.*

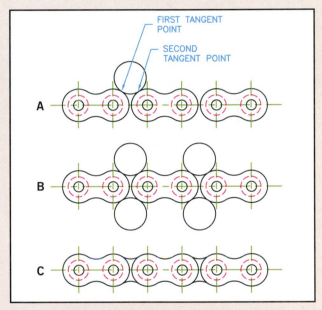

Fig. 2.5.16 *Procedure for connecting the links using tangent circles.*

5. Use arcs of the same radius you used within the link in step 3 to connect the three links, as shown in Fig. 2.5.16. The easiest way to create these arcs accurately is to create a circle tangent to the two existing curved surfaces. Enter the **CIRCLE** command and **Ttr** for the tangent, tangent, radius option. Select the two circular ends to be connected as shown in Fig. 2.5.16A. Then enter the radius of the arc you need. Repeat this for all four connections, as shown in Fig. 2.5.16B. Trim away the part of the circle that is not needed for this illustration to produce the result shown in Fig. 2.5.16C.

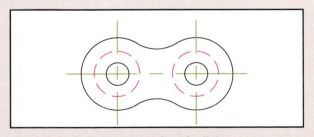

Fig. 2.5.14 *Basic representation of a link.*

Each hook is made of a 16″ length of ¼″ steel rod. The rod is bent at a 90° angle 1 ¼″ from each end. It is also bent in the middle so that it forms a double prong. Refer again to the photograph in Fig. 2.5.12 for the general shape of the hook. The hooks are placed at 36″ intervals.

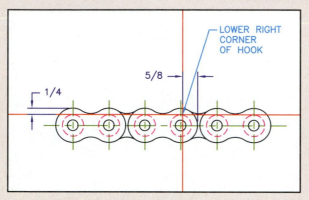

Fig. 2.5.17 *Location for the lower right corner of the hook.*

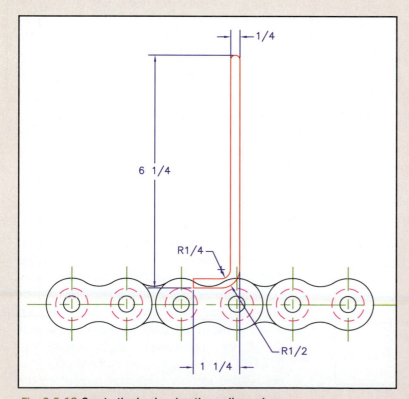

Fig. 2.5.18 *Create the hook using these dimensions.*

6. To locate the hook, first locate its lower right corner. (Do not worry yet about the radius of the bend.) Create a vertical construction line at the right quadrant of the right side of the center link. Offset this line **5/8″** to the left to find the right edge of the vertical portion of the hook. Create a horizontal construction line by snapping to the top quadrant of a link. Offset this line down by **1/4″** to locate the bottom edge of the hook. Erase the temporary lines. The intersection of the two remaining lines represents the lower right corner of the hook. See Fig. 2.5.17. *Note:* The lines in Fig. 2.5.17 and the hook in Fig. 2.5.18 are shown in red for clarity. You may place the hook temporarily on the red **Hatch** layer to reduce confusion as you work, but be sure to change the hook to the **Object** layer when you finish.

7. Complete the hook according to the dimensions shown in Fig. 2.5.18. Use the **FILLET** command to create the .25″ and .50″ radii at the bend.

8. A cross-piece made of a Ø1/2″ steel rod is welded to the hook at the curve in its base. Refer again to Fig. 2.5.12. The end view of this rod is visible in the side view. Create the rod, as shown in Fig. 2.5.19. Use the **CIRCLE** command with the **Ttr** option.

9. For clarity in the drawing detail, break away the part of the middle link that interferes with the lines of the hook, as shown in Fig. 2.5.20.

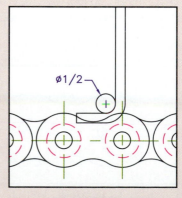

Fig. 2.5.19 *Location of the 1/2″ rod.*

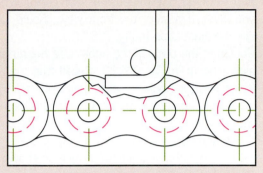

Fig. 2.5.20 *Break away a portion of the middle link to show the hook more clearly.*

10. Add the item identification numbers. Remember to place them on the **Text** layer.

11. Add a partial centerline for the vertical portion of the hook.

12. Add the dimensions, as shown in Fig. 2.5.21. Notice the 36″ dimension that has no left extension line. This dimension shows the space between the hooks on the conveyor. When you create this dimension, pick a point anywhere to the left of the side view for the second endpoint. Then use the **Properties** window to suppress the extension line so that the arrow appears to be pointing at the left side of the drawing sheet. Override the text as shown in Fig. 2.5.21. This completes the side view.

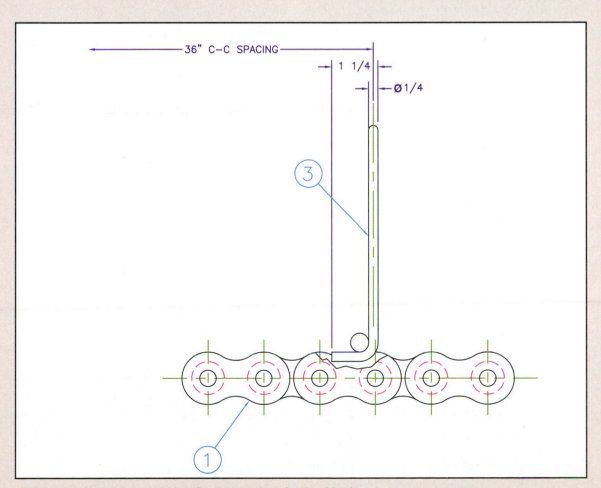

Fig. 2.5.21 *Dimensions and identification numbers for the side view.*

13. Zoom out to see the entire drawing area. Enter the **XLINE** command. Place horizontal construction lines at the top and bottom quadrants of the circle in the side view that represents the end view of the 1/2″ rod. To finish the rod, construct a vertical line where you want the left end of the rod to be. Use the **Nearest** and **Perpendicular** object snaps to ensure that the vertical line meets the horizontal lines exactly. Offset the vertical line to the right by **7**″ to establish the width of the 1/2″ rod. Trim the construction lines to finish the rod.

14. The hook is symmetrical around a vertical centerline. Create the vertical centerline using the midpoints of the upper and lower lines of the rod. With Ortho on, extend the centerline up about **6**″ and down about **1**″.

15. Create the front view of the links. Transfer relevant distances from the side view. Refer to the specification sheet in Appendix C as necessary, and estimate any dimensions not given. When you finish this step, the front view should look like the illustration in Fig. 2.5.22.

16. Construct the body of the hook. Use the dimensions shown in Fig. 2.5.23. Recall that the rod that forms the hook has a diameter of **1/4**″. Trim away the parts of the 1/2″ rod that are hidden by the hook.

17. Create the dimensions shown in Fig. 2.5.23.

18. Add the item identification number, as shown in Fig. 2.5.23.

19. Add the weld specification, as shown in Fig. 2.5.23.

All of the geometry for the chain detail is now complete. To finish the drawing, you will need to work in paper space.

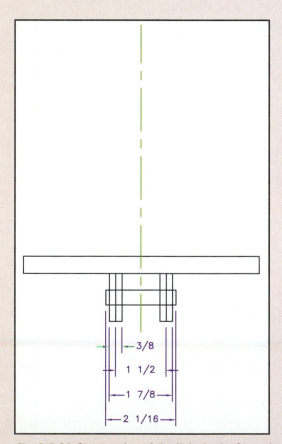

Fig. 2.5.22 *Construction of the links in the front view.*

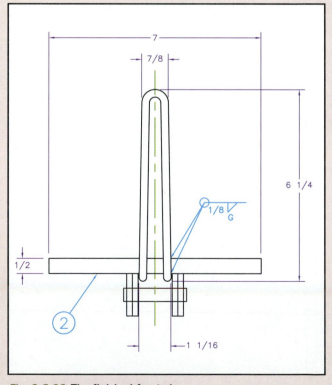

Fig. 2.5.23 *The finished front view.*

20. Zoom out to see the entire chain drive detail drawing.

21. Rightclick the **0023** tab and rename it **0024**.

Because you used drawing 0023 as a basis for this drawing, most of the work in paper space has already been done. However, you must adjust the drawing scale, edit the information in the title block, and add the partial parts list.

22. Doubleclick the viewport to see its properties, and change the drawing scale to **1:1**.

23. Click the **PAPER** button at the bottom of the drawing area to enter model space. Use the **PAN** command to place the drawing in the middle of the drawing area. Click the **MODEL** button to return to paper space.

24. Zoom in on the title block and edit the information for the current drawing as shown in Fig. 2.5.24.

25. Create the partial parts list as a separate spreadsheet file and embed it as an OLE object. The parts list is shown in Fig. 2.5.25. Notice that part number 1 has a multiline description. You can set this up automatically in Excel. Select **Cells...** from the **Format** menu, pick the **Alignment** tab, and pick the check box next to **Wrap text**.

26. When you finish inserting the OLE object, the drawing is complete. **ZOOM All** and save the drawing file.

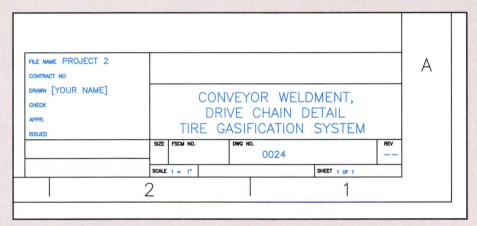

FILE NAME	PROJECT 2				A
CONTRACT NO					
DRAWN	[YOUR NAME]				
CHECK		CONVEYOR WELDMENT,			
APPR.		DRIVE CHAIN DETAIL			
ISSUED		TIRE GASIFICATION SYSTEM			

SIZE	FSCM NO.	DWG NO. 0024	REV --
SCALE 1 = 1"		SHEET 1 OF 1	

2   1

Fig. 2.5.24 *Title block for the drive chain detail. Be sure to remove the MATL.: C.R. STEEL 1018 line.*

ITEM #	PART/DRAWING #	QTY	DESCRIPTION
5			
4			
3	1/4" DIA X 16" LONG	31	MILD STEEL ROD
2	1/2" DIA X 7" LONG	31	MILD STEEL ROD
1	#120 X APPROX. 92' LONG	1	STD RIVETED ROLLER CHAIN W/CONNECTING LINK, BROWNING

Fig. 2.5.25 *Parts list for the chain drive detail.*

Cut Sheet

The framework that supports the tire conveyor consists mostly of welded steel, so it contains many steel parts that must be cut to shape. To document the various shapes that must be cut, the drafter provides a **cut sheet**. The cut sheet is similar to a detail in that it shows the exact dimensions of the various parts to be cut.

The finished cut sheet is shown in Fig. 2.5.26. With one exception, all of the parts are shown at the same scale. Part #21, however, is so small that it is shown at twice the scale of the other parts. It is very important to note the scale for part #21 to avoid confusion.

In Practice Exercise 2.5C, you will create the cut sheet, which is drawing 0025 in the series. Refer to Fig. 2.5.26 as necessary as you complete the exercise.

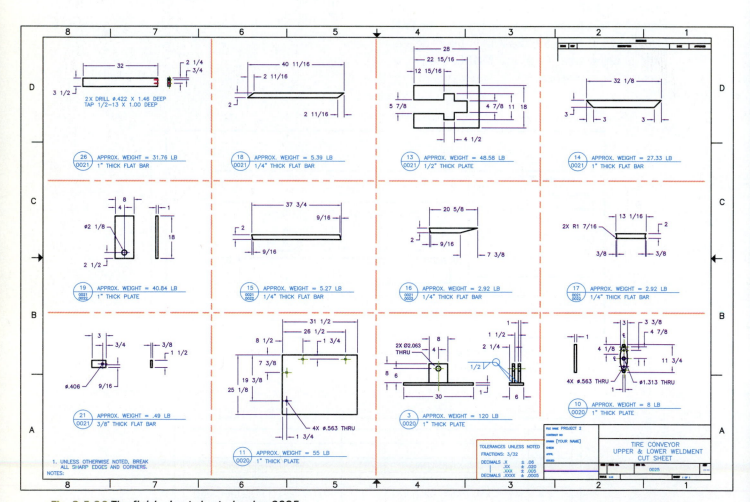

Fig. 2.5.26 *The finished cut sheet, drawing 0025.*

Practice Exercise

2.5C

1. Open **PROJECT 2-0024.dwg**. Use the **Save As...** option on the **File** menu to save the drawing with a new name of **PROJECT 2-0025.dwg**.
2. Prepare the file for drawing 0025 by doing the following:
   - Return to model space.
   - Erase all of the existing geometry.
   - Enter the **LIMITS** command and change the upper drawing limit to **285,190**.
   - Adjust the snap and grid if you intend to use them.
3. Before you begin drawing the parts, put the drawing space in perspective by dividing it into 12 segments. Make **Viewport** the current layer, and create a rectangle at the drawing limits. Then make **Phantom** the current layer. Enter **LTSCALE** and set a new linetype scale of **10**. Then place the horizontal and vertical division lines as shown in Fig. 2.5.27. Note that the boxes are not all exactly the same size. Some of the parts to be described require more space than others. Also, you may need to adjust the phantom lines slightly as you develop the parts.
4. Zoom in on the box in the upper left corner. This box will contain part #26. Complete the part drawing, including the dimensions and detail identification symbol, as shown in Fig. 2.5.28. *Hint:* Instead of redrawing the tapped holes, use **COPYCLIP** to copy them from drawing 0021. Note that the material from which the part is to be cut is specified below the horizontal line of the detail identification symbol. For the dimensions,

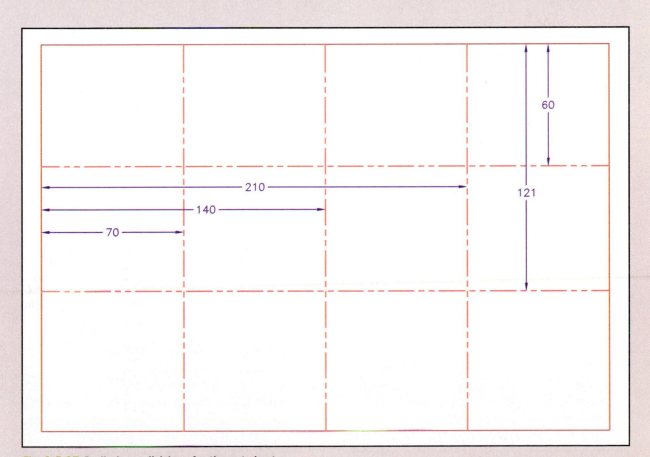

Fig. 2.5.27 Preliminary divisions for the cut sheet.

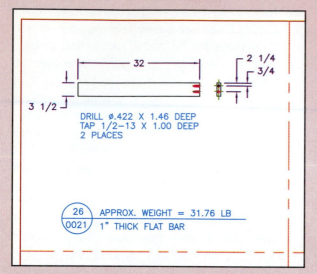

Fig. 2.5.28 *Part #26.*

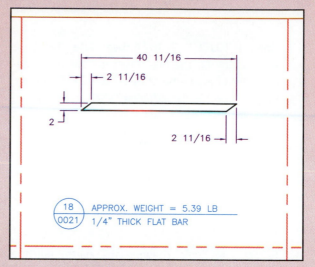

Fig. 2.5.29 *Part #18.*

use the **Detail** dimension style, but note that you may need to adjust the style somewhat for this drawing. Specifically, you may need to edit the style to make the text and arrows slightly smaller. The text and arrows in Fig. 2.5.28 are drawn at a size of **1.5**. Center the objects in the box, making sure to maintain the alignment between the front and side views.

5. Use **ZOOM Dynamic** to move the viewing window to the next box to the right (the second box on the top row). Create the detail for part #18, as shown in Fig. 2.5.29. Center the geometry, but align the detail identification symbol with the one for part #26. Aligning the part symbols across the page provides a better appearance for the drawing sheet.

6. Move the viewing window to the third box on the top row. This box contains a detail of the tire platform, part #13. Create the detail as shown in Fig. 2.5.30. *Hint:* Create a temporary horizontal centerline to make the geometry easier to draw using the dimensions shown.

7. Move the viewing window to the last box on the top row. This box contains a detail of one of the

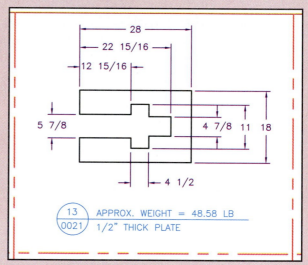

Fig. 2.5.30 *Part #13.*

support pieces at the lower end of the conveyor, part #14. Create the detail as shown in Fig. 2.5.31.

8. Move down to the first box on the left on the second row. This box contains a detail of part #19, the mounting feet for the lower weldment. Create the detail drawing as shown in Fig. 2.5.32. Notice that the detail identification symbol for

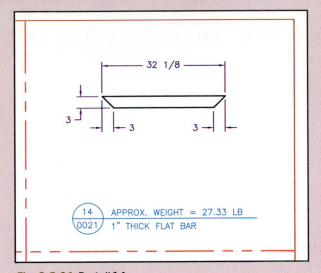

Fig. 2.5.31 *Part #14.*

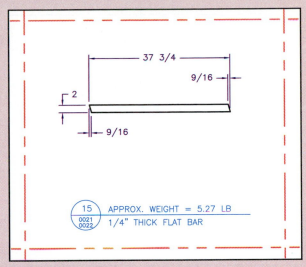

Fig. 2.5.33 *Part #15.*

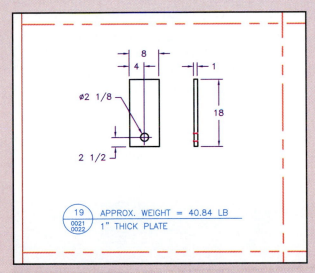

Fig. 2.5.32 *Part #19.*

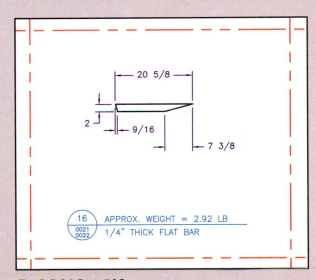

Fig. 2.5.34 *Part #16.*

this detail contains two drawing numbers. You will need to reduce the text size of the drawing numbers to **1.0** so that both numbers will fit. Also, align the detail identification symbol vertically with the one in the top row, and place it about the same distance from the bottom of the box as the symbols on the top row.

9. Move to the second box on the second row. This box contains a detail drawing of part #15. Create the detail drawing as shown in Fig. 2.5.33.

10. Move to the third box on the second row, which contains a detail of part #16. Create the detail drawing as shown in Fig. 2.5.34.

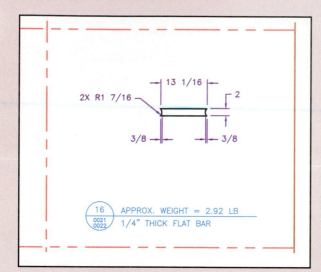

Fig. 2.5.35 Part #17.

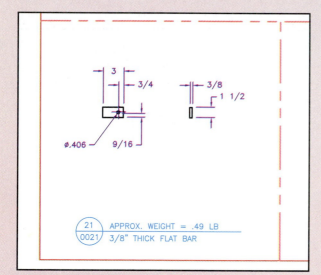

Fig. 2.5.36 Part #21.

11. Move to the last box on the second row, which contains a detail of part #17. Create the detail drawing as shown in Fig. 2.5.35.

12. Move to the first box on the left of the third row, which contains a detail of part #21. This is the mounting tab; it is so much smaller than the other parts that it must be shown at a larger scale. Refer to Fig. 2.5.36 as you develop this detail.

13. Create the geometry for both views at its true size, according to the dimensions in Fig. 2.5.36. Do not create the dimensions yet.

14. Enter the **SCALE** command, select both views, and enter a new scale of **2** to double their size.

15. Pick the **Dimension Style** button on the **Dimension** toolbar and create a new style based on the Detail style. Name the new style **Detail 2X**. Pick the **Primary Units** tab. In the Measurement Scale section (about halfway down the left side of the window), change the scale to **.500**. This tells AutoCAD that every object dimensioned with this dimension style is actually half of its measured size. Therefore, even though your geometry for the mounting tab is shown at twice its normal size, the dimensions will reflect its true size.

16. Make **Detail 2X** the current dimension style. Dimension the two views of the mounting tab. When you finish, be sure to change the dimension style back to **Detail** so that your dimensions for the rest of the parts will be correct.

17. Move to the second box on the third row. This box contains a detail of part #11, a side panel that will be attached to the lower weldment in the final assembly as an added safety precaution for workers. Fig. 2.5.37 shows the side panel in place on the operating tire conveyor. Create the detail drawing as shown in Fig. 2.5.38.

Fig. 2.5.37 Part #11 is the side panel (step 17), and part #3 refers to the mounting feet (step 18).

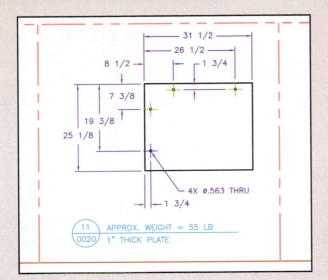

Fig. 2.5.38 *Part #11.*

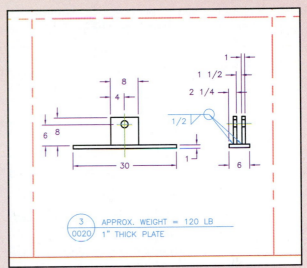

Fig. 2.5.39 *Part #3.*

18. Move to the third box on the third row. This box contains part #3, the mounting feet that will be used to bolt the tire conveyor permanently in place. Refer again to Fig. 2.5.37 to see them in use. Create the detail of the mounting feet, as shown in Fig. 2.5.39.

19. Finally, move to the last box on the third row. This box contains a detail drawing of part #10, the takeup access plate. Create the detail, as shown in Fig. 2.5.40. Place the geometry and the detail identification symbol as close to the top of the box as possible, because you will need room on the layout for the title block.

20. Zoom out to see the entire drawing. Erase the rectangle you created on the Viewport layer in step 3.

21. Pick the **0024** tab at the bottom of the drawing area. Change the name of the layout to **0025**.

22. Erase the parts list.

23. Doubleclick the viewport to display its properties, and change the scale to **1:10**. Close the Properties window.

24. Pick the **PAPER** button to move into model space. Use the **PAN** command to center the drawing on the drawing area. Then click **MODEL** to return to paper space.

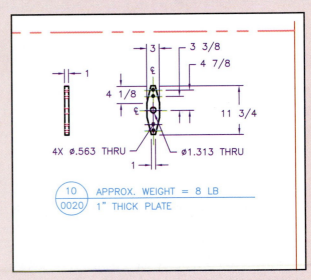

Fig. 2.5.40 *Part #10.*

25. Look carefully at the lower right corner of the layout. If any of the geometry in the last two boxes on the bottom row conflicts with the title and tolerance boxes, return to model space and adjust the position of the geometry. If you still can't make it work, you may need to adjust the lowest horizontal phantom line up to allow more space for the bottom row.

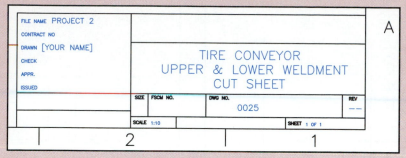

Fig. 2.5.41 Title block for drawing 0025.

26. Edit the text in the title box for the current drawing, as shown in Fig. 2.5.41.

27. Zoom in on the lower left corner of the drawing sheet. The note should already be in place, but check to make sure that the detail identification symbol does not interfere with it. If it does, go back into model space and move the identification symbol up to clear the note text.

28. The cut sheet, drawing 0025, is now complete. Save your work and close the drawing file.

Reference Assembly

The last drawing in the set is the reference assembly, which is drawing 0020. In field assembly situations such as this one, in which the drafter's job is to provide documentation of the finished prototype, the reference assembly is often drawn last. The drafter generally doesn't have enough information to complete the assembly until near the end of the actual construction. Another advantage of saving this drawing until last is that it will be much easier to complete now that the other drawings are finished.

The reference assembly is the most complex drawing in the set. It is the only drawing that shows the two weldments assembled. It contains both a left- and a right-side view, as well as a section and two details. The parts list is also much longer than those for the other drawings. For these reasons, it is shown at a smaller scale. For details, viewers must refer to the other drawings in the set. The reference assembly has only three purposes:

- show the relationships among all of the parts
- define the weld between the two weldments
- provide identification information for the major parts

The completed drawing is shown in Fig. 2.5.42. Notice that, in addition to the customary round "bubbles," this drawing contains triangular identification symbols. These symbols occur after the round bubbles, but they, too, contain part numbers. The purpose of the triangular symbol is to alert the reader that more information about this part is provided in the general drawing notes. This technique is often used on crowded drawings if there is not enough space on the drawing to place the note with the part or if the note is very long. In drawing 0020, both of these limitations apply. Both notes are also very long. They contain complex information that the drafter did not want to get "lost" in the drawing.

Left-Side View

In Practice Exercise 2.5D, you will complete the reference assembly drawing. You will begin with the PROJECT 2-0020.dwg file you started earlier in this project, but you will also draw in geometry from most of the other drawings in the set. Refer to the completed reference assembly drawing in Fig. 2.5.42 as necessary as you complete the exercise.

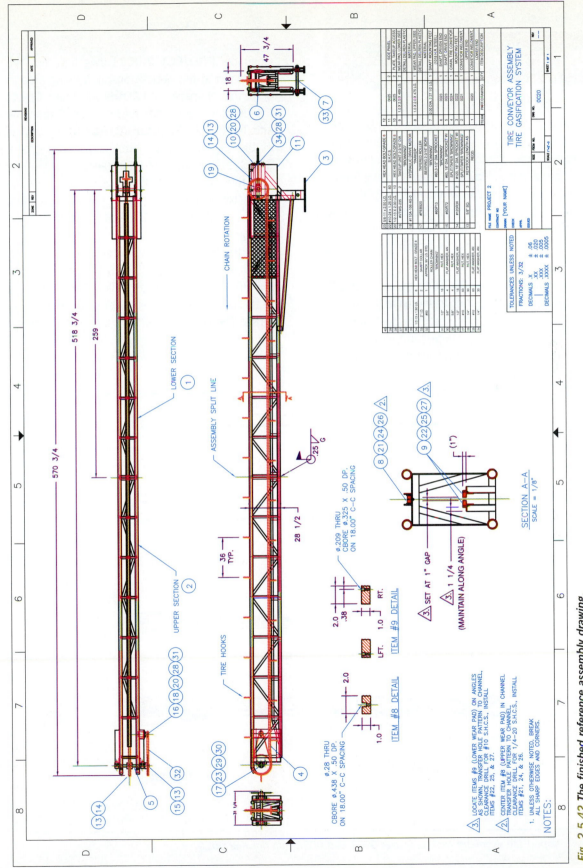

Fig. 2.5.42 The finished reference assembly drawing.

Practice Exercise

2.5D

The work you did on PROJECT 2-0020.dwg in sections 1 and 2 of this project provided much of the detail work on the shafts, roller bearings, sprockets, and other small parts. However, the structural framework was far from complete when you stopped working on this drawing. You will therefore need to incorporate much of your work from the upper and lower weldment drawings to complete drawing 0020.

1. Open the **PROJECT 2-0020.dwg** file and **ZOOM All**. Reset the upper limit to increase the size of the drawing area to accommodate the two weldment drawings. Move the existing views to the top left corner of the drawing area. Leave this drawing open.

2. Open the **PROJECT 2-0021.dwg** file. Pick the **Model** tab at the bottom of the drawing area to enter model space, and **ZOOM All**. Make **Object** the current layer, and freeze layers **Dimension** and **Text**. Use **COPYCLIP** (CTRL+C) to copy all of the geometry that remains on the screen. Switch to drawing 0020 and paste the geometry into the drawing. Paste it below the existing geometry on the right side of the drawing area. Return to drawing 0021 and close it *without* saving any changes.

3. Open the **PROJECT 2-0022.dwg** file and repeat the procedure in step 2 to copy the geometry into the PROJECT 2-0020.dwg file. Place the copy to the left of the geometry from drawing 0021. Close PROJECT 2-0022.dwg *without* saving your changes and return to PROJECT 2-0020.dwg, which should now look like Fig. 2.5.43.

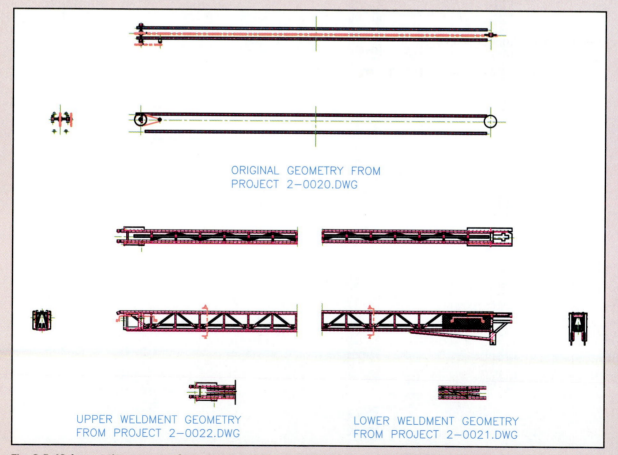

Fig. 2.5.43 **Import the geometry from drawings 0021 and 0022 into drawing 0020.**

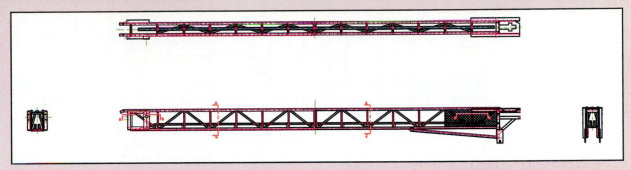

Fig. 2.5.44 The joined upper and lower weldments.

4. Erase the sections that appear at the bottom of the upper and lower weldment drawings.

5. Zoom in so that you can see all of the views of the upper weldment and the left end of the front and top views of the lower weldment. Carefully, using the object snaps, move the views of the upper weldment into position against the lower weldment. Be sure to keep the side view of the upper weldment aligned with the front view. The joined upper and lower weldment drawings should look like Fig. 2.5.44.

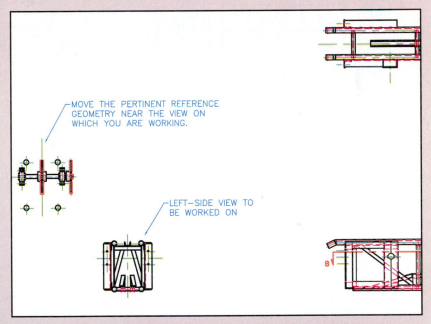

Fig. 2.5.45 Move the reference geometry for the side view close to the side view so that geometry is easier to transfer from one to the other.

Superimposing the finished structural framework on the shafts, bearings, and so on, of drawing 0020 would be difficult. It is easier to move the smaller parts from your original drawing 0020 into place on the finished structural framework from the weldment drawings. This is the approach we will use. From this point, we will work exclusively on the joined weldment drawings. Therefore, we will refer to the joined weldment drawings as "the drawing." We will refer to the original geometry in drawing 0020 as "reference geometry."

6. Concentrate first on the left-side view of the drawing. To make it easier to transfer items from the reference geometry to the drawing, move the left-side view of the reference geometry to a location near the left-side view, as shown in Fig. 2.5.45.

The finished left-side view for the reference assembly drawing is shown in Fig. 2.5.46. Refer to this drawing as you complete the steps to adapt the existing geometry.

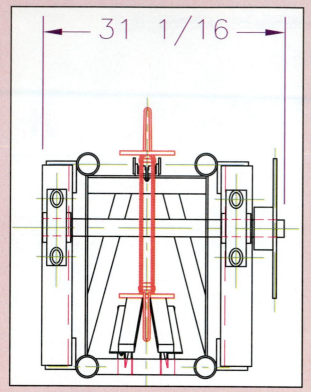

Fig. 2.5.46 *The finished side view for the reference assembly drawing. The dimension may appear too large, but it is actually at the correct size. Because of the size of the assembled tire conveyor, each view is very small.*

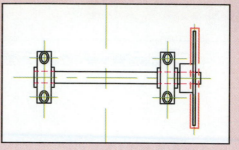

Fig. 2.5.47 *The reference geometry with the central sprocket and Schedule 40 pipes removed and the shaft lines mended.*

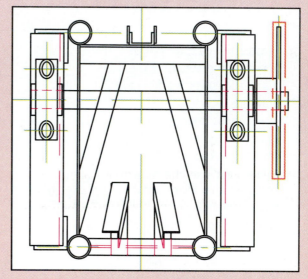

Fig. 2.5.48 *Shaft and bearings in place.*

7. In the left-side view, erase the four small holes in the rectangular framework structure (two on each side), along with their centerlines. Do not erase the horizontal centerline that represents the shaft.

8. Working on the reference geometry, erase the center sprocket. Be careful not to erase the main vertical centerline. Mend the shaft lines so that each line is a single object.

9. Still working on the reference geometry, erase the end views of the four Schedule 40 pipes and their centerlines. The reference geometry should now look like Fig. 2.5.47.

10. All of the objects that remain in the reference geometry need to be moved into place on the left-side view. Enter the **MOVE** command and use a window to select all of the objects in the left-side view of the reference geometry. Press **R** (for Remove), pick the vertical and horizontal centerlines to remove them from the selection set, and press **Enter**. For the base point, use the **Intersection** object snap to select the intersection of the vertical and horizontal centerlines of the reference geometry. For the second point of displacement, select the intersection of the vertical and horizontal centerlines of the left-side view. This places the shaft, sprockets, and bearings accurately into the left-side view, as shown in Fig. 2.5.48. Erase the remaining centerlines from the reference geometry.

11. Add the end view of the bolts and washers that attach the bearings in place.

Check your work carefully. If for any reason your left-side view does not look like the one in Fig. 2.5.48, use the UNDO command to undo the move and try again. It is very important to place the geometry accurately.

12. Trim away the portions of the diagonal supports that lie behind the shaft.

13. Erase the phantom line that represents the chain drive on the right-side sprocket.

The next task is to add the chain drive. Refer again to Fig. 2.5.46. In this drawing, the entire chain drive will be placed on the Phantom layer. You can pick up the hook from drawing 0024, but you will need to create the rest of the chain drive from scratch.

14. Zoom in on the top chain support. Add the upper wear pad to the inside of the chain support, as shown in Fig. 2.5.49.

15. Add the bolt, washer, and nut to fix the track in place. First use the dimensions shown in Fig. 2.5.50A to create them in a blank portion of the screen. You may wish to block them for use with other drawings. Then place the bolt and nut as shown in Fig. 2.5.50B.

For demonstration purposes, the side view will show two hooks on the drive chain: one at the top of the conveyor, and one at the bottom. The hook at the

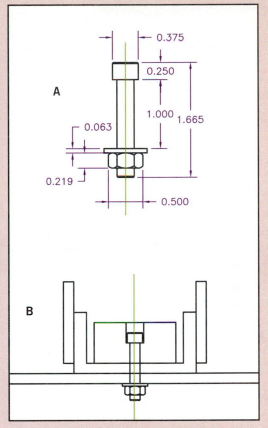

Fig. 2.5.50 (A) Dimensions for building the bolt, washer, and nut. (B) Position the fasteners as shown.

top of the assembly can be added now. The $\varnothing 1/2''$ steel rod of the hook assembly is located $2\,5/8''$ above the top of the steel angle cross-piece at the top of the framework.

16. Offset the top line of the steel angle up by **$2\,5/8''$** to locate the bottom line of the angle assembly.

17. Save your work, but do not exit the drawing.

18. Open **PROJECT 2-0024.dwg** and switch to model space. Freeze the **Dimensions** and **Text** layers. Then use **COPYCLIP** to copy the entire front view of the hook. Close the drawing without saving it. In PROJECT 2-0020.dwg, paste the hook somewhere near the left-side view.

19. Use a crossing window to select the entire hook assembly and move it to the **Phantom** layer. You may include even the centerline in this operation.

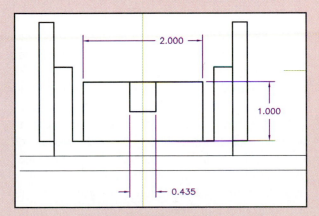

Fig. 2.5.49 Add the upper wear pad.

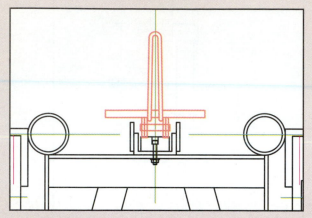

Fig. 2.5.51 **Hook assembly in place on the upper chain rail.**

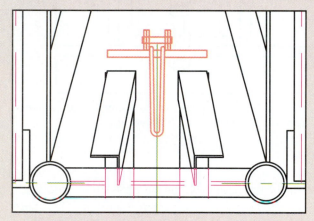

Fig. 2.5.52 **Hook assembly in place on the lower chain rail.**

20. The hook assembly is already in the correct orientation for the top of the chain drive, so you can now move it into place on the drawing. However, instead of using the MOVE command, use COPY. This leaves the original intact so that you can use it later at the bottom of the chain drive. Enter the **COPY** command and select the entire hook assembly. For the base point, select the intersection of the vertical centerline and the lower edge of the Ø1/2″ steel bar on the hook assembly. For

the second point of displacement, select the intersection of the vertical centerline on the left-side view and the line you offset in step 15. Erase both the offset line from step 15 and the vertical centerline that is on the Phantom layer. The hook assembly should be placed as shown in Fig. 2.5.51.

21. In this view, the hook is seen from the opposite side from the view in drawing 0024, so that the Ø1/2″ steel bar is in front of the hook itself. Trim away the parts of the hook that are hidden by the steel bar and mend the lines of the bar.

22. Move the other, unused hook assembly to a blank area near the lower chain support and zoom in on that area.

23. Because of the path of the drive chain, the lower hook is upside down. Before you can insert the hook into the left-side view, you will need to rotate it 180°. Enter the **ROTATE** command and select the entire hook assembly. Pick the bottom of the vertical centerline as the base point, and enter **180** for the angle of rotation.

24. The edge of the Ø1/2″ steel bar is located 7 3/4″ above the top edge of the bottom steel angle cross-piece. Offset the top line of the cross-piece up **7 3/4″** as a guide.

25. Move the entire hook assembly into place. For the base point, select the intersection of the centerline of the hook assembly and the lower edge of the Ø1/2″ steel bar. For the second point of displacement, select the intersection of the main vertical centerline on the left-side view and the line you offset in step 23. Erase the hook centerline (the one on the Phantom layer) and the temporary offset line from step 23. The hook should now be positioned as shown in Fig. 2.5.52. Note that, for now, the hook appears to be hanging in midair. You will add the wear pads later.

You may have noticed that we have routinely omitted many of the hidden lines in this view. Because of the size of the assembled tire conveyor, the individual views will be quite small on the drawing sheet. It is acceptable to remove hidden lines entirely in the assembly view to increase the clarity of the drawing.

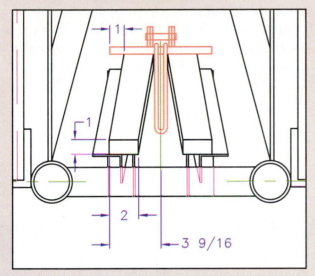

Fig. 2.5.53 *The wear pads rest on the rails of the lower chain support.*

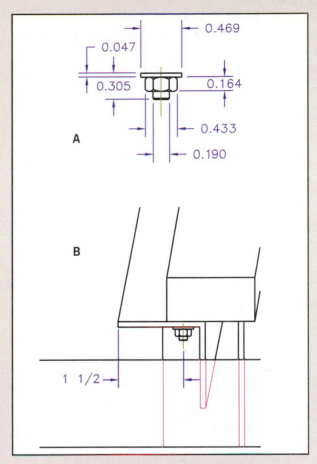

Fig. 2.5.54 *(A) Dimensions for the bolt, washer, and nut. (B) Location of the fasteners.*

26. Delete the hidden lines that represent the edge view of the lower 2″ flat-bar cross support.

27. The lower wear pads are 1″ × 2″ strips that rest on the rails of the lower chain support. The wear pads are symmetrical about the main vertical centerline and help guide the drive chain. Add the wear pads using the dimensions shown in Fig. 2.5.53.

28. The wear pads are bolted to the chain rail as shown in Fig. 2.5.54. Add the visible portion of the bolt, washer, and nut on the bottom of each rail of the lower chain support. The dimensions of the fasteners are shown in Fig. 2.5.54A. Place them as shown in Fig. 2.5.54B.

29. Zoom out to see the central area of the left-side view, including the upper and lower hooks. Use the **EXTEND** command to extend the three lines on each side of the vertical centerline that represent the drive chain. The lines should extend all the way from the top to the bottom of the drive chain assembly, as shown in Fig. 2.5.55.

30. Add the overall horizontal dimension to finish the view. Refer again to Fig. 2.5.46.

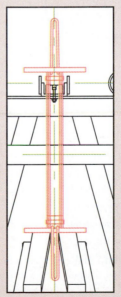

Fig. 2.5.55 *Extend the roller chain lines to complete the drive chain.*

Top View

The structural framework of the tire conveyor needs little work in the top view. However, the conveyor itself needs significant work. You must add the sprockets, shafts, bearings, hydraulic motor, and so on. You will not need to alter anything in the center of the top view, but you will need to work on both ends. Much of the geometry to be added already exists in the reference geometry, so you can save time by using it. After you have finished the geometry, you will add the overall dimensions, notes, and identification numbers.

In Practice Exercise 2.5E, you will complete the top view of the reference assembly. Refer again to the completed reference assembly in Fig. 2.5.42 (page 231) as necessary.

Practice Exercise 2.5E

1. Zoom out to see the entire drawing area. Move the top view of the reference geometry down to a spot just above the top view so that it is easily available for copying.
2. Zoom in so that you can see the left end of both the top view and its reference geometry.
3. The first task will be to move the sprockets and shaft from the reference geometry to the top view. To provide a point of reference, first create the vertical centerline for the main shaft. This centerline is located **7³/4″** from the vertical centerline of the left-most cross support, as shown in Fig. 2.5.56.
4. Enter the **MOVE** command and use a crossing window to select the entire left end of the top view of the reference geometry, including all of the shafts, sprockets, and the hydraulic motor. Enter **R** and remove the main horizontal centerline and the lines that represent the Schedule 40 pipe from the selection set. For the base point of displacement, pick the intersection of the main horizontal centerline of the reference geometry and the vertical centerline of the shaft. For the second point of displacement, pick the same intersection in the top view of the drawing. When you finish the move, the left end of the top view should look like Fig. 2.5.57.

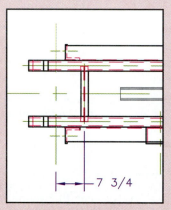

Fig. 2.5.56 *Location of the centerline of the shaft.*

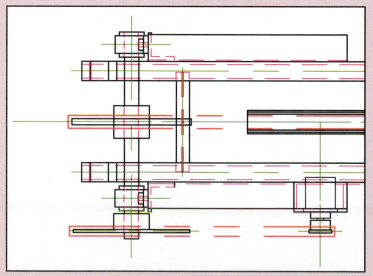

Fig. 2.5.57 *Correct position of the shafts, sprockets, and hydraulic motor at the left end of the top view.*

Again, it is very important to check your work carefully for accuracy. The parts should be centered around their centerlines and the hydraulic motor should be centered inside the motor housing. If your drawing differs, undo the move and try again.

5. Zoom in on the top bearing on the main shaft. Add the bolt, washer, and nut as follows:

   • Washers: Ø1 3/4″ by 1/8″ thick
   • Nuts: 15/16 hex (1.083 across corners) × 9/16″
   • Bolt: Head diameter is 15/16 hex by 3/8″; shaft is Ø5/8″; total length 3 7/16″

   Place the fasteners on the horizontal centerline of the bearing, as shown in Fig. 2.5.58.

6. Mirror the fasteners from the top bearing about the main horizontal centerline of the top view to place them easily onto the bottom bearing.

7. Zoom in on the hydraulic motor and motor housing. The geometry is all in place, but notice that the the motor is now hidden by the motor housing. Change the interior lines of the hydraulic motor to the **Hidden** layer.

8. The motor must also be fastened to the motor housing. Add the fasteners, as shown in Fig. 2.5.59. Use 1/2–13 × 2 1/4 bolts with Ø1 3/8″ washers.

This completes your work on the geometry at the left end of the top view. You will add dimensions and identification numbers later, after you complete the right end.

9. Zoom in on the right end of the top view. The first task is to add the shaft and sprocket. Begin by creating a vertical centerline for the sprocket 22″ from the right side of the tire platform.

10. Create the Ø2 1/4″ shaft around the centerline. Use hidden lines because the shaft is below the tire platform. For now, trim the shaft to the edges of the tire platform.

11. Zoom so that you can see the right end of the top view as well as the right end of the reference geometry, including the sprocket. Create a temporary horizontal centerline through the center of the sprocket. Then move the sprocket into place on the top view. Use the intersection of the vertical and horizontal centerlines of the sprocket for the first point of displacement. Use the intersection of the main horizontal centerline and the vertical sprocket centerline on the top view as the second point of displacement. The sprocket should fit as shown in Fig. 2.5.60.

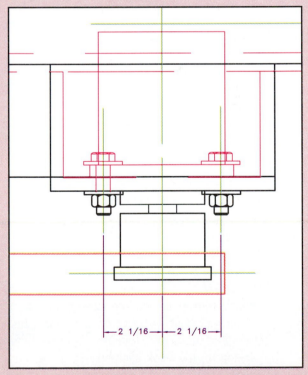

Fig. 2.5.59 *Add the fasteners to mount the motor in the motor housing.*

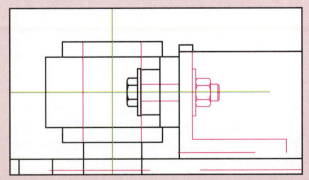

Fig. 2.5.58 *Place the bolt, washers, and nut on the horizontal centerline of the bearing.*

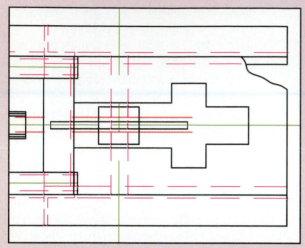

Fig. 2.5.60 *Place the sprocket on the shaft in the top view as shown here.*

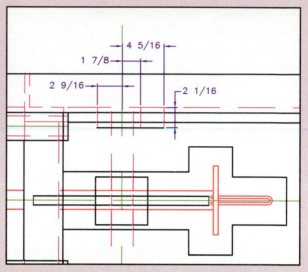

Fig. 2.5.62 *The upper bearing in the top view.*

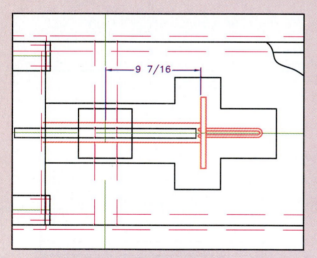

Fig. 2.5.61 *Add a hook at the right end of the conveyor drive chain.*

12. Add the hook at the end of the drive chain, as shown in Fig. 2.5.61. Place the left edge of the Ø1/2″ bar **9⁷/₁₆″** to the right of the centerline of the shaft. Trim or extend the existing phantom lines that represent the drive chain to meet the right edge of the hook. *Hint:* You can save time by picking up the hook geometry from the left-side view.

13. Remove the break in the tire platform and mend the tire platform lines.

14. Add the upper bearing on the shaft, placing it as shown in Fig. 2.5.62.

15. Mirror the upper bearing about the main horizontal centerline of the top view to create the lower bearing.

Next, refer to part #10 on the cut sheet (drawing 0025). This part is fastened to the ends of the flat-bar extensions at the lower end of the tire conveyor. It acts as an interface between the flat bar and two 1¹/₄″ steel rods that are part of the takeup unit, which you will add in the front view.

16. The flat bars to which part #10 is fastened share a centerline with the 1¹/₄″ rods. Add a centerline for the upper flat bar for reference. Then create part #10 to the top view, as shown in Fig. 2.5.63.

You could mirror or copy the part to create its lower counterpart, but in this case, it is probably more efficient to create the rest of the connecting geometry first, since it is the same for the upper and lower flat-bar pieces.

17. Add the two identical nuts to the right side of part #10. Use the dimensions shown in Fig. 2.5.64.

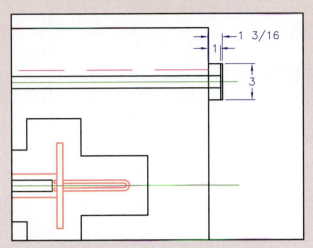

Fig. 2.5.63 *Dimensions for part #10 in the top view.*

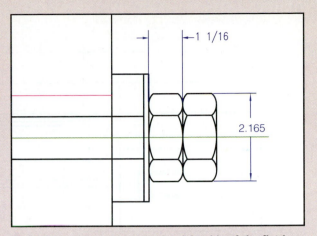

Fig. 2.5.64 *Add two nuts at the right side of the flat-bar extensions.*

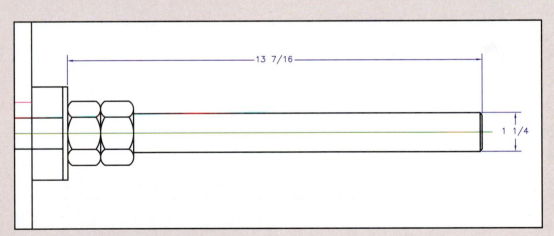

Fig. 2.5.65 *Dimensions for the threaded rod.*

18. Add the **1 1/4″** threaded rod. The rod extends **13 7/16″** to the right of the left edge of the bracket (part #10), as shown in Fig. 2.5.65.

19. Mirror the bracket (part #10), the two nuts, and the steel rod about the main horizontal centerline to add them accurately to the other flat-bar extension. When you finish this step, the right end of the top view should look like Fig. 2.5.66.

20. Add the mounting tabs to the upper rail in the top view, as shown in Fig. 2.5.67. Include the centerlines for the fasteners.

21. Add a washer and bolt head at each of the mounting tabs, centering them on the vertical centerlines you created in the previous step. A detail of the appropriate bolt and washer is shown in Fig. 2.5.68.

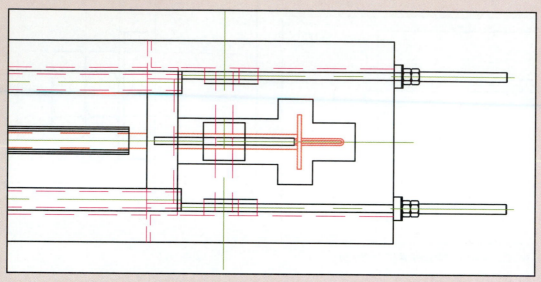

Fig. 2.5.66 *The right end of the top view after step 19.*

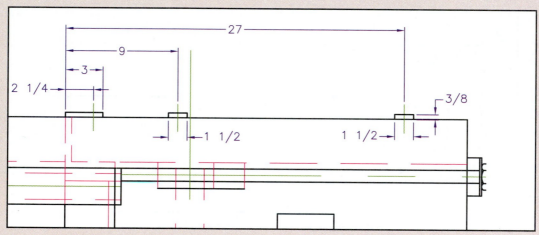

Fig. 2.5.67 *Mounting tabs and fastener centerlines for the top rail.*

22. Mirror the mounting tabs and fasteners about the main horizontal centerline to create the tabs and fasteners for the lower rail.

This finishes the geometry for the top view. The right end of the top view should look like Fig. 2.5.69.

23. Zoom out to see the entire top view. Erase the remaining reference geometry for the top view. (Do not erase the reference geometry for the front view yet.)

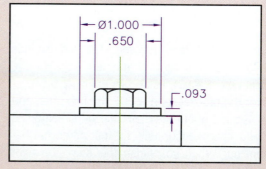

Fig. 2.5.68 *Fastener detail.*

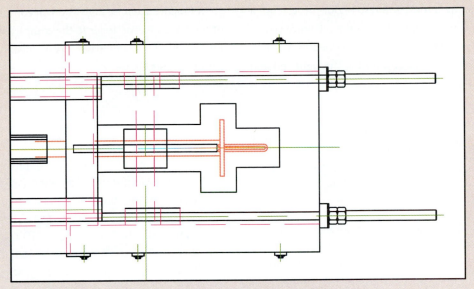

Fig. 2.5.69 *Finished geometry for the right end of the top view.*

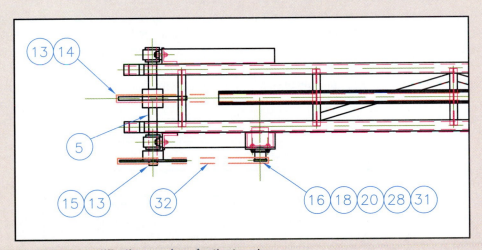

Fig. 2.5.70 *Identification numbers for the top view.*

24. Zoom in on the left end of the top view once more. Add the identification numbers, as shown in Fig. 2.5.70.

25. Add the remaining two identification numbers near the middle of the top view to identify the upper and lower weldments. Refer to Fig. 2.5.42 on page 231 for placement.

26. Add the overall dimensions for the top view. Refer again to Fig. 2.5.42 on page 231 for the dimensions to use and their placement.

Front View

Several objects are still missing from the front view. You have already drawn the sprockets and so on for the left end of the view, but they must be moved from the reference geometry to the front view. They must also be added to the right end. You must also add the takeup unit at the right end and complete the representation of the drive chain. Because this is the reference assembly, the drive chain is shown in the front view with all of the hooks in place at 36″ intervals.

In Practice Exercise 2.5F, you will complete the front view of the tire conveyor. Refer to the completed reference assembly in Fig. 2.5.42 on page 231 as necessary as you work through this exercise.

Practice Exercise 2.5F

Before you begin working on the front view, zoom out to see the entire drawing file. Some of the geometry you need to add to the front view exists in the reference geometry. You should still have the front view of the reference geometry in the drawing file.

1. Move the reference geometry for the front view to a point below the front view of the "real" front view, but near it, to make it easier to move the necessary geometry.
2. Zoom in on the left end of the front view so that you can see both the front view and its associated reference geometry.
3. Erase the cutting-plane line for Section B-B in the front view.

4. Still working in the front view, offset the vertical centerline of the motor housing **28 1/2″** to the left to locate the vertical centerline for the main shaft. With Ortho on, use grips to extend the line up through the top Schedule 40 pipe.
5. Now shift your attention to the reference geometry. Erase all of the lines that make up both of the Schedule 40 pipes. This will make it easier to select the geometry to be moved to the front view. Be careful not to erase the horizontal centerline, however. The sprockets, bearings, shaft, hydraulic motor, and motor drive chain should be left, along with the horizontal and vertical centerlines for the main shaft.
6. Move the sprockets, bearings, shaft, hydraulic motor, and motor drive chain from the reference geometry to the front view. Use the points of displacement shown in Fig. 2.5.71.

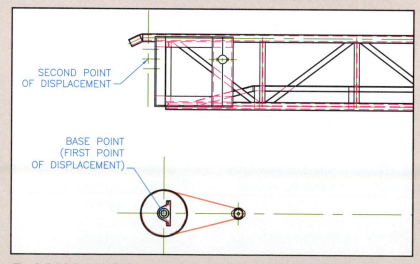

Fig. 2.5.71 **Points of displacement for moving geometry to the front view.**

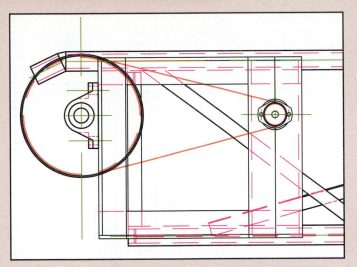

Fig. 2.5.72 *Sprockets, etc., in place in the front view.*

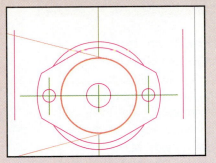

Fig. 2.5.73 *The motor and motor housing.*

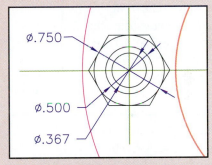

Fig. 2.5.74 *Fasteners for the motor housing.*

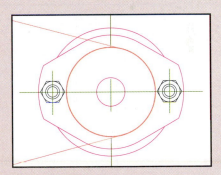

Fig. 2.5.75 *The finished motor housing.*

7. The result of the move operation is shown in Fig. 2.5.72. If the objects did not center correctly in your move, undo it and try again. Proper placement is very important. The motor should fit within the motor housing exactly, for example.

8. Zoom out temporarily and erase the remaining pieces of reference geometry. You will not need it further.

9. Zoom in on the left end of the front view once more. Then zoom in further to work with the hydraulic motor and motor housing. For clarity, erase the two inner circles that appear on the object layer. Then change the rest of the motor to the **Hidden** layer. Leave the small sprocket circle on the phantom layer, and attach the drive-chain lines directly to it using the **Tangent** or **Nearest** object snap. The motor should now look like Fig. 2.5.73.

10. Erase the Ø.531 holes and, in their place, add the fasteners to attach the motor to the motor housing. See Fig. 2.5.74. *Hint:* Fasteners should appear on the Object layer. However, recall that the Object layer is set to print with thick lines, which may make smaller objects such as fasteners hard to see. You may want to add a layer to the drawing named Object 2. Make the color for this layer white also, but assign the **THIN** linewidth to it. Place all of the fasteners and other very small objects on layer Object 2. After you have added the fasteners, the motor housing should look like Fig. 2.5.75.

11. Zoom in on the large sprockets. Change both to the **Phantom** layer.

12. Add two phantom circles to represent the inner and outer edges of the drive chain. Make the circles **Ø16 1/4″** and **18 15/16″**, respectively.

13. Trim the circle that represents the outer edge of the drive chain at the point it intersects the upper Schedule 40 pipe and at the bottom quadrant, leaving the left segment. Trim the circle that represents the inner edge in a similar manner. See Fig. 2.5.76.

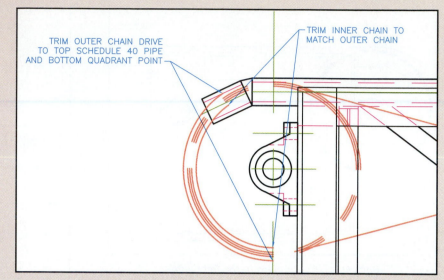

TRIM OUTER CHAIN DRIVE
TO TOP SCHEDULE 40 PIPE
AND BOTTOM QUADRANT POINT

TRIM INNER CHAIN TO
MATCH OUTER CHAIN

*Fig. 2.5.76 **Trim points for the outer edge of the drive chain.***

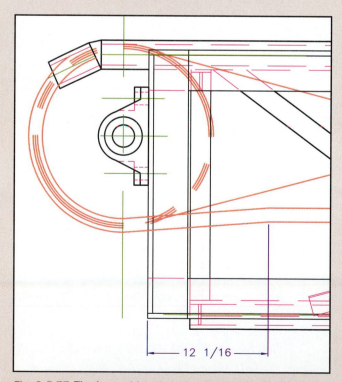

12 1/16

*Fig. 2.5.77 **The lower drive chain in place in the front view.***

14. To locate the lower drive chain, offset the bottom line of the lower Schedule 40 pipe up **12 1/16″**. Change the offset line to the **Phantom** layer. Then offset the line down **1 3/8″** to create the lower edge of the drive chain.

15. Offset the leftmost line of the rectangular framework **12 1/16″** to the right and trim away the part of the drive chain lines to the left of the offset line. Erase this last offset line. Enter the **LINE** command and connect the lower drive chain lines to the arcs that represent the drive chain at the sprockets. When you finish, the left end of the front view should look like Fig. 2.5.77.

16. Zoom in on the bearing and add the fasteners as shown in Fig. 2.5.78. *Hint:* These are the same size as those used on the left side of the top view. You can copy them to save time. Be sure to change all of the objects to the **Object 2** layer.

This completes all of the work on the left side of the front view, with the exception of the hooks. We will do the hooks later.

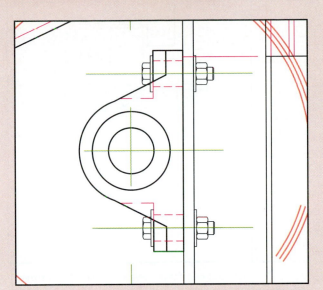

Fig. 2.5.78 *Add the fasteners to hold the bearing in place.*

17. Zoom out to see the entire front view. Notice that there are two cutting-plane lines for Section A-A. Erase the one on the upper weldment. Leave the one on the lower weldment in place.

18. Zoom in on the right end of the front view. Erase the cutting-plane line for Section B-B.

19. Add the side panel. This panel fits over the flat-bar extensions and drive shaft both to protect them and to increase the safety of the people who work near the drive chain. Use the dimensions from the cut sheet (drawing 0025) and place it as shown in Fig. 2.5.79.

20. Mend the breakout line of the top steel angle. Then break the lines as necessary to place the objects that are now hidden by the side panel on the **Hidden** layer. Note that the mounting tabs are on the outside of the side panel, so they are not hidden.

21. To locate the vertical centerline of the shaft at the lower end of the tire conveyor, offset the right side of the side panel to the left by **22"**. Change the offset line to the **Center** layer.

22. The easiest way to create the sprockets on the right end of the front view is to mirror those on the left end. However, you can't use the junction of the two weldments as the mirror line because the weldments are slightly unequal in length. To create the mirror line, first use grips to extend the vertical centerline of the sprockets and shaft on the left side of the view upward beyond the rest of the view. (Be sure Ortho is on to keep the line vertical.) Do the same for the vertical centerline you created in step 21 for the right end. Make sure this one extends even higher than the one on the left side. With Ortho still on, reenter the **LINE** command. Use the **Endpoint** object snap to snap to the top of the left centerline. Use the **Perpendicular** object snap to place the other end of the line perpendicular to the right centerline. Then create a vertical line starting at the midpoint of the line you just created. Extend the line through the front view. (Ortho should still be on.) This line is the mirror line.

23. Mirror the large sprockets, shaft, and drive chain lines from the left end of the front view to the right end. Use the mirror line you created the previous step. After the mirror operation, the right end of the front view should look like Fig. 2.5.79.

24. Delete the temporary mirror line.

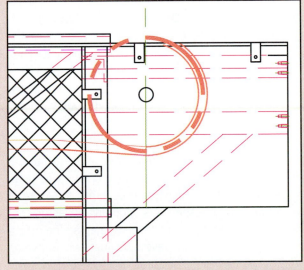

Fig. 2.5.79 *Right side of the front view after mirroring the large sprockets.*

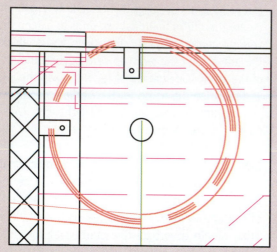

Fig. 2.5.80 *Add the fasteners to hold the bearing in place.*

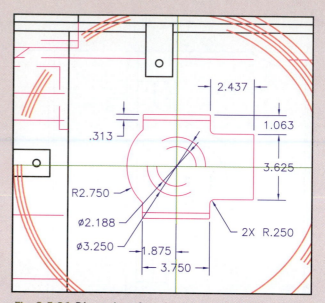

Fig. 2.5.81 *Dimensions for a basic representation of the takeup unit.*

Notice that the outside edge of the drive chain stops short of the top of the sprocket. This is because the pipes have a different configuration at the other end of the assembly. On this end, the upper edge needs to curve around the sprocket to the top quadrant point and then extend in a horizontal line to the left to represent the upper drive chain. This is easy to achieve.

25. Enter the **EXTEND** command and select the vertical centerline as the boundary edge. Pick the outside arc to extend it to the centerline. Then use the **LINE** command to create a line from the top quadrant point left to the edge of the Schedule 40 pipe. The intended result is shown in Fig. 2.5.80.

26. Add the horizontal centerline for the shaft.

The purpose of the **takeup unit** is to maintain a fairly constant tension in the drive chain. This is an important feature. If the chain becomes too loose, it might jump off the sprockets, stopping the entire conveyor system and possibly even damaging the mechanism. The engineers have specified two Browning #VTWS-235 Ø2 3/16" takeup units, one on each side of the lower sprocket.

In the front view, the takeup units are directly in line with each other, so you need only draw the closer one. Recall that because this is a standard purchased part, you do not need to create a detailed drawing of it. You only need a basic representation.

27. Create the body of the takeup unit representation using the dimensions shown in Fig. 2.5.81. Center it on the centerlines of the shaft. Assume that the part is symmetrical about the horizontal centerline. All fillets are 1/16" unless otherwise specified in Fig. 2.5.81. Place the entire takeup unit on the **Hidden** layer. Also, change the circle that represents the shaft to the **Hidden** layer. *Note:* In Fig. 2.5.81, other hidden lines in the area have been removed to allow you to see the dimensions clearly. Because there are other hidden lines in the area, you may choose to create the takeup unit on a different layer, such as **Hatch**, that is represented by a different color. When you finish the construction, move the entire takeup unit to the **Hidden** layer.

28. Add the adjustment nut on the left side using the dimensions shown in Fig. 2.5.82. Estimate any sizes not shown.

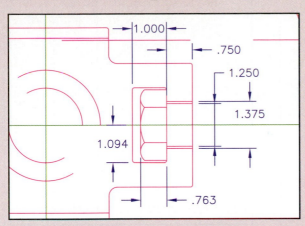

Fig. 2.5.82 *Dimensions for the adjustment nut.*

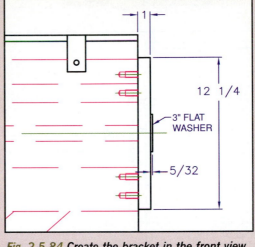

Fig. 2.5.84 *Create the bracket in the front view.*

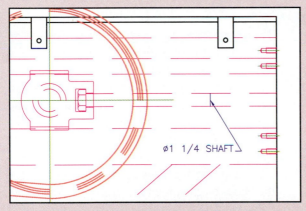

Fig. 2.5.83 *Add the portion of the threaded rod that runs from the adjustment nut to the right side of the side panel.*

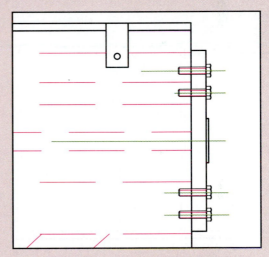

Fig. 2.5.85 *Replace the bolt holes with bolts.*

29. Extend the 1 1/4″ threaded rod that passes through the #10 bracket, as shown in Fig. 2.5.83.

30. Zoom in on the holes at the right end of the flat-bar extensions. The bracket (part #10) will be bolted into these holes. Add the bracket and flat washer as shown in Fig. 2.5.84.

31. With Ortho on, use grips to extend the centerline of each bolt hole to the right so that it extends through the bracket. Erase the bolt holes, but do not erase their centerlines.

32. Add the bolts and washers on the bolt-hole centerlines, as shown in Fig. 2.5.85. Use **Ø.50 ×** **1.868** bolts.

33. The size and shape of the exterior bolts and rod extensions for the takeup units are the same in the top and the front views. Copy them from the top view and insert them in the front view as shown in Fig. 2.5.86.

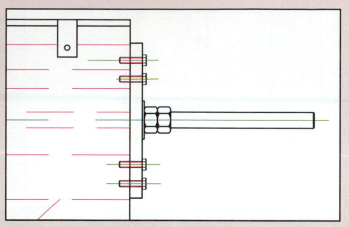

Fig. 2.5.86 *Copy the rod and connectors from the top view.*

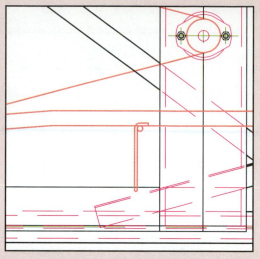

Fig. 2.5.88 *Add a hook on the lower drive chain just to the left of the motor housing.*

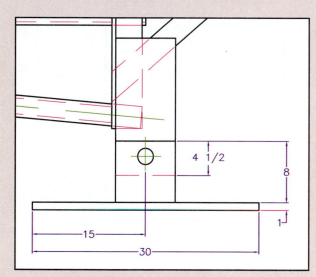

Fig. 2.5.87 *Create the mounting feet in the front view.*

34. Add the mounting feet at the bottom of the right end of the front view. Use the dimensions shown in Fig. 2.5.87. Notice that the mounting feet overlap the mounting plate by 4 1/2″, so you will need to change the bottom line of the mounting plate to the **Hidden** layer.

35. Zoom out to see the entire front view. Then zoom in on the left end of the front view. It is time to add the hooks at 36″ intervals along the drive chain. We will begin with the hooks on the bottom part of the chain. Create the first hook and place it as shown in Fig. 2.5.88. *Hint:* Copy the geometry from drawing 0024. You may also want to block it so that it is easier to manipulate and insert into the drawing. Remember to place the hooks on the phantom layer for this drawing.

36. Enter the **ARRAY** command. There are a total of **14** hooks along the bottom of the drive chain. Remember that the hooks are placed at **36″** intervals. Enter the appropriate specifications and array the hook to create the remaining hooks on the bottom of the drive chain. Check your work by using the **Distance** command to measure the distance between two consecutive hooks. It should be 36″.

37. Zoom in again on the left end of the front view and insert a hook approximately at the left quadrant of the arc that forms the left end of the drive chain. See Fig. 2.5.89.

38. The hook needs to be modified for use along the top of the drive chain because part of it is hidden by the top Schedule 40 pipe. Insert a hook in approximately the location shown in Fig. 2.5.90.

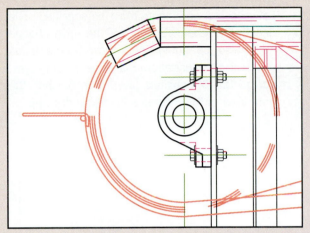

Fig. 2.5.89 *Add a hook approximately at the left quadrant of the circular portion of the drive chain.*

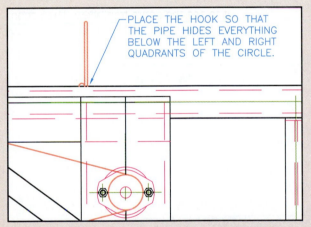

PLACE THE HOOK SO THAT THE PIPE HIDES EVERYTHING BELOW THE LEFT AND RIGHT QUADRANTS OF THE CIRCLE.

Fig. 2.5.90 *Place the first hook on the upper drive chain as shown here.*

Then explode it if necessary and trim away the parts that are hidden by the pipe so that it looks like the hook in Fig. 2.5.90. Array it to create the **14** hooks along the top of the drive chain, remembering to keep the copies at **36**″ intervals.

39. Add the identification numbers for the front view. These numbers are clustered at the left and right ends of the front view. Refer to Fig. 2.5.91 for those on the left end and to Fig. 2.5.92 for those on the right end. *Note:* You may need to move

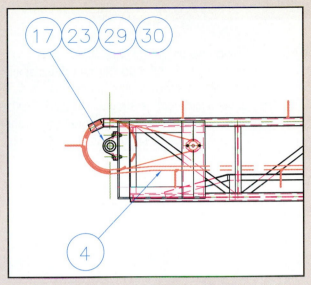

Fig. 2.5.91 *Identification numbers on the left end of the front view.*

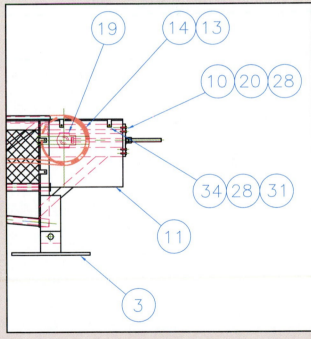

Fig. 2.5.92 *Identification numbers on the right end of the front view.*

the right-side view slightly to the right to fit the identification bubbles on the right end of the front view. If you do so, be sure Ortho is on so that you maintain the alignment between the front and side views.

40. The upper weldment of the tire conveyor is shown in Fig. 2.5.93. Add the two dimensions and the note as shown in the illustration.

41. The lower weldment is shown in Fig. 2.5.94. There are no dimensions on this end of the front view. However, because this is the assembly drawing, the assembly split line between the two weldments must be called out and the weld between the weldments must be specified. The direction of chain rotation for the drive change must also be noted. Add these elements as shown in the illustration. Place the welding symbol on the **Dimensions** layer. This finishes the front view of the reference assembly drawing.

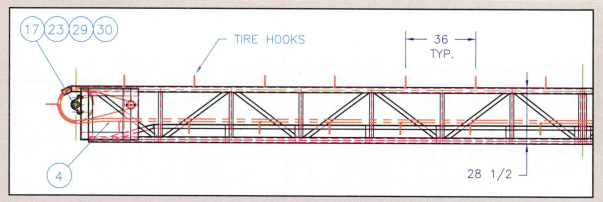

Fig. 2.5.93 *Dimensions and note for the upper weldment on the reference assembly drawing.*

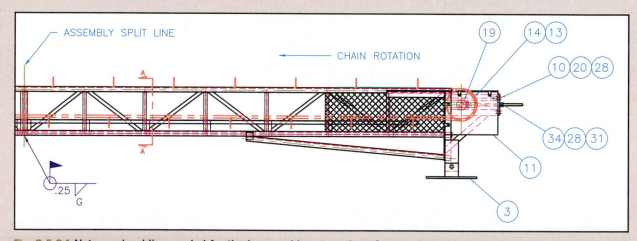

Fig. 2.5.94 *Notes and welding symbol for the lower weldment on the reference assembly drawing.*

Right-Side View

Only a few modifications need to be made to the existing right-side view. The drive chain must be added, but it looks exactly the same in the right-side view as it does in the left-side view. You can therefore simply copy the geometry from the left-side view. You must also add the side panels, the takeup unit, and the mounting feet, along with the fasteners. While the visual impact of these changes is major, the actual modifications are fairly easy to accomplish.

In Practice Exercise 2.5G, you will finish the right-side view. Refer to the finished assembly drawing in Fig. 2.5.42 on page 231 as necessary while you complete this exercise.

Practice Exercise 2.5G

1. Zoom in on the right-side view.
2. Erase the lines that represent the tire platform. We are removing the tire platform in this view so that the drive chain and sprockets can be seen more clearly.
3. Focus your attention on the bolt holes on the left side of the view. These holes are for the bolts that fasten the #10 bracket to the lower weldment. Erase the four holes, but be sure to leave their horizontal and vertical centerlines in place.
4. You could recreate the geometry for part #10, but recall that you have already drawn it for the cut sheet. Open **PROJECT 2 - 0025.dwg**, switch to model space if necessary, and zoom in on part #10 in the lower right corner. Freeze the Dimensions layer so that the front view of the bracket and its centerlines are easier to select. Use **COPYCLIP** (**CTRL+C**) to copy the bracket and its centerlines to the Windows® clipboard. Close **PROJECT 2 - 0025.dwg** without saving. Paste the bracket in an empty space near the left side of the right-side view. Then move it into place, using the intersection of the vertical and top horizontal centerlines as the points of displacement. When you finish, the bracket should be placed as shown in Fig. 2.5.95.
5. Change the object lines that are now hidden by the bracket to the **Hidden** layer, breaking lines as necessary. For clarity in this view, erase the lines that represent the ends of the flat bars.

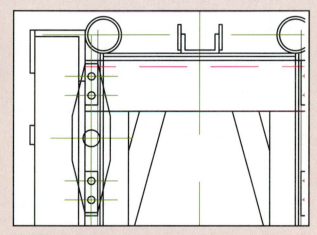

Fig. 2.5.95 *Correct position for the bracket.*

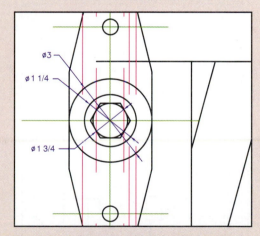

Fig. 2.5.96 *Correct position for the bracket.*

6. Erase the center hole in the bracket, but do not erase its centerlines. Replace the center hole with the end view of the rod, washer, and nut. Use the dimensions shown in Fig. 2.5.96.

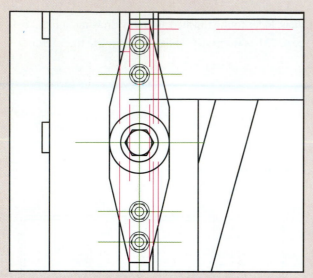

Fig. 2.5.97 *Fasteners in place on the bracket.*

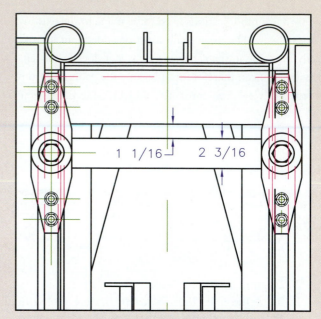

Fig. 2.5.99 *The central shaft for the takeup units.*

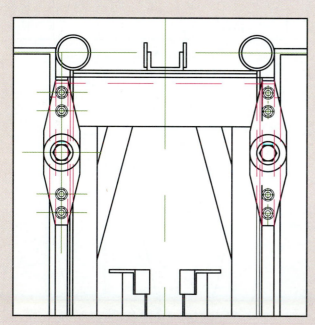

Fig. 2.5.98 *The finished bracket assemblies.*

7. In place of the bolt holes, add the four fasteners, as shown in Fig. 2.5.97. The washer diameter is **1.000**, and the bolt head is **.750 hex**.

8. You will copy the left bracket to create the right bracket. Before you do so, however, prepare the location for the right bracket by deleting the bolt holes and endlines for the flat-bar extensions.

9. Copy the entire bracket, including the left end of the uptake unit and the fasteners, to the appropriate location on the right side of the view. Do not include the object lines that were changed to hidden lines in the mirroring operation. After the mirror is complete, break and change the existing object lines to the **Hidden** layer as necessary.

10. For clarity trim away the portion of the lower edge of the steel angle that protrudes into the brackets on each side. When you finish, the brackets should look as shown in Fig. 2.5.98.

11. Add the central shaft for the takeup units. Trim away the parts of the diagonal supports and vertical steel angles that fall behind the shaft. See Fig. 2.5.99.

To add the drive chain assembly to the right-side view, you need only copy the geometry from the left-side view. Even though this is a copy within a single drawing, it is better to use COPYCLIP instead of the COPY command in this situation. COPYCLIP retains the copied objects on the clipboard until you replace them by issuing another COPYCLIP command. This is helpful when, as in this case, you need to copy small objects from one part of a large drawing to another.

12. This copy is most easily done using two separate operations. For the first operation, zoom in on the upper chain support in the completed left-side view. Select the wear pad and its fasteners and press **CTRL+C** to copy the selected items to the clipboard. Zoom in on the right-side view and paste them in place, being careful to center them on the main vertical centerline.

The rest of the drive chain manually would be tedious to select for copying from among its surrounding geometry in the left-side view. Fortunately, there is an easier way to do it because the entire drive chain assembly resides on the Phantom layer.

13. Make **Phantom** the current layer. Then freeze all of the layers except Phantom so that only the drive chain assembly remains in the left-side view. Use a window to select the entire drive chain assembly, including the sprockets. Press **CTRL+C** to copy the selected items to the clipboard. Then thaw all of the other layers in the drawing. Zoom in on the right-side view and insert the copy, being careful to place it precisely. When you finish, the right-side view should look like Fig. 2.5.100.

14. Add the lower wear pads and their fasteners, as shown in Fig. 2.5.101. *Hint:* Copy the fasteners from the left-side view.

15. Add the end view of both side panels. Recall that the side panels are $1/4''$ thick and **$28\,1/2''$** long. Place them inside the mounting tabs, as shown in Fig. 2.5.102. Add the fasteners as shown. A detail of the fasteners is shown in Fig. 2.5.103.

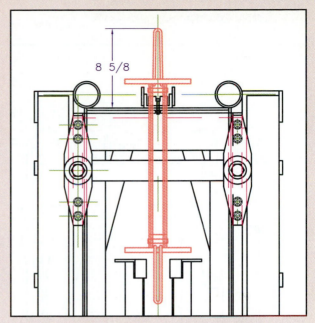

Fig. 2.5.100 **Correct position for the drive chain assembly.**

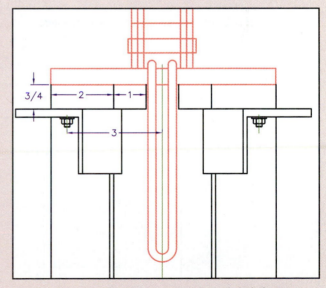

Fig. 2.5.101 **Add the lower wear pads and their fasteners.**

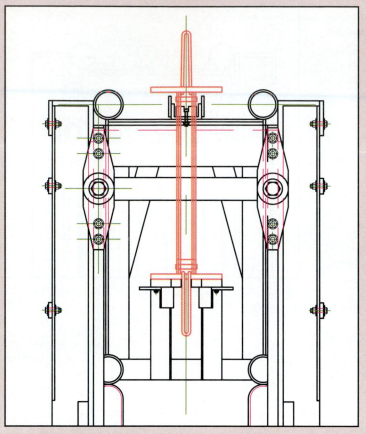

Fig. 2.5.102 *Add the side panels and their fasteners.*

16. Add the mounting feet and their shaft as shown in Fig. 2.5.104. Recall that the mounting feet are constructed of 1″ plates. *Hint:* The mounting feet are located symmetrically about the main vertical centerline. You can create one of the mounting feet and then mirror it to create the other.

17. Zoom out to see the entire right-side view. Add the dimensions and identification numbers as shown in Fig. 2.5.105 to finish the view.

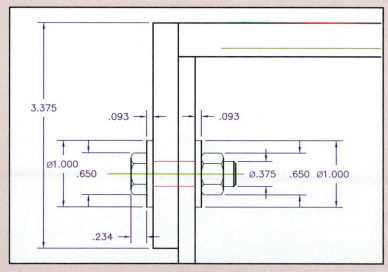

Fig. 2.5.103 *Fasteners to fasten the side panel onto the mounting tabs.*

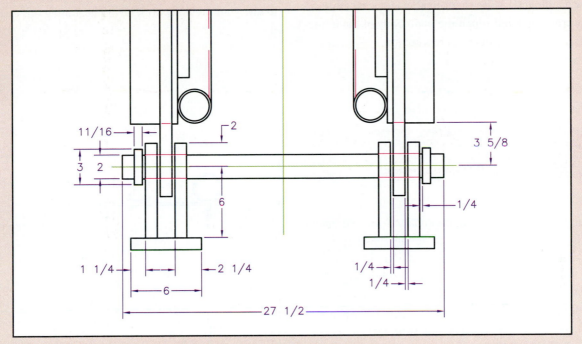

Fig. 2.5.104 *Dimensions for adding the mounting feet and shaft.*

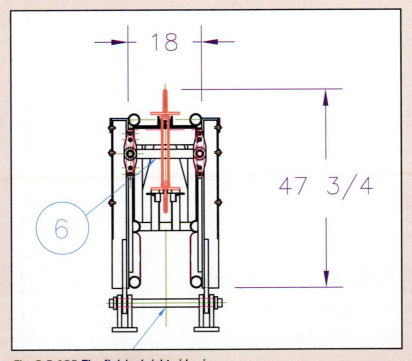

Fig. 2.5.105 *The finished right-side view.*

Section A-A

Because both ends of the tire conveyor contain modifications for handling the drive chain, neither end view shows an accurate cross section through the main body of the framework. Section A-A provides this cross section.

In Practice Exercise 2.5H, you will use the right-side view as a basis for creating Section A-A. In this case, instead of adding geometry, you will need to erase much of it. Be sure to save your previous work before beginning this exercise.

Practice Exercise

Before you begin working on Section A-A, refer again to the completed reference assembly drawing in Fig. 2.5.42 (page 231). Note the location of the section on the drawing. Also note the similarities and differences between Section A-A and the right-side view.

1. Zoom out to see the entire drawing. Temporarily freeze the **Dimensions** and **Text** layers.

2. Enter the **COPY** command and use a window to select the entire right-side view. Place the copy in its approximate location as demonstrated in Fig. 2.5.42. Then thaw the **Dimensions** and **Text** layers.

3. Zoom in on the copy, which will now be referred to as Section A-A.

4. Remove the items that are not needed for the sectional view:

   - The drive chain assembly. (*Note:* Do not erase the upper wear pad and its fastener.)
   - Both side panels and their fasteners.
   - The 4″ × 4″ steel angles that support the pieces at each end of the conveyor assembly and the mounting tabs attached to them.
   - The brackets (item #10) and their fasteners and centerlines.
   - The central shaft for the takeup unit.
   - The mounting feet and shaft at the bottom of the view, along with the mating pieces on the lower end of the conveyor and the vertical supports for them.
   - The additional Schedule 40 support pipes near the bottom of the view.

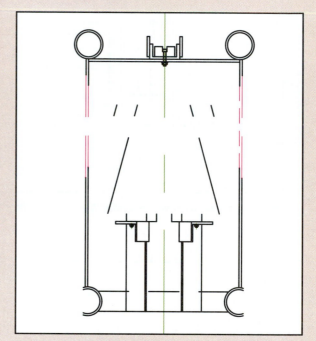

Fig. 2.5.106 *Section A-A after unneeded geometry has been removed.*

For now, leave the lower chain rail as it is. When you finish the deletions, Section A-A should look like Fig. 2.5.106.

5. Mend the broken lines and circles so that the view looks as shown in Fig. 2.5.107. *Hint:* You may need to recreate the lower Schedule 40 pipes. If so, simply establish their center points and copy the upper pipes. Be sure to place the copies accurately on the centerlines you have established.

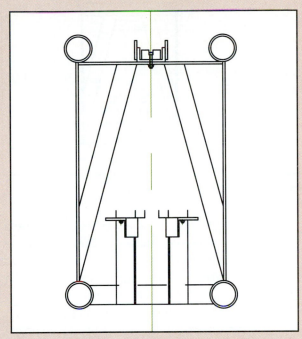

Fig. 2.5.107 *Mend the major lines as shown here.*

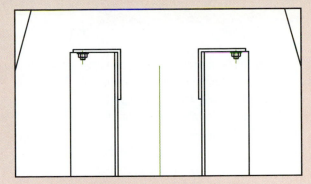

Fig. 2.5.108 *Add the lower chain rails.*

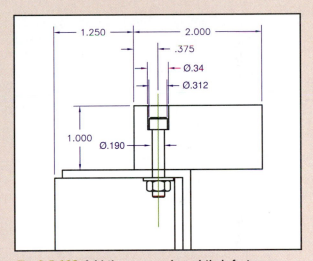

Fig. 2.5.109 *Add the wear pads and their fasteners.*

6. Offset the bottom line of the top vertical cross piece down by **2″** to finish the **2″ × 2″** steel angle at the top of the view. Trim the diagonal supports away from the interior of the steel angle.

7. Zoom in on the rails of the lower chain support. Erase all of the lines that represent the wear pads and the flanged portion of the rails. Leave the support posts and the fasteners intact.

8. Mend the existing lines of the support posts. Add the **2″ × 2″ × 1/4″** steel angles that form the rails of the lower chain support, as shown in Fig. 2.5.108.

9. Recreate the wear pads and the upper portion of their fasteners. The dimensions for the left wear pad are shown in Fig. 2.5.109. *Important:* The existing fasteners are no longer in the correct location because you have removed the flanges. You will need to move both sets of fasteners into place. Refer to Fig. 2.5.109 for correct placement. *Hint:* Construct the top part of the bolt

first, along with its centerline. Then simply move the nut and the existing lower part of the bolt into place along the centerline. When you have finished the left wear pad, mirror it to create the right wear pad.

10. Since this is a sectional view, you will need to hatch the cut surfaces. These surfaces include:

   • walls of the Schedule 40 pipe
   • 2″ × 2″ × 1/4″ steel angles
   • support block on the upper chain support
   • wear pads

Use the **ANSI31** pattern with a scale of **2**. For pieces that lie adjacent to each other, such as the wear pads and the lower chain rails, specify different hatch angles to distinguish the parts. Remember to place the hatches on the **Hatch** layer.

11. For clarity, Section A-A will be displayed at a larger scale than the main views on the reference assembly. Enter the **SCALE** command and use a window to select all of Section A-A. Enter a new scale of **3**.

12. Since there are only three dimensions to be added to the section, you don't need to spend time creating a new dimension style for the enlarged scale. Instead, you can simply change the scale individually for the dimensions. Add the dimensions as shown in Fig. 2.5.110. The default dimensions will not match those shown in Fig. 2.5.110. Doubleclick each of the two dimensions that show measurements to display

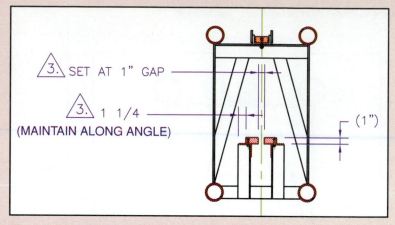

Fig. 2.5.110 Dimensions for Section A-A.

its properties. In the **Primary Units** portion of the Properties window, change the **Dim scale linear** value from 1.0000 to **.3333**. This will change the dimensions to their correct values. Add the extra text and triangles as shown.

13. Add the identification numbers and the section title, as shown in Fig. 2.5.111, to complete Section A-A.

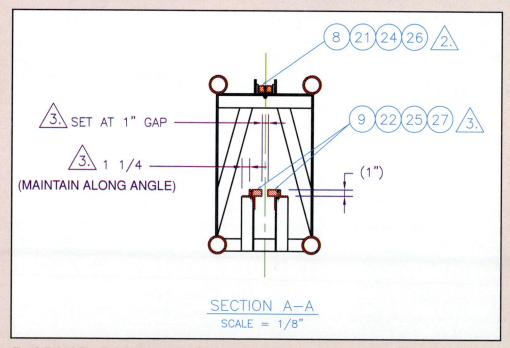

Fig. 2.5.111 The finished Section A-A.

Details of the Wear Pads

The wear pads are small enough that they are not easily seen in place on the reference assembly. Because fastening locations and holes are created at the assembly stage, details of both the upper and lower wear pads are shown on the reference assembly drawing. Complete Practice Exercise 2.5I to add the wear pad details to the drawing.

You could recreate the geometry for the details of the upper and lower wear pads; the geometry is simple enough. However, there is really no need. The wear pads already exist both in the side views and in Section A-A. In fact, in Section A-A, the wear pads are already shown in section, as needed for the details. However, recall that Section A-A is already shown at a larger scale than the rest of the drawing. The scale will need to be even larger for the details. To make scale calculations easier, it is better to start with a view drawn at the base scale.

1. Zoom in on the upper wear pad in the right-side view. Make sure there is enough clear space in the zoom window to place a copy of the wear pad in.

2. Enter the **COPY** command. Select all of the lines that make up the upper wear pad, including the bolt hole, but not the bolt itself. *Hint:* You may want to freeze the Phantom layer temporarily to make the wear pad lines easier to select. Place the copy in a clear area near the right-side view. Some of the lines that make up Section A-A are shared with other geometry, so you will need to trim the lines of the copy. When you finish, the copy should look like Fig. 2.5.112. From now on, the copy will be known as the upper wear pad detail.

3. Zoom out and move the detail to a location above and to the left of Section A-A. Refer to the completed reference assembly drawing on page 231 for guidance.

4. Use the **SCALE** command to enlarge the detail to a scale of 8.

5. Zoom in on the detail and hatch the cut surfaces using **ANSI31** and a hatch scale of **16**.

6. Add the centerline, note, dimensions, and detail title as shown in Fig. 2.5.113. In the **Properties** window, change **Dim scale linear** for the dimensions to $1/8$. Change the **Dim units** to **Decimal** with a **Precision** of **0.0**. *Note:* You may need to adjust the position of both the detail and Section A-A in the drawing to avoid interference. You can place them more exactly later.

Turn your attention to the lower wear pads. You cannot use the same procedure of copying them from

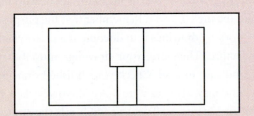

Fig. 2.5.112 Geometry for the detail of the upper wear pad.

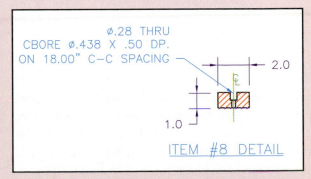

Fig. 2.5.113 Completed detail for item #8, the upper wear pad.

a side view because the side views show the flanged ends of the lower chain rails. You could copy the wear pads from the sectional view, but again, you would need to perform additional calculations to achieve the proper scale. Since the geometry is so basic, the easiest way to create the lower wear pads is simply to draw the left wear pad and then mirror it to create the right one.

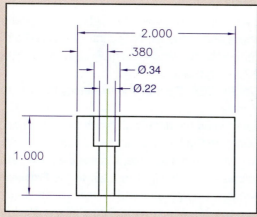

Fig. 2.5.114 *Dimensions for the left lower wear pad.*

7. Zoom in on the space just to the right of the upper wear pad detail. Create the left lower wear pad and the centerline for the bolt hole. The dimensions are given in Fig. 2.5.114 for reference. Create the geometry at its actual size.
8. Enter the **SCALE** command and change the scale of the left lower wear pad to **8**.
9. With Ortho on, create a temporary vertical line slightly to the right of the left wear pad. Use this line as a mirror line to mirror the left wear pad, thus creating the right wear pad. Erase the vertical line after the mirror operation is complete.
10. Hatch the cut surfaces of both of the lower wear pads using **ANSI31** and a hatch scale of **16**.
11. Add the dimensions, note, and detail title to finish the detail, as shown in Fig. 2.5.115. Use the **Properties** window to change the characteristics and scale of the dimensions as described in step 6. If necessary, adjust the spacing of the views so that the lower wear pads can be placed in line with and to the right of the upper wear pad on the drawing.

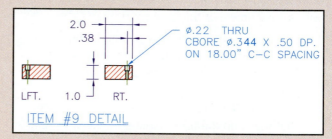

Fig. 2.5.115 *The finished detail for the lower wear pads, item #9.*

Parts List and General Drawing Notes

The reference assembly drawing is now complete except for placing it on the layout sheet and adding the parts list and general drawing notes. If you refer once more to the completed drawing on page 231, you will notice that the parts list for the reference assembly is very long. You will again create the parts list in a spreadsheet, but you will need to alter your methods slightly to include all three segments of the parts list in a single spreadsheet file.

In Practice Exercise 2.5.J, you will complete the reference assembly drawing, as well as the entire set of working drawings for the tire conveyor. Remember that the purpose of this set of working drawings is to document a prototype for later manufacture. Therefore, your drawings must be accurate, clear, and easy to read. Check your finished drawings carefully. You may also want to have them checked by your instructor. On an actual field assembly job, the drafter's work would most likely be checked by his or her supervisor as well as the chief engineer for the prototype.

Practice Exercise 2.5J

1. Zoom out to see the entire reference assembly drawing.

2. Pick the **ANSI D Title Block** tab to switch to the layout view in paper space. Change the name of the tab to **0020**.

3. Erase any existing viewports and create a new one for the reference assembly drawing, as described in earlier sections.

4. Doubleclick the edge of the viewport to show its properties. Change the scale to **1/32**.

5. The geometry for this drawing will be a tight fit at the declared scale. You may need to adjust all of the views to get them to fit within the drawing area of the layout sheet. Be very careful to keep the front, top, and side views in alignment with each other. You have more flexibility with the details and Section A-A, but try to place them roughly as shown on page 231.

6. Add the drawing notes on the **Text** layer, as shown in Fig. 2.5.116. Use the **POLYGON** command to create the triangles.

Next, turn your attention to the parts list. To fit on the drawing, this parts list will be divided into three segments. The following steps are based on the Microsoft Excel spreadsheet software. If you are using a different spreadsheet, you will need to customize the steps accordingly.

7. Open Excel and begin a new spreadsheet. Copy the parts list as shown in Fig. 2.5.117, formatting the columns approximately as shown in the illustration. Notice that one narrow column has been left between each segment of the parts list.

8. On the **Tools** dropdown menu, select **Options**. Pick the **View** tab in the Options window. Near the bottom of the window, pick the checkbox next to **Gridlines** to clear (uncheck) it and pick **OK**. This removes the background grid from the drawing. Now use a window to select the first segment of the parts list and change the border to **All Borders**. Do the same for the other two segments. When you finish, the Excel file should look like Fig. 2.5.117. Save and close the file.

9. In the AutoCAD drawing, insert the file as an OLE object. You may want to start with a height of 50% instead of 100%. Then size and position it as shown in Fig. 2.5.42 on page 231 to finish the parts list.

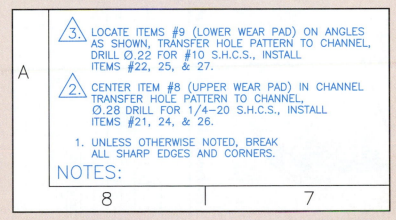

Fig. 2.5.116 *General drawing notes for the reference assembly drawing.*

10. Finish the title block. If you want, you can copy the title block information from one of the other drawings in the set to use as a basis. Change the scale to **1″ = 2′-0** and the drawing number to **0020**. The reference assembly takes only one sheet, so this will be sheet 1 of 1. Change the title of the drawing to:

TIRE CONVEYOR ASSEMBLY
TIRE GASIFICATION SYSTEM

11. Add the tolerance block to complete the drawing. The tolerances are the same as those for the rest of the drawings in the set, so you can copy the tolerance block from one of the other drawings.

This completes the reference assembly drawing as well as the documentation for the prototype of the tire conveyor. You should check your drawings carefully. Make sure there are no inconsistencies among the various drawings in the set. In the field, you would also check and cross-check your dimensions against the actual prototype. The finished, operating tire conveyor is shown in Fig. 2.5.118.

Fig. 2.5.118 The finished tire conveyor in operation.

Fig. 2.5.117 Parts list for the reference assembly drawing.

ITEM#	QTY	PART/DRAWING	ITEM DESCRIPTION
40			
39			
38			
37			
36			
35			
34	8	1/2-13 x 1.50 LG.	HEX HEAD BOLT - GRADE 8
33	2	2" I.D.	SHAFT COLLAR
32	1	#60 APPROX. 95" LG. STD.	ROLLER CHAIN "BROWNING"
31	18	1/2"	NUT, HEX
30	4	5/8"	FLAT WASHER, AN
29	4	5/8"	NUT, HEX
28	18	1/2"	FLAT WASHER, AN
27	60	#10	NUT, HEX
26	30	1/4"	NUT, HEX
25	60	#10	FLAT WASHER, AN
24	30	1/4"	FLAT WASHER, AN
23	4	5/8-11 x 2.50 LG.	HEX HEAD BOLT-GRADE 8
22	60	#10-24 x 1.25 LG.	S.H.C.S.
20	10	1/2-13 X 2.00 LG.	HEX HEAD BOLT-GRADE 8
19	2	#VTWS-235	TAKEUP UNIT 2-3/16" DIA "BROWNING"
18	1	#110A-106-AS-0	HYDRAULIC DRIVE MOTOR "PARKER"
17	2	#PBE920	TAPERED ROLLER BEARING 2-3/16" BORE "BROWNING"
16	1	#60P13	#60,3-1/2" DIA. SPROCKET "BROWNING"
15	1	#60R72	#60, 18" DIA. SPROCKET W/ SPLIT/TAPER "BROWNING"
14	2	#120R36	#120, 18" DIA. SPROCKET W/ SPLIT/TAPER "BROWNING"
13	3	5/8" SQ.	KEYSTOCK LENGTH AS REQD.
12	2	0025	SIDE PANEL
11	2	0025	PLATE, TAKEUP ACCESS
10	2	1.0 X 2.0 X 489 LG.	WEAR PAD, LOWER (SEE DETAIL) DELREN PLASTIC MATERIAL
8	1	1.0 X 2.0 X 475 LG.	WEAR PAD, UPPER (SEE DETAIL) DELREN PLASTIC MATERIAL
7	1	2.00 DIA. 2 27 1/2 LG.	SHAFT MOUNTING FEET (1018 H.R. STEEL)
6	1	0023	SHAFT, DRIVEN END
5	1	0023	SHAFT, DRIVE END
4	1	0024	CHAIN, TIRE CONVEYOR
3	2	0022	MOUNTING FEET
2	1	0021	CONVEYOR WELDMENT, UPPER END
1	1	0020	CONVEYOR WELDMENT, LOWER END

Review Questions

1. Why is the reference assembly drawing often left for last in creating a set of working drawings to document a field assembly?
2. Why do items such as the shafts and the drive chain need special detail drawings?
3. Explain the purpose of a key and keyseat. Why is accuracy so important in machining these items?
4. What information can be obtained from the detail identification symbol on a detail drawing?
5. What is a cut sheet? Why is it an important part of the documentation for a project like the tire conveyor?
6. What is the significance of a triangular "bubble" on a drawing?
7. What term is used to describe the distance between the centers of the two circular ends of each link in a roller chain?
8. What is the purpose of the takeup unit?
9. What is the easiest way to create a chamfer in AutoCAD?
10. In the extended parts list for the reference assembly drawing, what is the purpose of hiding the gridlines in the Excel file before inserting it into the AutoCAD file as an OLE object?

Portfolio Project

Biomass Conveyor (continued)

At this time, you must complete your work on the set of working drawings for the biomass conveyor you have designed.

1. Create a reference assembly drawing, if you have not done so already.
2. Add the main parts list to the reference assembly drawing.
3. Add any detail drawings or cut sheets needed to describe your design completely.
4. Review all of your drawings carefully to be sure they are consistent and accurate.
5. Save your work.
6. Print a copy of the entire set of working drawings to include in your portfolio.

Practice Problems

In Practice Problems 1 through 4, you will design each device to your own specifications. Illustrations have purposely not been provided to encourage you to use your own creative abilities.

1. Create a design for a simple toy truck to be made of wood. Make assembly and detail drawings to describe the truck completely. Include a cut sheet to show how the various pieces of wood should be cut.
2. Create a design for a simple cuckoo clock, complete with a moving cuckoo mechanism. Make a set of working drawings to describe the clock completely. Include a large-scale detail showing the cuckoo mechanism. Use your imagination—the cuckoo does not have to be a bird. You may want to research clocks that are already on the market, but do not copy an existing design.

3. Create a design for an up-scale desk tray to hold pencils and pens, paper clips, rubber bands, and other typical desk accessories. Design the tray to be made of polished wood. Include a creative emblem or rosette on the front or side of the tray. You may want to go to an office supply store and research desk trays that are already on the market, but do not copy an existing design. Make a set of working drawings to describe the desk tray completely. Include a cut sheet to show how the various pieces of wood should be cut, as well as a large-scale detail drawing to document the emblem or rosette.
4. Create a flat-screen monitor design with a tilt/swivel mechanism to allow the screen to be placed at various angles to increase its ergonomic usefulness. Make a set of working drawings for the monitor case. (Do not include the internal electronics.) Include a large-scale detail of the tilt/swivel mechanism.

Summary

- The drafter is an important member of the team that builds or manufactures a prototype.
- The drafter's role in a field assembly is different from his or her role in a normal drafting job.
- Parts of an assembly that are too small to be dimensioned accurately on the larger drawings should be drawn as a large-scale detail on one or more separate drawing sheets.
- For projects such as the tire conveyor assembly, which consists largely of welded metal, a cut sheet must be included to document the size of the finished metal pieces.

Project 2 Analysis

1. In what ways was working on this project different from working on a typical mechanical drafting project? Do you think it would have been different still if you had actually been out in the field during the field assembly of the tire conveyor? Explain.
2. What extra steps or tasks are involved in documenting a field assembly?
3. In Project 1, you placed the entire set of working drawings into a single AutoCAD file. In Project 2, however, each drawing was given its own dedicated AutoCAD file. Discuss the pros and cons of both methods.
4. Three-dimensional models are used in several places in the text for this project. Explain why 3D models are sometimes more useful than photographs in a field assembly project.
5. On a reference assembly drawing for a new network server design, the drafter has decided to include a detail of one of the connection devices. The drafter plans to show the detail on the same sheet as the reference assembly, but at a scale that is four times larger than the actual size. First, he draws it to actual size. What value should he use with the SCALE command to scale up the geometry? How can he change the dimensions to reflect the new scale automatically?

Synthesis Projects

Apply the skills and techniques you have practiced in this project to these special follow-up projects.

1. Actually participating in a field assembly project is different from just reading about one. In this problem, you will simulate an actual field assembly project. In a team of three to five people, decide on an item for which you will create a prototype. The project must be something your team can actually construct given your team's time, material, and equipment restraints. In other words, you must be able to build the actual prototype, so a proposed project that includes welding is unacceptable unless your team has access to structural steel members, welding equipment, and a knowledgeable welder. Organize the team so that every member is responsible for at least part of the documentation. Document your field assembly.
2. Consider the field assembly simulation you performed in the previous problem. Discuss any problems that occurred. How did you handle them? Why is involvement of the entire team required?

Compensator Swivel Bracket: 3D CAD/CAM

When you have completed this project, you should be able to:

- Create simple solid models using AutoCAD's predefined primitives.
- Revolve and extrude closed polylines to create complex solid models.
- Use Boolean operations to refine a solid model.
- Apply appropriate tolerances so that components are manufacturable using a 3D solid model as a guide.
- Follow accepted design guidelines when building a 3D solid model.
- Create a 3D solid model of an injection mold for use in manufacturing a component for which a 3D solid model already exists.
- Create 2D profile drawings directly from a 3D solid model.

Fig. 3.1 CNC endmill following a tool path while cutting a cavity in a bottom mold half.

Introduction

In recent years, drafters have been called upon to do more than just create sets of 2D working drawings for parts and assemblies to be manufactured. Three-dimensional modeling has advanced to the point that computers can be used not only to generate 3D models of proposed parts, but also to control machine tools to manufacture the parts directly.

In this project, you will create a 3D model of the compensator swivel bracket manufactured by Browning Component, Inc., for its SmartShift® high-performance bicycle. You will then use this model to investigate CAD/CAM techniques and numerical control methods.

Some manufacturers prefer to work through the design and approval stages using an accurate 3D model, but then they require drafters or engineers to produce 2D detail drawings for the final working drawings. Therefore, you will also experiment with extracting 2D views from a 3D model for the purpose of detail generation.

Introduction to Solid Modeling

Modeling in 3D requires many of the same skills needed for 2D drafting. However, the focus of 3D modeling is slightly different. Instead of preparing a set of drawings that thoroughly describe the part or assembly to be manufactured, the 3D modeler actually creates a virtual representation of the part itself. This has many advantages. It allows engineers to test parts of an assembly to make sure they don't interfere with each other, for example. Perhaps most importantly, the 3D models can be used for structural and other forms of simulated testing before a physical prototype is built. More and more companies are using 3D to design their parts.

This project will introduce you to 3D modeling concepts. You will translate a 2D drawing into a 3D solid model using AutoCAD. The basic concepts of 3D modeling are discussed in this section. The section also provides background information about the compensator swivel bracket that comprises this project.

Section Objectives
- Create simple solid models using AutoCAD primitives.
- Create solid models using extrusion and revolution techniques.
- View solid models as wireframes, hidden images, and shaded images.

Key Terms
- analysis
- CAD/CAM
- computer-aided manufacturing (CAM)
- extrusion
- primitive
- profile
- revolution
- solid modeling
- UCS icon

David Parker, 600 Group Fanuc/Photo Researchers/SPL

3D Design

The computer affects most aspects of life in modern society. There are computers in everything from the automobile to the refrigerator. See Fig. 3.1.1. One of the areas that has been vastly affected is the area of manufacturing. Manufacturing has been affected by the advent of the computer on two levels. The first is the use of computers to aid in the design of manufactured components and the use of the computer to assist in the manufacture of the components, and the second is the use of computers to guide the machine tools that actually create the components.

Fig. 3.1.1 An automobile assembly line is one example of the integration of computers into the manufacturing process. This assembly is completely computer-controlled.

Terminology

Computers have several acronyms associated when they are used in industry. Two of the most common are CAD and CAM. CAD commonly refers to computer-aided design or computer-aided drafting, depending on the context. This term refers to the use of a computer during the design portion of the product or component's life cycle.

CAM refers to **computer-aided manufacturing**. This encompasses how computers are used in the manufacturing environment. Computers are used for equipment control, computer-numerical control machining, tracking systems, environmental controls, and machine vision inspection. They are also used in many other aspects of manufacturing. When a manufacturer uses both CAD and CAM in the design and manufacture of a product, the process is commonly known as **CAD/CAM**.

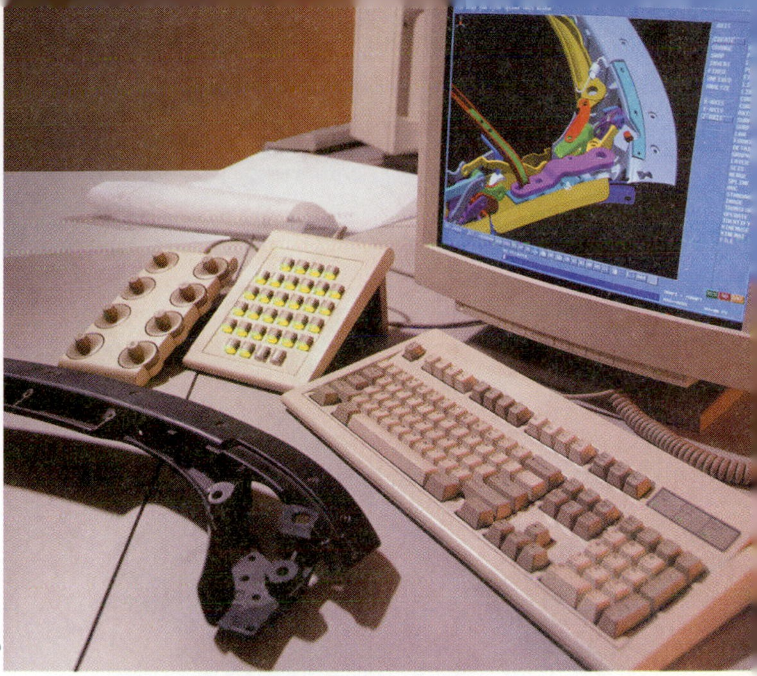

Roger Ball/Corbis

Fig. 3.1.2 *Solid models can be developed and tested virtually using 3D CAD systems.*

Solid Modeling

Computers have advanced over the past 20 years with tremendous speed. The software programs associated with computers, including CAD/CAM systems, have advanced at a similar rate.

The first **solid modeling** CAD system was developed in the early 1970s. Up to that point, all modeling on the computer was done using lines, points, arcs, and circles. This is an effective way to draw, but it had only one distinct advantage over paper: the ability to edit the drawing easily. A drafter working on paper racing against an equally skilled computer drafter might be in an even tie until it came time to change a line. Then the computer drafter would have a distinct advantage; erasing and changing items on the CAD drawing is a much simpler process. Two-dimensional CAD systems have the ability to show distance and area; a 3D system can show area, distance, and volume.

When solid models are used, the operator can determine weights, volume, surface area, and other key features of a design. See Fig. 3.1.2. The design phase of a component will determine 70% of the production costs. These costs are incurred by the specification of materials, dimensions, tolerances, surface finishes, and other parameters that specify the component. Therefore, it is important to be able to obtain as much information as possible during the design phase of a component with the use of both CAD and CAM systems.

Autodesk has developed many CAD products based around solid modeling, each with different features and options. This project will focus on the basic AutoCAD program, which is capable of performing 3D functions. These functions can be found in AutoCAD versions 13 through 2002 and beyond. Note that the LT versions of AutoCAD do not have the features and functions required to complete this project. The Mechanical Desktop and Inventor programs are capable of performing these functions, but the procedure required to complete the project utilizes different steps and functions.

Boolean Modeling

The type of solid modeling performed in this chapter is referred to as *Boolean modeling*. This style of modeling uses a boundary representation (B-Rep) to store and display the modeling information. The biggest advantage of Boolean modeling is that it minimizes the information that the computer must process and store about each model. The major disadvantage is that once the model is drawn, it cannot be edited or modified easily.

The name of the modeling comes from the types of operations that are used to perform each one of the functions. These operations are referred to as *Boolean operations*, after

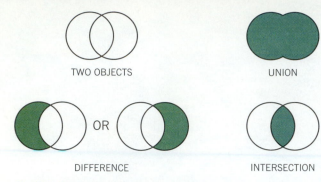

TWO OBJECTS

UNION

OR

DIFFERENCE

INTERSECTION

Fig. 3.1.3 *Boolean operations.*

George Boole, the mathematician who originally developed the concept. The operations and symbols are:

- Union: $\cup$
- Difference (Subtraction): $-$
- Intersection: $\cap$

These symbols are also commonly used in the study of geometry. In addition, Boolean math is the basis for all modern computer calculations. The basic operations in Boolean math are shown in Fig. 3.1.3.

Bicycle Shifting System

The first bicycle was invented as a walking machine in 1817 to assist Baron von Drais around the royal gardens. In the almost 200 years since its inception, the bicycle has seen some very dramatic improvements. The bicycle is now widely accepted as a source of both transportation and entertainment. Bicycles have evolved from simple one-speed cruising bicycles to multiple-speed mountain bikes and street racing bikes. See Fig. 3.1.4. Some of these improvements have been made possible by the inception and incorporation of the modern computer and computer-aided design.

Browning Component, Inc., has developed an automatic shifting system known as the Browning SmartShift®. This system is unlike any other system currently available on the market. It allows the rider to shift automatically at any time, even under a heavy load, with the push of a button. This shifting is done with the help of an onboard computer and an advanced shifting device that allows the chain to be fully engaged at all times. The shifting system can be customized to the operator's preference of shift settings or can be switched over to a button-activated manual shifting mode.

Courtesy of Browning Component, Inc./Patented by Browning Automatic Transmission, LP

Fig. 3.1.4 *Browning's SmartShift bicycle is an example of today's high-performance bikes.*

Most bicycles that have multiple speeds use a derailleur system to shift the chain from sprocket to sprocket to change the gear ratio. Changing the gear ratio changes the wheel-to-pedal rotation ratio. The higher the wheel-to-pedal rotation ratio, the faster the bicycle will go. The system used in the Browning SmartShift is highly specialized, but the hardware of the system is designed in a way that is similar to a typical derailleur system.

One of the assemblies that is used on the Browning SmartShift bicycle is called the *compensator assembly* or *chain-length compensator*. This assembly looks almost like a derailleur on a manual-shift bicycle. The compensator, shown in Fig. 3.1.5, does not move or force the chain. Instead, it acts as a guide and maintains the chain tension. The compensator keeps the chain wrapped around the rear sprockets by approximately 180°.

In this project, you will convert a 2D manufacturing drawing into a 3D solid model. The project will focus on the swivel bracket component of the compensator assembly. The 2D manufacturing drawing will provide all of the necessary information to create the swivel bracket in a 3D solid model.

Courtesy of Browning Component, Inc./Patented by Browning Automatic Transmission, LP

SWIVEL BRACKET

COMPENSATOR

Fig. 3.1.5 *The compensator and its associated swivel bracket.*

Some of these features and considerations include gates, sprues, runners, shrink factors, draft angles, alignment pins, cooling lines, surface coatings, and injection pressures. The student will create a basic mold from the 3D solid that is created in this section.

Swivel Bracket

The purpose of the swivel bracket is to interface with and mount the compensator assembly to the bicycle frame. The back of the bicycle frame has a flat plate that contains two threaded (tapped) mounting holes. The compensator assembly mounts to these tapped holes through mating holes.

The swivel bracket is constructed of black nylon. The bracket is manufactured using a plastic injection molding process. The mold for the component is constructed by machining a negative image of the component into two blocks of specially treated stainless steel. Depending on the material and other factors, this cavity tends to be slightly larger than the finished part to compensate for shrinkage.

The mold will be made up of two blocks, one for the top section and one for the bottom section, as shown in Fig. 3.1.6. Each block will contain half of the component, making a complete component cavity when the two sides are assembled together. Many of the more complex features and design considerations built into a mold will not be considered here.

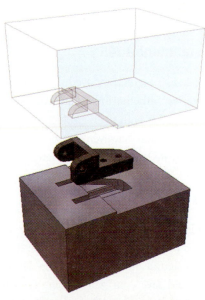

Fig. 3.1.6 *The mold is made up of two stainless steel halves in which a negative image of the component is embedded.*

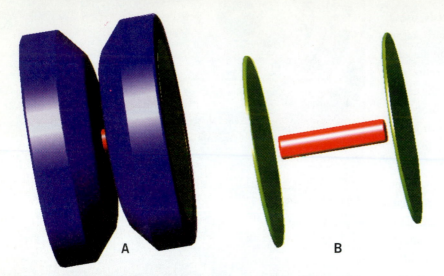

A   B

Fig. 3.1.7 Because each component of the yo-yo shown in (A) is on a separate layer, the components can be displayed or frozen independently. In (B), the body has been frozen so that the shaft and endcaps are more readily visible.

It is important to model these components in 3D. The 3D model supplies a wide variety of information about the components and the process. It will be used to calculate mass, volume, interferences, and key assembly techniques. A solid model can take more time to draw than a traditional 2D drawing, but time is saved during the drafting, manufacturing, and assembly of the device. It has been shown that 3D modeling reduces errors in the components that fit into an assembly.

Technical Documentation

The technical documentation needed for a CAD/CAM project often varies with the individual component to be manufactured, as well as with the manufacturer's preference. In addition to 2D views, however, components for which solid models are created also require 3D views.

Drawing Files

This project is different from the other two projects developed so far. This project will be based on one drawing file. You will spend a significant amount of time developing the one file and should save your work often. Typically, when designers work with AutoCAD 3D solids, they open one file containing several solid models (components); this file is an assembly or layout file. Each component will be placed on a different layer. To make it easier to navigate the drawing, each layer name describes the component on that layer. For example, a solid model of a yo-yo is shown in Fig. 3.1.7. The yo-yo consists of three elements: a shaft, endcaps, and the body. Each element exists on a separate layer, and each layer can be displayed or frozen independently of the others.

The advantage of designing with all of the components in one drawing file is that the designer can see how the component fits into the assembly and can design the interface between components properly. Once the components have been completed, they are often saved in their own files and cross-referenced into an assembly file. This project will be focused around one component, but we will be using that component to produce the negative impression in a mold later in the project.

Cross-References in AutoCAD

In AutoCAD, cross-referenced drawings appear as part of the drawing into which you cross-reference them. Cross references, which are often simply called *Xrefs*, are generally used for reference only. If you want to change the contents of a cross-referenced file, you typically open the original file to do so. AutoCAD allows you to cross-reference drawings by using the XREF command.

AutoCAD cross references are a variation on a block. As you may recall, a block is a collection of geometry that is identified by a unique name. It is stored in the drawing's symbol table or database and essentially behaves as if it is a single entity. Xrefs share many of the characteristics of a block, and they are similarly defined in a symbol table. However, unlike blocks, Xref definitions are not stored in the current drawing. An Xref is stored in an entirely separate drawing file. Like a block, an Xref can be inserted many times in a drawing but has only one definition. Therefore, using Xrefs is a very good way to achieve smaller file sizes.

The advantage of using Xrefs is that every time a drawing containing Xrefs is opened, the most recently saved versions of those externally referenced drawings are loaded into the current drawing. This is also true if the Xrefs are reloaded or the current drawing is plotted. Using Xrefs instead of blocks saves disk space, because there are not numerous copies of the reference drawing's geometry stored locally in all of the drawings that may need to use that data.

Typically, Xrefs are used to display the geometry of a common base drawing in the current drawing without bloating the size of the current drawing, allowing changes to that reference drawing to be reflected in all host drawings.

There are two methods of defining an Xref in a drawing: overlaying or attaching it. Overlays are typically used when you need to view another drawing's geometry temporarily but do not plan to plot using that data. Attaching a drawing is typically for permanent use and plotting in the host drawing. The only behavioral difference between overlays and attachments is how nested references are handled. Nesting occurs when an Xref drawing contains an Xref to another drawing. An attachment to an attached Xref is always visible in the current drawing, whereas an overlay to an overlaid Xref is not.

Drawing Numbers

This project does not require the use of formal drawing numbers. There is only one component, which will simplify data storage and tracking. Create separate files as directed in the Practice Exercises, but do not be too concerned with drawing numbers.

Model Viewing Techniques

There are four ways to view a model: 3D wireframe, hidden image, shaded image, or rendered image. These techniques do not alter the model. They only affect the way that the model appears on the screen. Fig. 3.1.8 shows an example of a simple solid model using each of these techniques.

A 3D wireframe represents an object using lines and curves to represent its boundaries, as shown in Fig. 3.1.8A. The advantage of using wireframes is that, for complex objects, they do not take as much internal calculation time to display on the screen as hidden, shaded, or rendered representations. However, wireframes are often visually confusing.

A hidden image like the one in Fig. 3.1.8B regenerates a three-dimensional model with hidden lines suppressed. This image appears realistic but is still a line style representation.

For more realistic-looking objects, shading or rendering is used. Fig. 3.1.8C shows an example of shading applied to the object. Shading is often used on models because it creates fairly realistic highlights and shadows, but does not take long to generate on the screen. Shading is a rather limited method because it is restricted to the object color, among other things.

Rendered objects can be assigned actual material bitmaps so that they look more real. For example, in Fig. 3.1.8D, a brass bitmap has been applied. You can also create special spotlights to highlight rendered objects at key points. However, for a complex object or assembly, rendering is often not practical because the mapping process is very math-intensive, so it takes the object a long time to appear.

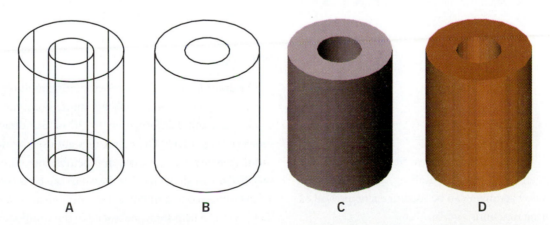

Fig. 3.1.8 *Display methods: (A) 3D wireframe; (B) hidden; (C) shaded; (D) rendered with attached material.*

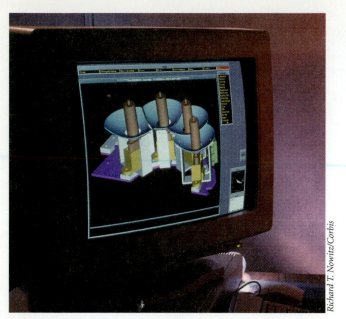

Fig. 3.1.9 *CAD can be used in the design of many different items. Here, it is being used to design computer chips and assemblies.*

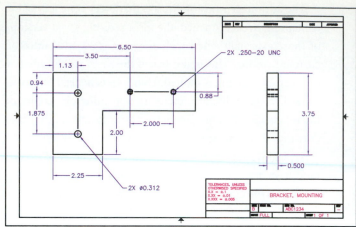

Fig. 3.1.11 *The drafted version of the model in Fig. 3.1.10.*

Richard T. Nowitz/Corbis

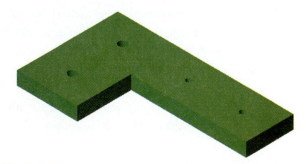

Fig. 3.1.10 *A simple reference model is often made of each part in an assembly.*

Fundamentals of 3D CAD

As we have discussed, CAD is a term that can mean many things: computer-aided design, drafting, and sometimes analysis. The CAD acronym can be used to encompass all of these engineering functions.

Design

Designing is the process of creating 2D or 3D geometry that fully defines the component or assembly to be built. The designer must take into account all of the interactions of each component with the other components in the assembly. These interactions include clearances, fits, kinetic movements, structural considerations, and manufacturability of the component. The design process is an iterative process that changes as the overall design matures and develops.

CAD software is now powerful enough that designers can complete a design for a new product using only the computer and the appropriate solid modeling and analysis software. See Fig. 3.1.9. A simple solid model is often constructed of each component at the design stage, as shown in Fig. 3.1.10, for reference as the drafter works through the design stage. Components are designed, and then they are drafted.

Drafting

The drafting process includes generating multiple isometric drawing views and applying dimensions, notes, title blocks, and other information to 2D or 3D drawings, as shown in Fig. 3.1.11. These drawings are used by manufacturing personnel to produce or assemble the components specified on the drawings. The drawings can be as complex as a fully dimensioned, multifaceted component to a diagram that is used to help someone assemble the components.

Analysis

The process of applying theoretical loads and restraints to a component, processing the model, and presenting the results is known as **analysis**. Analysis and simulation can be loosely defined as a component of a CAD system due to the recent advances and integration of analysis programs. Analysis programs work with both 2D and 3D geometries, including both solid models and surface models (3D models that do not contain internal characteristics or parameters). Often the analysis program is either embedded in the CAD system or runs with some translation between the models in a separate program.

An analysis program divides the model or geometry up into very tiny pieces, applies loads and restraints, solves the equations, and links the pieces together using nodes. The linked results can then be used to represent the solution for the entire model. Many different types of analyses can be run, but typically, they include structural analysis, thermal analysis, vibration (frequency analysis), fluid flow, motion simulation, and electromagnetic simulation. See Fig. 3.1.12.

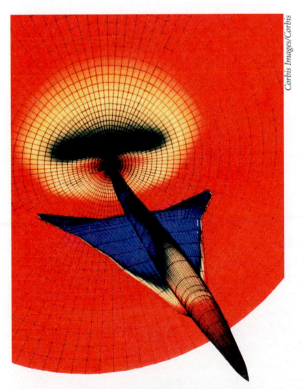

Corbis Images/Corbis

Fig. 3.1.12 Computational fluid dynamics analysis showing the predicted pressure field on an aircraft in flight.

Types of Solid Models

Solid models fall into two basic categories: primitives and generated objects. **Primitives** consist of the box, sphere, cylinder, cone, wedge and torus, as shown in Fig. 3.1.13. These objects can be combined or used as forming tools with Boolean operations. Primitives can be used to make a majority of components.

More complex components can be generated using extrusions or revolutions. These are more complicated and require the operator to provide more input to their creation. A set of geometry must be defined for each object before it can be extruded or revolved. There may be several ways to make a solid object; the operator typically follows his or her own preference.

Solid Modeling Practice

It is worth becoming familiar with the creation of simple solid models before you embark on the formal swivel bracket project. In Practice Exercise 3.1A, you will create a series of practice files in which you can practice creating solid primitives as well as more complex objects using extrusion and revolution techniques.

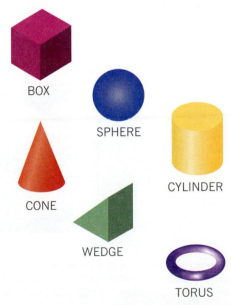

Fig. 3.1.13 Solid primitives that are predefined in AutoCAD.

Practice Exercise

3.1A

The AutoCAD commands that create the predefined primitive solid models are named according to the shape of the solids they create. For example, the SPHERE command creates a sphere, the CYLINDER command creates a cylinder, and so on. Follow the steps below to create an example of each predefined primitive.

1. Create a new drawing file and name it **SOLID PRIMITIVES.dwg**. Then **ZOOM All**.
2. Rightclick anywhere in the menu area at the top of the drawing area. From the shortcut menu, pick **ACAD** and then **Solids** to display the Solids toolbar. Become familiar with the buttons on the toolbar by moving the cursor slowly over the buttons to show the tooltips that identify them.
3. Pick the **Box** button. For the first corner, pick a point near the lower left corner of the drawing area. At the second corner prompt, notice that a box begins to form on the screen. The two corners that you specify establish the base of the box. Pick the second corner to create a base approximately **1** square. Specify a height of **2**.

The finished box looks like a simple square in this view, as shown in Fig. 3.1.14A. By default, solids are viewed as 3D wireframes in the top, or plan, view. To see the box in three dimensions, you will need to change the view.

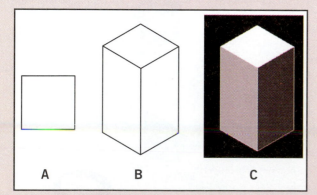

Fig. 3.1.14 *A 3D box primitive: (A) plan view; (B) SE isometric view, hidden image; (C) SE isometric view, shaded.*

Fig. 3.1.15 *The UCS (user coordinate system) icon helps orient drafters to the current view.*

4. From the **View** menu at the top of the screen, pick **3D Views** and then **SE Isometric** to see the box from the southeast isometric view. Enter **ZOOM 1**. Then enter the **HIDE** command at the keyboard. The box should appear similar to the one shown in Fig. 3.1.14B. Enter the **SHADE** command to shade the box, giving it a 3D appearance. The shade color is taken automatically from the object color.

Notice the **UCS icon** (XYZ axes) that appears near the box. See Fig. 3.1.15. (If it is not present, enter **UCSICON** and specify **On**.) It is possible to view a model from any point in 3D space. This UCS icon helps you keep track of the position from which you are viewing the model. In the plan view, the Y (vertical) axis points straight up, the X (horizontal) axis points to the right, and the Z axis points back toward the back of the computer, away from the operator.

5. To return to the wireframe view, choose **Shade** and **3D Wireframe** from the **View** menu.
6. To return to the plan view, enter the **PLAN** command. Press **Enter** to accept the Current default. **ZOOM All**.
7. Pick the **Sphere** button on the Solids toolbar. Place the center of the sphere in the bottom center part of the drawing area and specify a radius of **.5**.
8. Pick the **Cylinder** button on the Solids toolbar. Place the center point of the cylinder in the lower right part of the drawing area. This center point is actually the central axis of the cylinder. Specify a radius of **.5**. For the cylinder height, enter **1.5**.
9. Pick the **Cone** button on the Solids toolbar. Place the center point for the base of the cone in the upper left part of the drawing area. Enter a base radius of **.75** and a cone height of **1.25**.

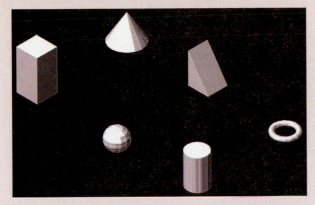

Fig. 3.1.16 *The completed, shaded solid primitives.*

10. Pick the **Wedge** button on the Solids toolbar. Place the first corner of the base in the upper center part of the drawing area. Use the cursor to place the second corner to make the base approximately **1** square. Specify a height of **1.5**.

11. Pick the **Torus** button on the Solids toolbar. Specify a center point in the upper right part of the drawing area. Make the radius of the torus **.5** and the radius of the tube **.1**.

12. Change the current view to see the solids from the southeast isometric point of view.

Notice that only four defining lines, called *isolines*, are used for each object. This is the default in AutoCAD, but you can change the number of lines using the ISOLINES command so that the objects are more clearly defined without using the HIDE or SHADE command.

13. Enter the **ISOLINES** command and enter a new value of **10**. Enter **REGEN** to regenerate the drawing and show the additional isolines.

14. For practice, view the solid primitives as a hidden image and as a shaded image. They should appear as shown in Fig. 3.1.16.

15. Save and close the drawing file.

Extrusion is very similar to a manufacturing extrusion process in which material is pushed through a predefined opening (a die) for a specific length. The geometry profile defines the shape, and the operator defines the length of the extrusion. The extrusion path can be a straight line or a combination of joined 3D curves and lines. To practice creating a solid model using the extrusion method, you will create a simplified version of the bracket in Fig. 3.1.10. You will use the front view as the basis for the extrusion.

16. In a new drawing, create the front view of the bracket using the dimensions in Fig. 3.1.17. Do not add dimensions or centerlines.

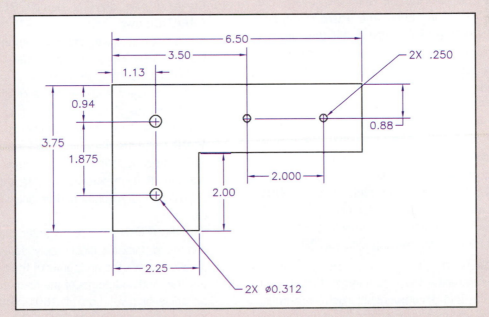

Fig. 3.1.17 *Bracket dimensions.*

17. Save the file as **EXTRUSION PRACTICE.dwg**.
18. Before you can extrude the bracket, you must change the lines you used to create the front view into a single polyline. Enter the **PEDIT** command and select one of the lines that makes up the bracket. When prompted, enter **Y** to turn the line into a polyline. Then enter **J** to activate the Join subcommand. Pick the remaining line segments in order, either clockwise or counterclockwise. When all of the segments have been selected, press **Enter** to create the polyline. End the PEDIT command.
19. Pick the **Extrude** button on the Solids toolbar and select the object. For the height of extrusion, enter **.5**, which is the thickness of the bracket. Press **Enter** to accept the taper angle of 0.
20. Change the view to **SE Isometric** to see the extrusion. At first, it may look okay, but enter the **HIDE** command. Notice that the holes do not appear. They must be extruded separately. (They could also be created as primitive cylinders, but in this case, it is easier to extrude them.)
21. Enter **PLAN** to return to the plan view. Enter the **EXTRUDE** command and select all four holes. Extrude them to a height of **.5** with a taper of **0**.
22. Change the view to **SE Isometric** to see the revised extrusion. Enter the **HIDE** command again. Now the holes appear, but they are not true holes. They are simply solid cylinders that interfere with the bracket. To finish the model, you need to subtract these cylinders from the bracket using Boolean math.
23. Enter the **SUBTRACT** command at the keyboard. (You do not need to return to the plan view to perform the subtraction.) Pick the bracket as the object to subtract from, and pick all four cylinders as the objects to subtract. Enter the **HIDE** command and notice the difference in the model. The holes in the bracket are now true holes.
24. Save and close the drawing file.

Revolution is the process of extruding a profile around a central axis. The revolution method is best used to make round components such as o-rings, washers, pipe sections, or even some primitives such

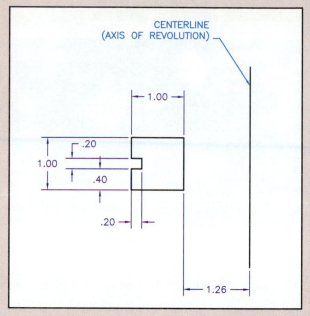

Fig. 3.1.18 *Dimensions for the object to be revolved.*

as cylinders, cones, tori, or spheres. A good rule of thumb to keep in mind is to keep the creation operations to the absolute minimum. This will reduce the number of steps that the program must follow to create an object. Follow the steps below to explore the REVOLVE command.

25. Create a new drawing named **REVOLUTION PRACTICE.dwg**. Pick the **Layers** button and create layers called **Objects** and **Center**. Make the Objects layer red and the Center layer green. Make **Objects** the current layer.
26. Draw the object and centerline on their respective layers, as shown in Fig. 3.1.18. It doesn't matter how long the centerline is. Either use **PLINE** to create the object as a polyline or use **PEDIT** to convert the individual lines to a closed polyline. It is important to note that for both extrusion and revolution, the polyline must be closed.
27. Pick the **Revolve** button to enter the REVOLVE command. Pick the closed polyline as the object to revolve. Pick the endpoints of the centerline as the start and end points of the revolution. Specify an angle of revolution of **360°**. The revolution should appear as shown in Fig. 3.1.19.

28. View the object from the SW Isometric (southwest isometric) viewpoint and shade it. The completed object should look like the one in Fig. 3.1.20A.

You can also specify an angle of revolution that is less than 360°. Fig. 3.1.20B shows what the result would be if an angle of 270° had been specified for the object in this exercise.

29. Save and close the drawing file.

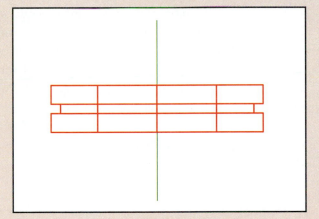

Fig. 3.1.19 *Plan view of the revolution.*

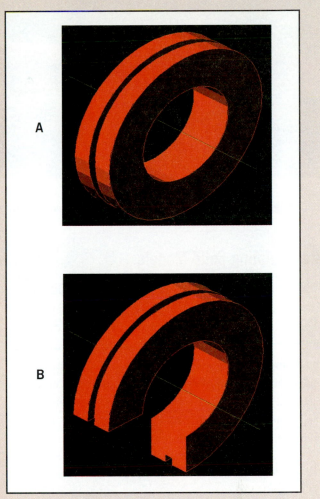

Fig. 3.1.20 *Shaded view of the object revolved 360° (A) and 270° (B).*

User Coordinate Systems

In Practice Exercise 3.1A, the user coordinate system (UCS) icon was introduced as a tool to help drafters keep track of the current point of view. The UCS system in AutoCAD is actually much more than that. It provides a movable coordinate system for coordinate entry, planes of operation, and viewing. Drafters can create their own UCSs based on the geometry in the current drawing.

Most AutoCAD geometric editing commands are dependent on the location and orientation of the UCS. AutoCAD creates new 2D geometry on the XY plane of the current UCS. Therefore, solid modeling requires mastery of UCS orientation and manipulation. To define other orientations, the user must manipulate and move the UCS to new locations to create geometries on other surfaces of the model. To refresh your knowledge of UCS manipulation, refer to the *CAD Reference* at the end of this book.

Fig. 3.1.21 *The complex curvatures on the body of a sports car require advanced techniques in CAD/CAM technology. Many manufacturers have developed their own proprietary software to meet this need.*

Manipulating Solid Models

Many 3D operations can be performed on solid models. These include rotation, scaling, copying and arraying, and aligning. These functions, combined with Boolean union, difference, and intersection operations, can be used to create most simple geometries. Some industries, such as automotive and nautical design, need to produce complex curves. See Fig. 3.1.21. These industries need advanced methods to shape the contours of components. Basic AutoCAD does not include these advanced features.

Create the Base Model

Now that you are more familiar with the concepts and basic techniques of solid modeling, it is time to begin work on the swivel bracket by creating the base model. The images shown in Practice Exercise 3.1B are displayed as 3D wireframes, hidden, and shaded images. The image display was selected to provide the most appropriate view for you to see the key aspects of the process. You can use any style of viewing that you prefer as you complete the exercise; you may choose to use several depending on the circumstances.

Practice Exercise

A 2D drawing of the swivel bracket is provided for reference in Fig. 3.1.22. Examine the geometry carefully before proceeding with this exercise. The geometry in the view labeled 2D Profile #1 will be used as a basis for the 3D model.

1. Open a new drawing file from scratch and specify **English (feet and inches)** for the units of measurement. Then **ZOOM All**. Name the drawing **PROFILE 1.dwg**.
2. Using the dimensions shown in Fig. 3.1.23, create the 2D geometry from Profile #1. Do not add the dimensions or text.

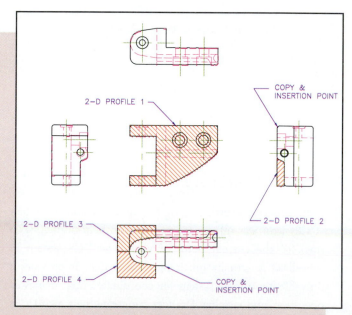

Fig. 3.1.22 *The 2D geometry for the swivel bracket.*

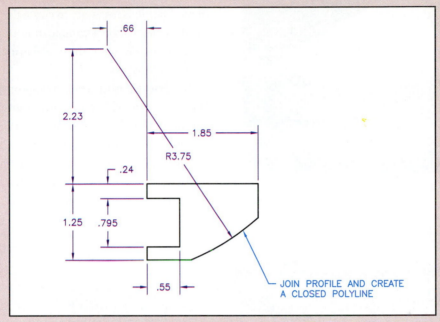

Fig. 3.1.23 Dimensions for the base geometry.

3. Use the **PEDIT** command to convert the lines that make up the profile into a single polyline.

4. Without closing the PROFILE 1.dwg file, create another new drawing from scratch. Name it **3D BRACKET MODEL.dwg**. Then **ZOOM All**.

5. Make **PROFILE 1.dwg** the current file. From the **Edit** menu, select **Copy with base point**. Pick the lower left corner of the 2D Profile 1 bracket and select the entire polyline to copy.

6. Make **3D BRACKET MODEL.dwg** the current file. Paste Profile 1 at point **0,0,0** (the origin).

7. Change the view to **SE Isometric**.

8. From the command line, type **UCSICON** and set it to **ON** to display the UCS icon. Enter **OR** (origin) to move the UCS icon to the origin (0,0,0). The base geometry, or **profile**, should now look like Fig. 3.1.24.

The next step is to extrude the profile into a basic 3D solid model. If the Solids toolbar is not currently displayed, you may want to display it now.

9. Extrude the profile to a height of **–.75** with a taper of **0**. The base model (shaded) should look like the one in Fig. 3.1.25.

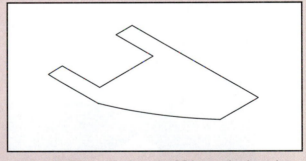

Fig. 3.1.24 The profile from the SE isometric viewpoint.

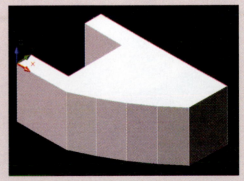

Fig. 3.1.25 The base model after step 9. (Note that the model has been shaded for this illustration.)

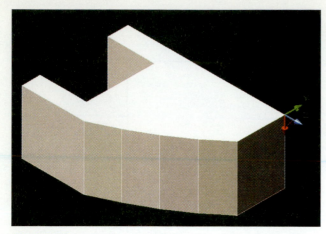

Fig. 3.1.26 *The origin and UCS for the solid model of the swivel bracket.*

Initial Part Configuration

Now that the basic model exists, you can begin the series of modifications necessary to create the final model. The finished model will be a combination of multiple operations. These operations shape the model in a manner similar to that of an artist shaping clay.

The first step is to create an appropriate UCS. To do this, you will use the UCS command to move the origin and reorient the *X*, *Y*, and *Z* axes as shown in Fig. 3.1.26. In Practice Exercise 3.1C, you will create a new UCS to achieve this. You will also make the first two cuts to refine the solid model of the swivel bracket. Refer to the *CAD Reference* at the end of this textbook if you need to review the UCS command.

Practice Exercise

The UCS command allows you to specify any point in 3D space as the origin of the drawing (the point at which the *X*, *Y*, and *Z* axes intersect). In addition, the command allows you to specify the orientation of the axes to create a true "user" coordinate system.

1. Enter the **SHADE** command to shade the solid model in 3D BRACKET MODEL.dwg. This will make it easier to see the required points in the following steps.
2. Enter the **UCS** command and create a **New** UCS using the **3Point** option. Place the origin as shown in Fig. 3.1.27. Choose points on the positive *X* and *Y* axes as shown in Fig. 3.1.27. The UCS appears as shown in Fig. 3.1.26.

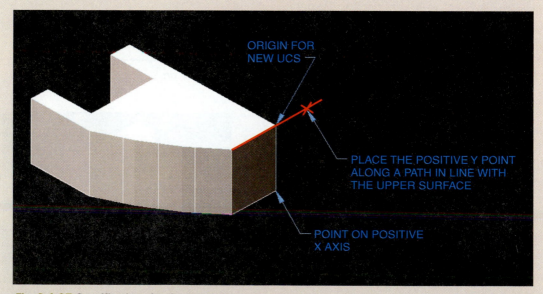

ORIGIN FOR NEW UCS

PLACE THE POSITIVE Y POINT ALONG A PATH IN LINE WITH THE UPPER SURFACE

POINT ON POSITIVE X AXIS

Fig. 3.1.27 *Specifications for the new UCS.*

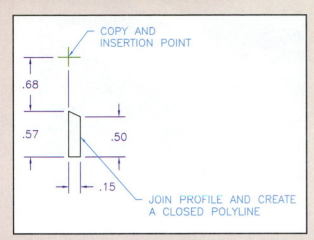

Fig. 3.1.28 *Dimensions for Profile 2.*

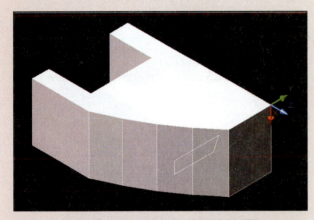

Fig. 3.1.29 *Profile 2 inserted into 3D BRACKET.dwg.*

Fig. 3.1.30 *Profile 2 extruded.*

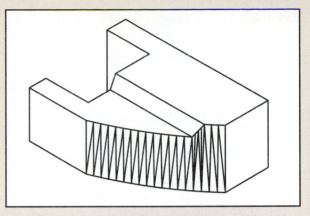

Fig. 3.1.31 *Result of the Boolean subtraction. (The hidden method shows this better than shading.)*

3. Create a new drawing file to hold Profile 2. Name the file **PROFILE 2.dwg**.

4. Create the geometry for Profile 2 using the dimensions shown in Fig. 3.1.28. Also create intersecting lines as shown for the base point for copying and inserting the profile into the model drawing, but do not add the dimensions and notes. Use **PEDIT** to convert the individual lines into a single polyline.

5. From the **Edit** menu, select **Copy with base point**. Pick the intersection of the two lines that define the base point and select the entire polyline to copy. Save and close PROFILE 2.dwg.

6. Make **3D BRACKET.dwg** the current drawing. Paste the profile into the drawing using a base point of **0,0,0**. See Fig. 3.1.29.

7. Extrude Profile 2 to a height of **–1.90**. The result is shown in Fig. 3.1.30. *Note:* The color of the extruded profile has been changed in Fig. 3.1.30 for purposes of illustration only. Your extrusion will be white.

The area in which the two colors interfere indicate interference between the two solids. This is exactly what is needed. You can now subtract Profile 2 from the bracket to refine the shape of the bracket.

8. Enter the **SUBTRACT** command to perform the Boolean subtraction operation. Select the model of the bracket (from Profile 1) as the solid to subtract from. Select the solid created from Profile 2 as the solid to subtract. The refined model is shown in Fig. 3.1.31.

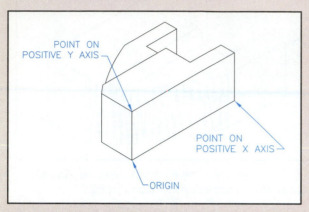

Fig. 3.1.32 *The three points needed to create the new UCS in step 10.*

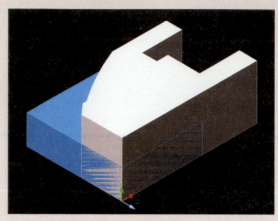

Fig. 3.1.33 *Expected interference between the box and the bracket model.*

Solid primitives can be subtracted in the same manner as extruded solids. For the second cut, you will use a solid box primitive to remove additional material from the bracket model.

9. Change the current view to **NE Isometric**.

10. Create a new UCS using the **3Point** option. Position the origin and axes as shown in Fig. 3.1.32.

11. Create a solid box primitive. Specify the default location of **0,0,0** for the first corner of the box. Enter **L** to activate the Length option. Specify a length of **1.10**, a width of **.4**, and a height of **−1.3**. The box should interfere with the bracket model as shown in Fig. 3.1.33. (The box is shown in blue for illustration purposes only.)

12. Subtract the box from the bracket model. The result is shown in Fig. 3.1.34.

13. Save the drawing file.

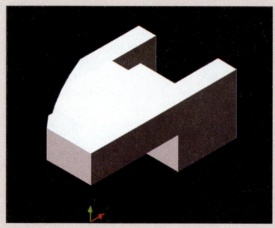

Fig. 3.1.34 *The result of the second cut on the bracket model.*

Review Questions

1. What is CAD/CAM?
2. Explain the advantages of solid modeling over 2D drafting. Are there any disadvantages?
3. What information can be obtained from a solid model that is not available from other types of models, such as surface models?
4. How are Boolean operations used in solid modeling? Describe the function of each operation.
5. What is an Xref?
6. Describe four ways of viewing a solid model.
7. In AutoCAD, what is a solid primitive?
8. Briefly explain the process of extrusion in AutoCAD.
9. For what kinds of solid models is the revolution method commonly used?
10. Why are UCSs used frequently in the process of creating a solid model?

Practice Problems

1. On the Internet, research computer systems that are dedicated to or set up for solid modeling. What components are important?
2. Develop a matrix to compare the features and cost of various types of 3D-capable CAD systems that are available on the market today.
3. Using *only* AutoCAD's primitives and Boolean operations, create the solid models shown in Fig. 3.1.35. Estimate the relative sizes.

Portfolio Project

Bicycle Component

Occasionally, when paper drawings have been lost or are incomplete, a manufacturer may need to *reverse-engineer* a part, or create new drawings based on the part. For this project, you will reverse-engineer a component on a bicycle to create a solid model of the part.

1. Using a ruler, a set of measuring calipers, and some triangles reverse engineer another component on your bicycle. Take the component off and measure it completely. **Safety precaution:** Do not ride the bike until the component can be replaced, and make sure the removed component has been replaced properly.
2. Measure all of the holes and features, including the outside perimeter of each face of the component. Transfer all of the dimensions and features into the computer using the techniques that were described in this section.
3. Create any necessary profiles and develop a simple solid model. (You will refine the model later in this project.)

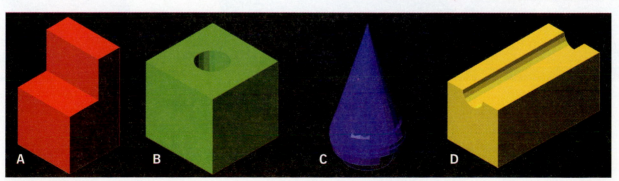

Fig. 3.1.35

4. Use the extrusion method to create each of the objects shown in Fig. 3.1.36. Estimate all dimensions.

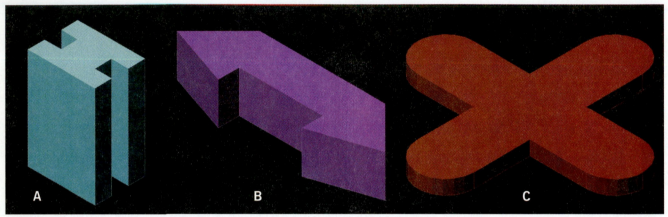

Fig. 3.1.36

5. Use the revolution method to create each of the objects shown in Fig. 3.1.37. Estimate all dimensions.

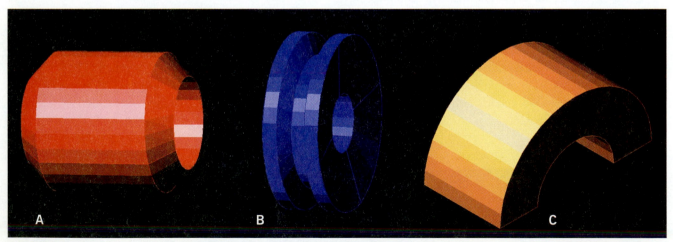

Fig. 3.1.37

Industry Practices and Principles

AutoCAD models use a boundary representation scheme (B-Rep) that saves only the final image of the part. The individual steps that were required to achieve the final part are not saved. The geometry that is used in the B-Rep must be defined fully before it can be made into a solid.

Some CAD packages allow you to modify and constrain a sketch drawing after it has been made. In AutoCAD, however, the geometry needs to be drawn to scale and closed before it can be used. As part of the design process, it is important to review dimensioning guidelines, tolerancing, and standard practices as part of the learning process.

This section reviews the dimensioning and tolerancing standards that relate to the creation of 2D drawings and solid models for use with CAD/CAM. It also provides an overview of best practices for working with solid models.

Section Objectives
- Review general dimensioning and tolerancing guidelines as they apply to solid models.
- Apply MMC and LMC conditions to tolerancing applications.
- Plan a solid model using best-practice techniques.

Key Terms
- bilateral tolerance
- Least Material Condition
- Maximum Material Condition
- parametric tolerancing
- Regardless of Feature Size
- unilateral tolerance

Dimensioning Guidelines

Developing a multiview drawing is the art of translating a designer's ideas into a universal language that can be understood anywhere in the world. The use of standard guidelines help the operator achieve better drawings. Dimensioning is a part of the overall drawing; you should pay close attention to it. The American National Standards Institute (ANSI) has specified the following overall guidelines to dimensioning:

1. Show enough dimensions so that the intended sizes and shapes can be determined without calculating or assuming any distances.
2. State each dimension clearly so that it can be interpreted in only one way.
3. Show the dimensions between points, lines, or surfaces that have a necessary and specific relation to each other or that control the location of other components or mating parts.
4. Select and arrange dimensions to avoid accumulations of tolerances that may permit various interpretations and cause unsatisfactory mating of parts and/or failure in use.
5. Show each dimension only once.
6. Where possible, dimension each feature in the view where it appears in profile and where its true shape appears.
7. Where possible, specify dimensions to make use of readily available materials, parts, tools, and gauges. For example, significant savings are often possible when drawings specify:
   - Commonly used materials in stock sizes.
   - Parts that are generally recognized as commercially standard.
   - Sizes that can be produced with standard tools and inspected with standard gauges.
   - Tolerances from acceptable published standards.

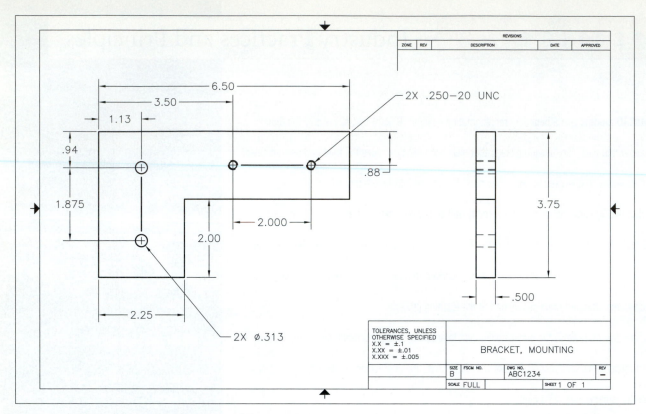

Fig. 3.2.1 Drawing dimensioned according to ASME Y14.5M.

The American Society of Mechanical Engineers (ASME) developed the ANSI standards for dimensioning and tolerancing a drawing. The drawing in Fig. 3.2.1 has been dimensioned according to ASME Y14.5M. Look carefully at the characteristics of the dimensions on this drawing. Notice the following items in particular:

- Decimal fractions on drawings dimensioned using U.S. Customary (English) units do not have leading zeros. (Drawings dimensioned using metric units do have leading zeros, however.)
- When a dimension refers to a feature, such as a hole, that occurs in more than one place, the proper style is to precede the dimension with 2X, 3X, etc. Do not use the older style of adding the text (2 PLACES) after the dimension.
- Two linewidths are used: thick lines for object lines, and thin lines for dimensions, centerlines, hidden lines, text, and hatches.

If you have any questions about specific guidelines, consult the latest version of ASME Y14.5M.

Tolerancing Guidelines

It is unreasonable to assume that a component can be manufactured to exactly the same dimensions every time. There must be some allowable variation in each component. The variations can be attributed to machine tool variation, stock variation, and operator error. Components must contain sufficient tolerances to ensure that the components will still perform the function for which they were designed. As tolerances become tighter (closer), the production methods used must be more precise and careful, often costing substantially more money.

Parametric Tolerancing

Basic tolerancing, called **parametric tolerancing**, is widely used to define allowable dimensions for components. Parametric tolerancing provides an upper and lower limit for each dimension that is specified on the drawing. There are two types of parametric tolerancing: unilateral and bilateral.

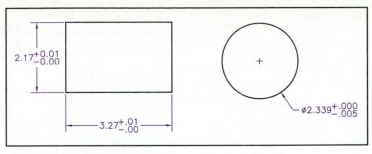

Fig. 3.2.2 *Unilateral tolerances.*

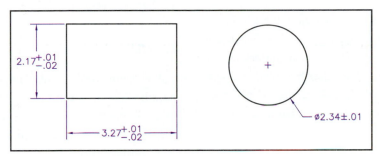

Fig. 3.2.3 *Bilateral tolerances.*

When variation can be allowed in only one direction, such as in a close-fitting shaft, a **unilateral tolerance** is used. Fig. 3.2.2 shows a typical unilateral tolerance.

In many cases, variation can be allowed in both directions, including most locating dimensions. Fig. 3.2.3 shows two examples of **bilateral tolerances**. The dimensions on the rectangle show how different (nonsymmetrical) tolerances are shown. The dimension on the circle shows how symmetrical tolerances are treated. See Project 1, Section 3 for more information about tolerancing in general and about symmetrical and nonsymmetrical tolerances.

Geometric Dimensioning & Tolerancing

Geometric dimensioning & tolerancing (GD&T) can be used in addition to parametric tolerancing to define critical geometric characteristics of a component. GD&T uses symbols to show control tolerances. These symbols and their meanings were presented in Project 1, Section 3. The techniques and symbols are controlled by the ASME Y14.5M specification. The more familiar aspects that are controlled are shown in Fig. 1.3.17 on page 86.

The ASME Y14.5M specification relies on another important feature to help control tolerances and tie in the geometric conditions. This feature is known as *material condition.*

- Maximum Material Condition (MMC, Ⓜ)
- Least Material Condition (LMC, Ⓛ)
- Regardless of Feature Size (RFS, Ⓢ)

These material conditions change the tolerances with respect to how much material is present in a geometric feature. A hole at **Maximum Material Condition** is a hole drilled in metal that is at the smallest value allowed with the tolerance. Because the hole is small, there is the maximum amount of metal present. Likewise, a hole at **Least Material Condition** is the largest, when there is the least amount of metal left around the feature. The designer can change the tolerances of other geometric features based on the existing conditions. **Regardless of Feature Size** means that the geometric tolerance in question is not affected by any change in the material conditions.

Again, these are merely the basics. If you are interested in more specific or detailed information about GD&T standards, refer to ASME Y14.5M.

Boolean Solids: Design Guidelines

The following is a brief summary of some guidelines that can be used to model AutoCAD solids. If you use a different modeling package, these rules may vary slightly depending on the software that is used. However, these guidelines will be useful to keep in mind while you are designing models regardless of the software used.

Layout File

As mentioned in Section 1 of this project, it is good practice to use one file for your design. Assign a different layer to each component of the assembly that you are modeling. This will allow you to freeze the components that you do not wish to see. Back the file up regularly and use the PURGE command to eliminate any extra information that may cause the system to slow down.

Planning

It is important to understand what you intend to model before you start. Thinking about it will help you plan for it. Think before you model, use scrap paper for brainstorming, and then start to model the components.

Manufacturability

Just because you can model it does not mean that you can manufacture it. Make sure that the wonderful design you came up with can actually be produced. When in doubt, call a machine shop and see if they can offer any suggestions. Machinists are often the best reference for manufacturability.

Tolerancing

Tolerance everything. Typically, all parts are designed on an average dimension with sufficient tolerance so that it will fit with the mating components. For example, holes are always larger than the shafts that are meant to spin in them. If the dimension of a hole and a shaft are the same, they will have a line-to-line fit and the shaft will not spin. There are some instances in which you would want to have a shaft stuck in a hole, but there are others in which you would want it to move freely.

It is important to keep in mind the proper tolerance for each component. As a general rule of thumb, when the tolerance is halved, the cost of that operation quadruples.

Alignment

AutoCAD does not contain any provisions to mate components together. This means that the components may be in a different plane than you anticipated. Always look at the drawing from several views to make sure that the component is where you want it.

Materials

Use commercially available materials. Materials shown in a textbook or on a website may not be available in your area. When in doubt, call a machine shop and see what materials they have on hand or work with regularly. If the material suits your purpose, specify it.

Taps and Drills

When you model a tapped hole in AutoCAD, it is a good idea to make the hole the same size as the tap drill. AutoCAD does not retain enough information to remember that the hole was to be tapped and not drilled. When you use the tap drill size, it can be easily reworked if you neglect the tap call-out on the detail drawing.

Milling and Drilling

It is also a good idea to use only commercially available drills, end mills, and taps. You can obtain a list of preferred sizes from your local machine shop or tool supplier. If you need to, you can specify odd sizes, but doing so will increase the production cost of the component.

Welds

It can become difficult to model welds. Instead, it is considered good practice to model the various components and place them together in the solid assembly, but specify the weld on the detail drawing.

Review Questions

1. Name three standard guidelines for dimensioning a drawing.
2. Why should the drafter specify standard or readily available materials, parts, and tools whenever possible?
3. In the United States, what society is responsible for developing and maintaining dimensioning and tolerancing standards?
4. Under what circumstances should the drafter include a leading zero before decimal fractions?
5. What is parametric tolerancing?
6. What is the difference between unilateral and bilateral tolerances?
7. In addition to geometric dimensions and tolerances and parametric tolerances, what other specifications control tolerances?
8. How does MMC relate to hole size?
9. What is the significance of the symbol (S)?
10. Name three modeling guidelines and explain the importance of each.
11. Why is it important to use the tap drill size to make a tapped hole?
12. Why is it important to check the manufacturability of a product during the design stage?

Practice Problems

1. In the drawing in Fig. 3.2.4, each square represents .25″. Reproduce the drawing in AutoCAD. Then dimension it in accordance with ASME Y14.5M. Use U.S. Customary dimensions.
2. The part in Practice Problem 1 is a bracket that will hold brass rods in place in an assembly. The rods have a diameter of .485″. Specify a suitable material for the bracket and tolerances for the holes.

Portfolio Project

Bicycle Component (continued)

Take the time to create a multiview drawing of the bicycle component you have chosen. This will help you understand the requirements for the model.

Add dimensions and tolerances to the drawing in accordance with ASME Y14.5M. Note that to tolerance the drawing properly, you will need to gather information about other components on the bicycle that interface with your chosen component.

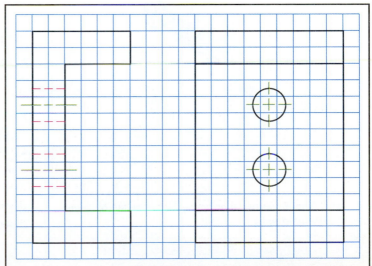

Fig. 3.2.4

Swivel Bracket Model

Now that you have practiced working with models in AutoCAD, it is time to concentrate on refining and finishing the solid model of the swivel bracket. To do so, you will need to work from several different points of view in 3D space.

AutoCAD provides a number of ways to move among viewpoints in a 3D drawing. The first part of this section provides a more in-depth discussion of 3D views and how to use them. Then you will continue to refine the swivel bracket by working on the outer construction. Finally, you will add the holes and fillets to complete the solid model.

Section Objectives

- View a solid model from various points in 3D space using AutoCAD's preset views.
- View a solid model from any point in 3D space using the 3D orbit feature.
- Complete a complex solid model by refining its outer construction and adding holes and counterbores.

Key Terms

- spotface
- 3D orbit

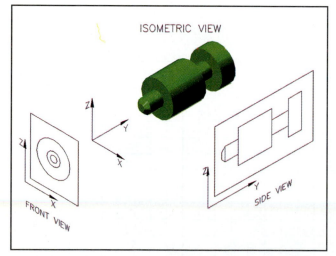

Fig. 3.3.1 *Two-dimensional orthographic views can be generated from a solid model.*

Working with Multiple 3D Views

Drawing components using 3D solid models is different from drawing in 2D, especially when it comes to the use of views. Instead of several views to show the various sides and features of the component, there is only one solid model that can be viewed from an infinite number of angles, much the same as an object in real life. Solid modeling packages have the ability to rotate components around on the screen to show any angle that the operator needs to identify.

The 3D solid model can also be used to generate multiple 2D views or projections before the files are detailed. Fig. 3.3.1 shows a front view and a side view as generated from a 3D model. Typically, only one view is used until the model is detailed. Then enough views are generated to ensure that the drawing dimensions can be fully defined.

Fig. 3.3.2 *The View toolbar allows you to select any of AutoCAD's preset views quickly.*

3D Orbit

Fig. 3.3.3 *The 3D Orbit button is located on the Standard toolbar.*

The View Toolbar

So far, you have used the View menu to access the NE Isometric and NW Isometric views. When you know that you will be changing the view frequently, however, there is an easier way to do so. AutoCAD has a View toolbar that includes buttons for several predefined views. See Fig. 3.3.2. These views provide different viewpoints from which you can look at a single model. This is very similar to moving a video camera around to see different parts of an object. You can cycle through each of the views until you find the ideal view for your current purpose. The preset views on the View toolbar include:

- Top
- Bottom
- Left
- Right
- Front
- Back
- SW Isometric
- SE Isometric
- NE Isometric
- NW Isometric

To display the View toolbar, move the cursor up to the menu portion of the screen (above the drawing area). Rightclick to display the shortcut menu. Choose **ACAD** and then **View** to display the toolbar. You may wish to dock the toolbar at the top or side of the screen to keep it out of the way, since you will be using it often. As you work through the

Practice Exercises in this section, try several of the views to understand what each one will show. The isometric views tend to be the most descriptive when you are working in 3D.

3D Orbit

Beginning with AutoCAD 2000, the program has included another view-setting feature known as **3D orbit**. This feature allows you to move the viewpoint quickly to any point in 3D space—not just the preset views. The 3DORBIT command can be accessed from the standard tool bar, as shown in Fig. 3.3.3. This command activates a 3D view in the current viewport. The 3D orbit view displays an arc ball, which is a circle divided into four quadrants by smaller circles. When 3DORBIT is active, the target of the view stays stationary and the point of view moves around the target. Keep in mind that you cannot edit objects while the 3DORBIT command is active. When you have positioned the view to your satisfaction, you must end the command before you can continue.

Swivel Bracket: Outer Construction

You may have noticed in previous sections that the outer construction of the finished swivel bracket includes several rounded features. In Practice Exercise 3.3A, you will shape the outer construction of the swivel bracket to include these rounded features. You will do this by creating the appropriate geometry, extruding it, and then subtracting the extrusion from the solid to remove the excess material.

Practice Exercise

3.3A

The procedure for shaping the outer surface of the swivel bracket is similar to that you used to form its basic shape in Section 2. Follow the steps below.

1. In **3D BRACKET.dwg**, change the view to **SE Isometric**.
2. Create a new UCS with the origin and axes oriented as shown in Fig. 3.3.4.
3. Create 2D Profile #3 in a separate drawing named **PROFILE 3.dwg**. Use **PEDIT** to join the individual lines into a single polyline. Use the dimensions shown in Fig. 3.3.5. Do not add the dimensions, but be sure to include the crossed lines that define the base point for copying.
4. Copy the entire profile (but not the crossed lines) using **Copy with base point**. Select the intersection of the crossed lines as the base point.
5. Open the **3D BRACKET.dwg** file. Insert 2D Profile #3 using an insertion point of **0,0,0**.
6. Create 2D Profile #4 in a separate drawing named **PROFILE 4.dwg**. Use **PEDIT** to join the individual lines into a single polyline. Use the dimensions shown in Fig. 3.3.6. Do not add the dimensions, but be sure to include the crossed lines that define the base point for copying.
7. Make **3D BRACKET.dwg** the current drawing. Insert 2D Profile #4 using an insertion point of **0,0,0**. The model should now look like the one in Fig. 3.3.7.

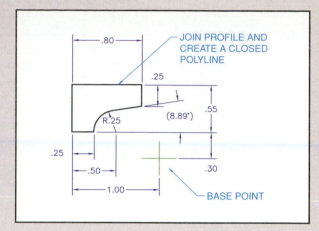

Fig. 3.3.5 *Dimensions for 2D Profile #3.*

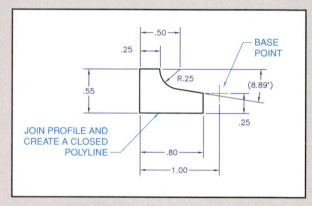

Fig. 3.3.6 *Dimensions for 2D Profile #4.*

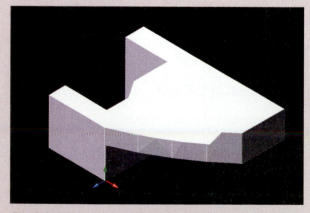

Fig. 3.3.4 *Orientation of the new UCS.*

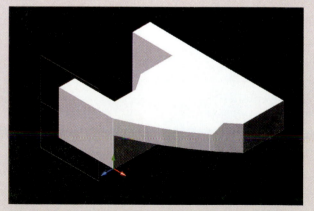

Fig. 3.3.7 *Profiles #3 and #4 inserted into the bracket model drawing.*

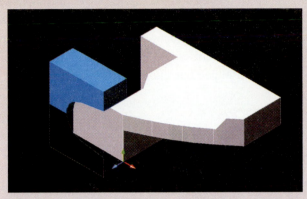

Fig. 3.3.8 *Profile #3 extruded.*

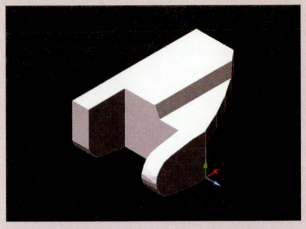

Fig. 3.3.11 *The result of subtracting the solid created from Profile #4.*

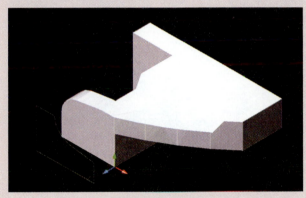

Fig. 3.3.9 *The result of subtracting the solid created from Profile #3.*

8. Save **PROFILE 3.dwg** and **PROFILE 4.dwg** and close them.
9. Enter the **MOVE** command and pick 2D profiles #3 and #4. Enter a first point of displacement of **0,0,0** and a second point of displacement of **0,0,.10** to move the profiles away from the model by a distance of **.10**.
10. Extrude Profile #3 to a length of **–.35** with a taper of **0**, as shown in Fig. 3.3.8.
11. Subtract the solid that resulted from extruding Profile #3 from the bracket solid, as shown in Fig. 3.3.9.
12. Change the view to **SW Isometric**.
13. Extrude Profile #4 to a length of **–1.4** with a taper of **0**, as shown in Fig. 3.3.10.
14. Subtract the solid that resulted from extruding Profile #4 from the bracket solid. The expected result is shown in Fig. 3.3.11.
15. Save the 3D BRACKET.dwg file.
16. Use **SAVEAS** to save another copy of the drawing as **BRACKET BASE.dwg**. This copy will be used later to create a mold.
17. Close the **BRACKET BASE.dwg** file and reopen **3D BRACKET.dwg**.

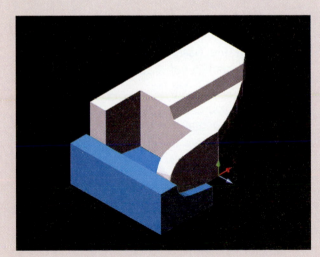

Fig. 3.3.10 *Profile #4 extruded.*

Swivel Bracket: Holes and Fillets

Although the basic contours of the swivel bracket are now complete, the bracket contains several holes on various faces. The next step is therefore to add these holes. In Practice Exercise 3.3B, you will first add the through hole and **spotface** (small counterbore) to the end of the bracket. Then you will add the counterbored holes to the top surface of the bracket and the tapered hole through the body of the bracket.

Finally, you will create the filleted edges to complete the bracket. The fillets are used to add radii to sharp corners and to add transition curves between features. Fillets are also used to remove sharp angles and edges that may either present a safety risk or impair operation of the component.

Practice Exercise 3.3B

As you may have guessed, each of the holes will be created by subtracting a series of cylinders from the swivel bracket model. The first cylinder will create the spotface.

1. Change to the **NE Isometric** view.
2. Create a new UCS. Place the origin at the center of the .25 radius and the axes as shown in Fig. 3.3.12. *Hint:* Use the **X** and **Z** options of the **UCS** command to rotate the icon into position. Be sure the axes are oriented as shown in Fig. 3.3.12 before you continue.
3. Create a solid cylinder. Place the center point for the base of the cylinder at **0,0,0**. Specify a diameter (*not* a radius) of **.25** and a height of **−.030**.

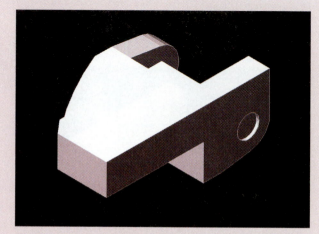

Fig. 3.3.13 *The spotface in place after step 4.*

4. Subtract the cylinder from the bracket model. The resulting spotface should look like the one in Fig. 3.3.13.
5. Create another solid cylinder. Place the center point for the base of the cylinder at **0,0,0**. Specify a diameter of **.126** and a height of **−1.4**.

Although you will only be able to see the end of this cylinder if you are working with the bracket in a shaded view, it actually extends all the way across and through the other leg of the bracket. You may wish to switch temporarily to a 3D wireframe view to verify this.

6. Subtract the cylinder from the bracket model. View the model from the NW Isometric view to verify that the hole extends through both legs of the bracket. See Fig. 3.3.14.

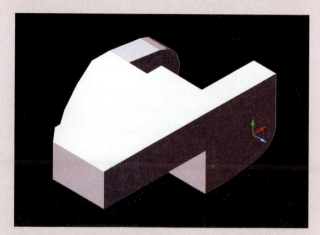

Fig. 3.3.12 *Place the new UCS as shown, with the origin at the center of the .25 radius.*

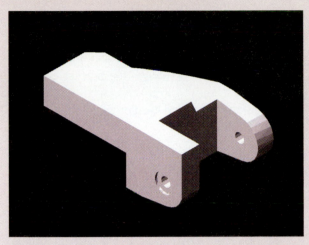

Fig. 3.3.14 *The first hole and its spotface completed.*

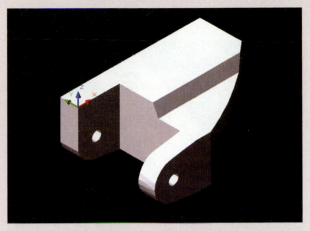

Fig. 3.3.15 *Location of the UCS in step 8.*

The next task is to add the counterbored holes to the top surface of the bracket. Again, you will add the holes by creating multiple solid cylinders that have the dimensions of the holes and then subtracting the cylinders from the bracket to create the holes and their counterbores.

7. Change to the **SW Isometric** view.
8. Create a new UCS. Position the origin and axes as shown in Fig. 3.3.15.
9. Create centerline location geometry as shown in Fig. 3.3.16. From the upper right corner, create a line **.30** long in the –*Y* direction. Then create two more lines in the –*X* direction a distance of **.30** and **.50**.
10. Begin by placing the counterbores. Create two cylinders of exactly the same size. Place their center points at the two intersections formed by the location geometry you created in the previous step. Specify a diameter of **.312** and a height of **–.175** for both cylinders.
11. Subtract the cylinders from the bracket model. The resulting counterbores are shown in Fig. 3.3.17.
12. Add the through holes. Create two additional cylinders with the same center points that you used in step 10. Specify a diameter of **.204** and a height of **–.40** for both cylinders.

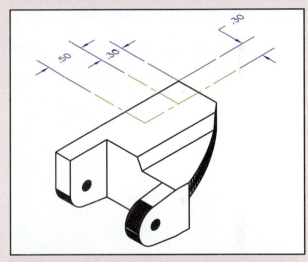

Fig. 3.3.16 *Location of centerlines on the top face for step 9.*

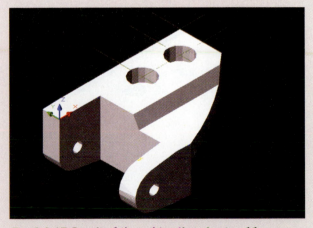

Fig. 3.3.17 *Result of the subtractions in step 11.*

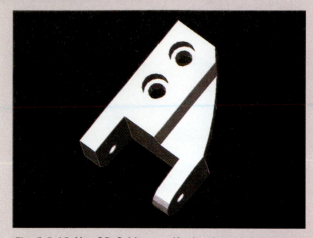

Fig. 3.3.18 Use 3D Orbit to verify the through holes.

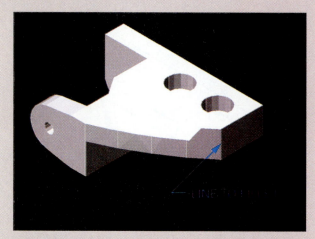

Fig. 3.3.19 Fillet the line shown here with a radius of .25.

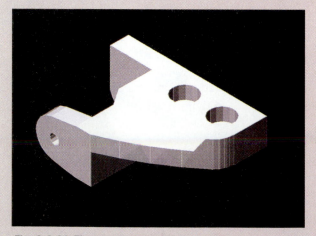

Fig. 3.3.20 The result of the fillet operation.

13. Subtract both cylinders from the bracket model to finish the through holes.

14. Enter the **3DORBIT** command and use it to change the point of view so that you can verify that the holes were created and that they do extend all the way through the bracket. See Fig. 3.3.18.

15. Delete the centerline location geometry you created in step 9.

Now add the fillet at the back of the swivel bracket. Follow these steps.

16. Change to the **SE Isometric** view.

17. Enter the **FILLET** command. Select the vertical line shown in Fig. 3.3.19 and enter a fillet radius of **.25**. Press **Enter** to accept the **Chain/Radius** default. The resulting fillet is shown in Fig. 3.3.20.

Next, add the tapered hole to the swivel bracket. This is done by positioning a tapered cylinder in the proper position and then subtracting it from the swivel bracket.

18. Rotate the *X* and *Y* axes to create a new UCS as shown in Fig. 3.3.21. Be sure the axes are facing as shown in the illustration.

19. Create centerline location geometry as shown in Fig. 3.3.22. by creating a line in the –*X* direction a distance of **.565** from the origin of 0,0,0. Create a second line in the +*Y* direction a distance of **.200**.

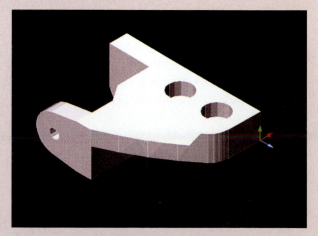

Fig. 3.3.21 Rotate the axes to create the new UCS.

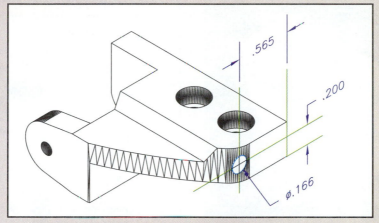

Fig. 3.3.22 *Location geometry for tapered hole.*

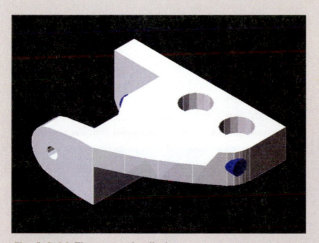

Fig. 3.3.23 *The tapered cylinder.*

Fig. 3.3.24 *The tapered hole after the subtraction.*

20. Create a circle where the two centerlines intersect. The circle diameter is **.166**.

21. The CYLINDER command does not have a taper option, so the tapered cylinder will be created by extrusion. Extrude the circle to a height of **–1.40** with a taper of **.67**. The cylinder should interfere with the bracket as shown in Fig. 3.3.23.

22. Subtract the cylinder from the bracket model to create the hole, as shown in Fig. 3.3.24. Notice that the hole looks distorted from this view because it opens onto a filleted surface.

23. Use **3D Orbit** to verify that the resulting hole extends all the way through the bracket. See Fig. 3.3.25.

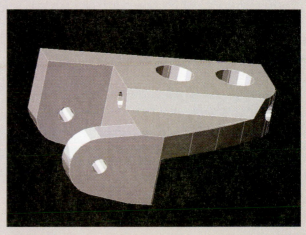

Fig. 3.3.25 *Verify the through hole.*

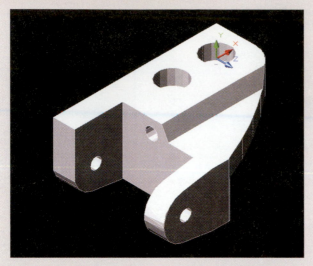

Fig. 3.3.26 *Rotate about the Y axis to create this UCS.*

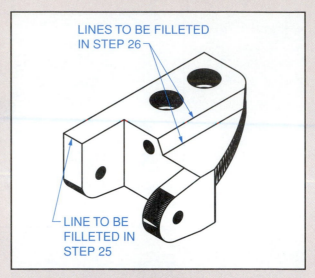

Fig. 3.3.27 *Line to fillet in step 25.*

All of the holes are now complete. To finish the swivel bracket, you need only add a few more fillets. First, you will fillet the left arm of the bracket. Then you will fillet the edges of the slanted surface on the top of the bracket to create a smooth transition between features.

24. Delete the reference geometry and change to the **SW Isometric** view.

25. Create a new UCS by rotating the current UCS about the **Y** axis by **270°**. Be sure that the new UCS is oriented as shown in Fig. 3.3.26.

26. Enter the **FILLET** command and pick the top edge of the arm to be filleted, as shown in Fig. 3.3.27. Enter a fillet radius of **.140** and press **Enter** to select the **Chain/Radius** default.

27. Reenter the **FILLET** command and pick one of the two horizontal lines on the top of the bracket, as shown in Fig. 3.3.27. Enter a fillet radius of **.05** and press **Enter** to select the **Chain/Radius** default. Repeat for the other line shown in Fig. 3.3.27. The finished bracket is shown in Fig. 3.3.28.

28. Save the drawing file.

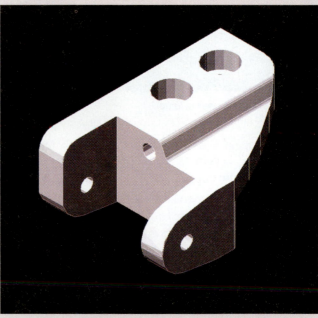

Fig. 3.3.28 *The finished bracket.*

Review Questions

1. How are views in 3D solid models different from views on 2D orthographic drawings?
2. What is the fastest way to change from one of AutoCAD's predefined viewpoints to another if you anticipate frequent view changes?
3. List the views that are predefined in AutoCAD.
4. What is the best way to move the viewpoint to any point in 3D space that is not predefined as a view in AutoCAD?
5. What is a spotface?
6. Explain why a new UCS was created for nearly every operation in working with the model. What is the advantage of moving the origin and axes in this manner?
7. Why might the method of creating a new UCS not be applicable in some cases?
8. Briefly explain how a tapered solid cylinder can be created in AutoCAD.
9. What is the purpose of the various profiles created in this section?
10. Why is it best to store the profiles for a solid model in one or more separate files?

Practice Problems

1. Create a solid cube in which each side is 2″ square. At the center of each of the six sides, create a hole, as follows:
   - Side 1: Ø.50, .12 deep
   - Side 2: Ø.25, .25 deep
   - Side 3: Ø.80, .20 deep
   - Side 4: Ø1.20, .10 deep
   - Side 5: Ø1.50, .05 deep
   - Side 6: Ø1.65, .15 deep

   Change the UCS as often as necessary to create and subtract the cylinders efficiently. When you have finished, view the cube from as many different points of view as necessary to see all six sides. Print or save the drawing with the views displayed. (You will need to save the drawing with two or more different file names.)

Portfolio Project

Bicycle Component (continued)

Refine the solid model of the bicycle component that you created in Section 1. Perform the following steps.

1. Refering to your orthographic drawings as necessary, create any profiles needed to refine the solid model. Be sure to create them in a separate drawing file.
2. Calculate the insertion point for each profile so that you can insert it at the origin in the model drawing file.
3. Import the profiles, extrude them, and use Boolean operations to refine the outside contours of the solid model.
4. Add any holes and fillets needed to complete the solid model.

2. Create a 3″ × 3″ × 4″ solid box like the one in Fig. 3.3.29. Create Ø1″ holes in the center of each side as shown. All of the holes should run through the center of the block, intersecting one another.

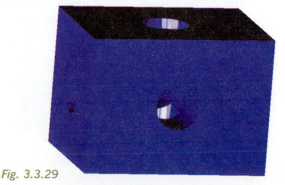

Fig. 3.3.29

3. Examine each solid model in Fig. 3.3.30 carefully. Determine the operations you would need to perform to create the models in AutoCAD. Then create the models, estimating all sizes.

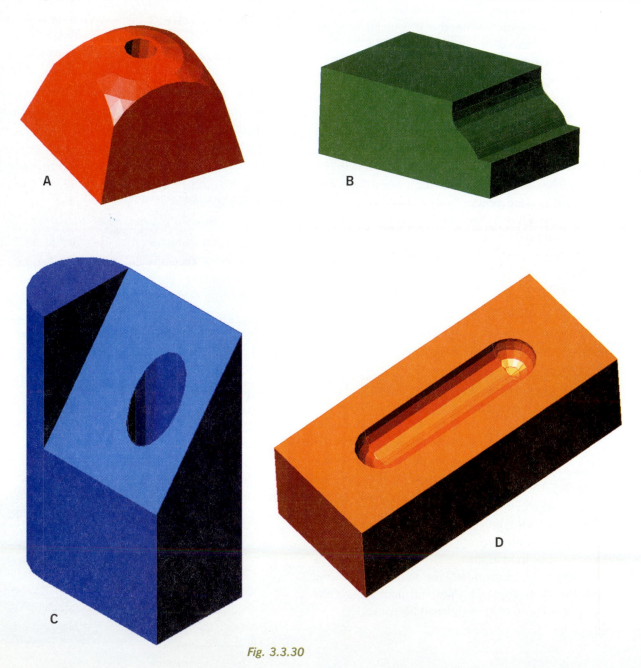

A

B

C

D

Fig. 3.3.30

CAD/CAM Techniques

Solid modeling packages today can store a vast amount of information that can be used for many different purposes. The resulting 3D models can therefore be used for many different things. Solid models can be used directly in applications such as simulation, analysis, creation of descriptive drawings, and creation of other components. They are also used indirectly in material resource planning (MRP), process planning, concurrent engineering, and tool design.

Solid models for commercial products can also be used as tools to subtract from an injection mold. In this section, the swivel bracket model that you created in previous sections of this project will be used to create a simulated injection mold. It will not be a complete mold, because some features have been simplified for the purpose of demonstration. However, it will give you an idea of the processes that can be used to create molds from solid models.

Section Objectives
- Set up a solid model for use in creating an injection mold model.
- Create the top and bottom halves of an injection mold model.

Key Terms
- assembly modeling
- injection mold

Fig. 3.4.1 Machining the base half of the injection mold.

Bracket Base Mold Setup

An **injection mold** is a "negative" of a component. Plastic polymers or other liquified material is poured into the mold to create the "positive." This is a viable manufacturing process for the swivel bracket, which will be molded of black nylon.

The injection mold that will be used to form the basic swivel bracket has two parts: a base and a top half. In Fig. 3.4.1, the bottom half of the mold is shown in a machining process. Solid models of the two pieces of the mold are created by subtracting parts of the 3D bracket model from a prepared box base.

In Practice Exercise 3.4A, you will prepare the model of the swivel bracket for use in a mold design. The holes will be filled and features will be simplified. The model created here will represent the component that would come from the mold before finish machining. Features such as holes and fillets will be added later using a machine tool.

3.4A

For this exercise, you will need BRACKET BASE.dwg, the file you created in Section 3.3 before you created the holes to finish the swivel bracket model. Because it doesn't contain the parts that will be added later by machining, BRACKET BASE.dwg is a good choice for creating the injection mold. However, you must add the fillets to the model in this drawing before you can use it.

1. Open the file named **BRACKET BASE.dwg**.
2. Enter the **FILLET** command and select the line shown in Fig. 3.4.2. Specify a radius of **.140** and press **Enter** to accept the **Chain/Radius** default.
3. Reenter the **FILLET** command and select one of the lines shown in Fig. 3.4.2. Specify a radius of **.05** and press **Enter** to accept the **Chain/Radius** default. Then select the other line shown in Fig. 3.4.2 and repeat the process. When you finish, the model should look like Fig. 3.4.3.
4. Change to the **SE Isometric view**.
5. Reenter the **FILLET** command and select the vertical line shown in Fig. 3.4.4. Specify a radius of **.25** and accept the **Chain/Radius** default. The result is shown in Fig. 3.4.5.

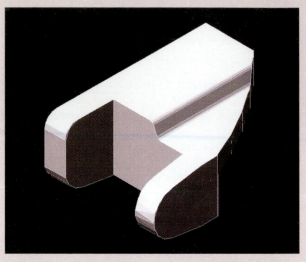

Fig. 3.4.3 The bracket base model after the filleting operations in steps 2 and 3.

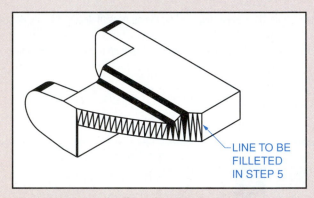

Fig. 3.4.4 Line to be filleted in step 5.

LINE TO BE FILLETED IN STEP 5

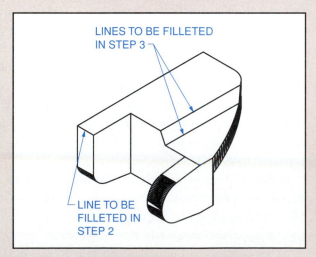

LINES TO BE FILLETED IN STEP 3

LINE TO BE FILLETED IN STEP 2

Fig. 3.4.2 Lines to be filleted in steps 2 and 3.

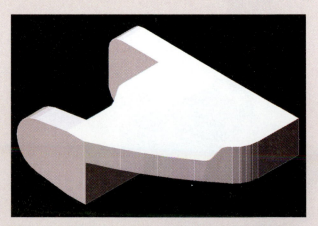

Fig. 3.4.5 The bracket base model after the filleting operations in steps 2, 3, and 5.

6. Change to the **NW Isometric** view.
7. Create a new UCS with its origin and axes positioned as shown in Fig. 3.4.6.
8. Use **SAVEAS** to save the file with a filename of **BRACKET BASE 4.dwg**.

Setup of the swivel bracket model for the injection mold is now complete. The BRACKET BASE 4.dwg file will be used directly to create the injection mold.

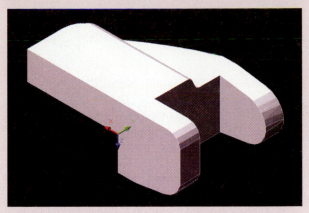

Fig. 3.4.6 Orientation of the new UCS for BRACKET BASE 4.dwg.

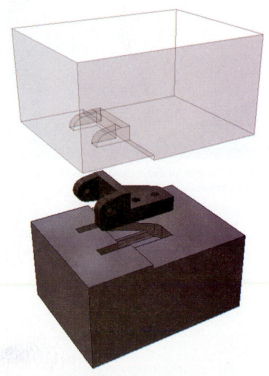

Fig. 3.4.7 Model showing how the two mold halves are used to create the swivel bracket.

Create the Mold Base

Now that you have completed the version of the solid model that you will use to create the injection mold, you can turn your attention to the mold itself. The mold has an upper half and a lower half. The bracket model will be positioned so that its upper half interferes with a solid that represents the lower half of the mold. The other half of the bracket model will interfere with the upper half of the mold. The bracket model will then be subtracted from the mold halves to leave a negative impression in the mold. See Fig. 3.4.7.

Liquified polymers—in this case, nylon—will be poured into the mold and allowed to harden to create the initial bracket. The mold must be created in two halves so that the bracket can be removed from the mold after it hardens. Further processing and machining will then be needed to complete the swivel bracket.

In Practice Exercise 3.4B, you will create a solid model of the base portion of the injection mold. Refer to Fig. 3.4.7 as necessary for orientation as you perform the steps in the Practice Exercise.

Practice Exercise

3.4B

Instead of being two simple boxes, the mold halves will be created so that each one has a lip or offset. This will facilitate the molding process and make it easier to align the halves exactly during the molding process.

1. Create a new AutoCAD drawing and name it **MOLD BASE.dwg**.

2. Create a solid box primitive. Place its first corner at **0,0,0**. Select the **Length** option and enter a length of **4.0**. At the remaining prompts, specify a width of **3.0** and a height of **−2.0** to create the box as shown in Fig. 3.4.8.

3. Create another, smaller box primitive. This one will rest on top of the first box. Place the first corner at **0,0,0**. Then select the Length option and enter a length of **1.50**, a width of **3.0**, and a height of **.10**. See Fig. 3.4.9.

4. Use AutoCAD's **UNION** command to combine the two boxes into a single solid.

5. Move the origin of the UCS to the *lower* edge of the lip on the solid, as shown in Fig. 3.4.10. *Note:* Do not change the viewpoint. The model is shown from a different point of view in Fig. 3.4.10 to clarify the position of the origin.

6. Create centerline location geometry **.90** in the +Y direction from the origin, as shown in Fig. 3.4.11.

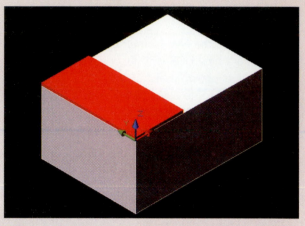

Fig. 3.4.9 *The second box primitive for the base mold.*

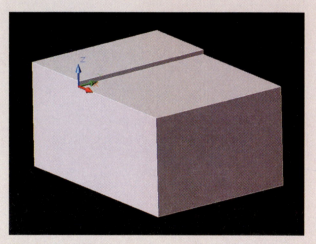

Fig. 3.4.10 *UCS origin for the combined solid that will make up the base mold.*

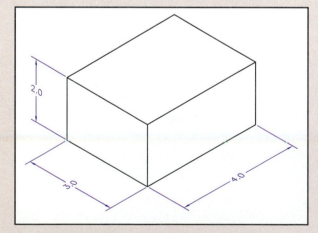

Fig. 3.4.8 *The first box primitive for the base mold.*

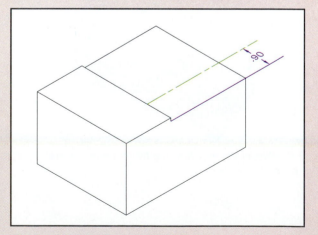

Fig. 3.4.11 *Centerline location geometry.*

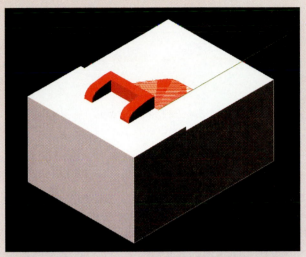

Fig. 3.4.12 *Insert the bracket model into the mold base so that the bottom half of the bracket interferes with the base model.*

Fig. 3.4.13 *Finished cavity in the base mold.*

7. Save your work in **MOLD BASE.dwg** and return to **BRACKET BASE 4.dwg**. (Do not close MOLD BASE.dwg.)

8. From the **Edit** menu, select **Copy with Base Point** and enter **0,0,0** as the base point. Select the model as the item to copy.

9. Return to **MOLD BASE.dwg** and paste the model into the drawing. Use the lower end of the centerline location geometry as the base point for insertion. The bracket model should now be positioned as shown in Fig. 3.4.12. *Hint:* If the model does not position properly in your drawing, there is a good chance that you did not define the UCS properly at the end of Practice Exercise 3.4A. Go back and review your work there to correct the problem.

10. Subtract the bracket from the base model to create the mold cavity in the lower half of the mold. Also delete the centerline location geometry at this time. The result is shown in Fig. 3.4.13.

11. Move the UCS origin back to the lower edge of the lip on the mold, as shown in Fig. 3.4.14. Be sure the axes are positioned correctly.

12. Save and close the MOLD BASE.dwg file.

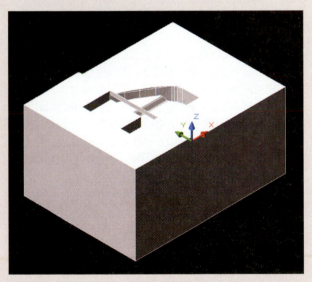

Fig. 3.4.14 *Locate the UCS origin on the lower edge of the lip on the mold base with the axes positioned as shown here.*

Create the Mold Top

The procedure for creating the mold top is very similar to that for creating the mold base. In Practice Exercise 3.4C, you will create the top of the mold, as shown in Fig. 3.4.15. You will then finish the injection mold by making some modifications to the mold top to allow it to interface smoothly with the mold base.

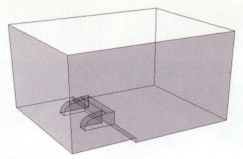

Fig. 3.4.15 *The mold top.*

Practice Exercise

For the injection mold to work correctly, the upper half of the mold must mate precisely with the lower half. Follow the steps below to create the top half of the mold.

1. Create a new AutoCAD drawing and name it **TOP MOLD.dwg**.

2. Create a box primitive as shown in Fig. 3.4.16. Place the first corner at **0,0,0**. Using the **Length** option, specify a length of **4.0**, a width of **3.0**, and a height of **–2.0**. Notice that this box has exactly the same size and orientation as the box primitive that formed the basis for the base mold.

3. Change to the **SW Isometric** view.

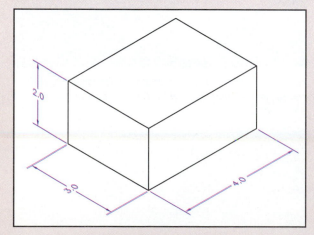

Fig. 3.4.16 *The first box primitive for the top mold.*

Fig. 3.4.17 *Add the second box primitive as shown here. (Red color is for instructional purposes only.)*

4. Create another box primitive, but this time, place the first corner at coordinates **4,0,0**. This places the corner at the back edge of the first box. Use the **Length** option to specify a length of **–2.50**, a width of **3.0**, and a height of **.10**. The second box appears, as shown in Fig. 3.4.17.

5. Use the Boolean **UNION** to transform the two box primitives into a single solid object. This forms the overall shape of the top mold.

6. Create a new UCS with its origin and axes oriented as shown in Fig. 3.4.18.

7. Create the centerline location geometry at the top of the far side of the lip, as shown in Fig. 3.4.19. Create the line in the –Y direction at a distance of **.90** from the UCS origin.

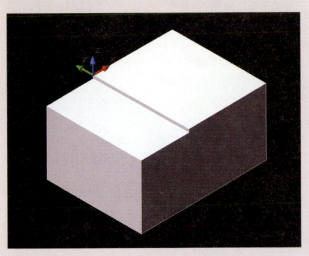

Fig. 3.4.18 *Origin and orientation of the new UCS for step 6.*

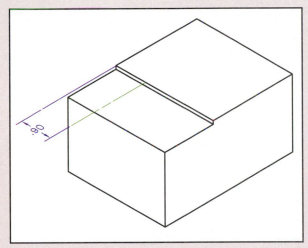

Fig. 3.4.19 *Centerline location geometry for inserting the bracket model.*

8. Create a new UCS with the same origin, but with the axes oriented as shown in Fig. 3.4.20. Note that the positive *Z* axis is now 180° from its previous position.

9. Save the **TOP MOLD.dwg** file, but do not close it. Return to the **BRACKET BASE 4.dwg** file.

10. From the **Edit** menu, select **Copy with Base Point** and enter **0,0,0** as the base point. Select the model as the item to copy.

11. Close the **BRACKET BASE 4.dwg** file and return to the **TOP MOLD.dwg** file.

12. Paste the bracket model into TOP MOLD.dwg. Specify the upper end of the centerline geometry as the insertion point. The bracket inserts as shown in Fig. 3.4.21.

13. Delete the centerline geometry.

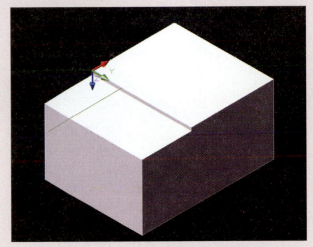

Fig. 3.4.20 *Origin and orientation of the new UCS for step 8.*

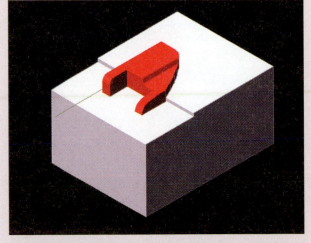

Fig. 3.4.21 *The bracket inserts and interferes with the mold as shown here.*

14. Subtract the bracket model from the mold top. The resulting cavity is shown in Fig. 3.4.22.
15. Create a new UCS with the same origin, but with the axes oriented as shown in Fig. 3.4.23.
16. Save and close the drawing file.

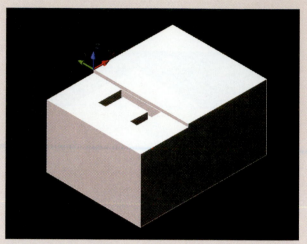

Fig. 3.4.23 The finished top mold with the UCS oriented correctly.

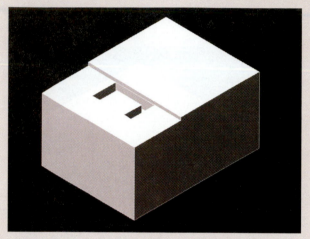

Fig. 3.4.22 The top mold cavity.

Fig. 3.4.24 Testing is done routinely on assembly models in a wide range of industries.

Ed Kashi/Corbis

Advantages of Assembly Modeling

A tremendous amount of information can be gained from assembling a model on the computer. As noted earlier, 3D solid models can be used by designers to detect interferences, motion simulation, FEA, and system level analysis. To achieve the maximum benefit, however, all of the components in an assembly should be modeled and placed together in a single compilation file, either physically, with each component on a separate layer, or by cross reference. This process is known as **assembly modeling**. See Fig. 3.4.24.

The use of assembly modeling can eliminate or drastically reduce the need to create prototypes and models. Solid assemblies can be created directly in the computer to examine and test the systems that are being designed using virtual simulation programs. These models and procedures provide data that previously could only be obtained through physical modeling and testing.

Review Questions

1. Briefly explain how components are created using an injection mold.
2. Why is the injection mold created in two separate halves?
3. What is the purpose of the lip on the mold halves?
4. Why was a special version of the swivel bracket—one without holes—needed to create the injection mold model?
5. What is assembly modeling?
6. If you were creating an assembly model that consists of five complex components, would you place all of the drawings in a single drawing file? If so, why? If not, how would you proceed?
7. Why are assembly models more useful than isolated solid models of individual components?

Practice Problems

1. The object in Fig. 3.4.25 consists of two 3″ cubes joined by a Ø1″ cylinder with endpoints at the center of the facing sides of the cubes.
   - Create the solid model.
   - Create an injection mold that could be used to manufacture the object.
2. The 2D drawing in Fig. 3.4.26 describes a part that will be created by the injection molding process.
   - Create a solid model using the 2D drawing as a guide.
   - Create an injection mold that could be used to manufacture the object.

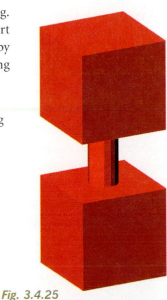

Fig. 3.4.25

Portfolio Project

Bicycle Component (continued)

Create a solid model of an injection mold that could be used to create the bicycle component. Keep in mind that the best approach might be to add features such as holes later in a machining process.

1. If modifications are needed to the basic model of the bicycle component (such as the removal of holes), make the modifications now.
2. Plan to create the mold as two separate halves.
3. Determine the UCSs and insertion points needed to place the bicycle component model correctly in the mold.
4. Create the two halves of the mold.

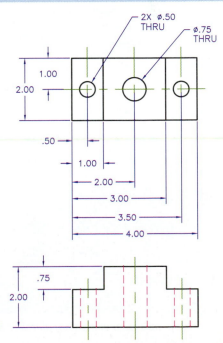

Fig. 3.4.26

| Machine Tools and Numerical Control

Various manufacturers use 3D solid models in different ways, depending on the type of components or assemblies they produce, the company's budget, and even the expertise of their employees. Some manufacturers use the models directly to drive computer-controlled machines. Others use the models for design and development, but prefer to move into the production stage with the traditional, fully documented set of 2D working drawings.

This section describes the development of computer-driven machine tools and explains how numerical control is used to drive these tools. In addition, the process of extracting a 2D drawing from a 3D solid model is demonstrated.

Section Objectives
- Understand the development of machine tools and CNC technology.
- Prepare a preliminary 2D view from a 3D solid model.

Key Terms
- CNC machine
- continuous control
- interpolation
- machine tool
- point-to-point control

Fig. 3.5.1 An industrial computer-controlled lathe.

David Parker, 600 Group/Photo Researchers/SPL

Machine Tools

A **machine tool** is a system that is used to manufacture an object. It can also be referred to as a "machine designed for shaping solid work."

The first machine tool was developed around 700 B.C. This machine was a lathe, which is a tool designed to turn components or tools and remove material. The first lathe was used to refine woodwork. The 15th century brought the addition of metal machining to the industry, and the first production metal machinery came into use in the 18th century. These humble beginnings, combined with the invention of the microprocessor, brought about one of the most influential developments of modern manufacturing: the development of the computer-controlled machine, commonly called a **CNC machine**. *CNC* stands for "computer numerically controlled." The CNC machine in Fig. 3.5.1 is an example of an industrial computer-controlled lathe.

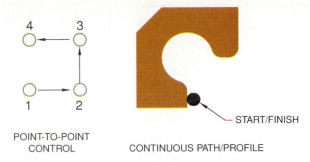

POINT-TO-POINT CONTROL

CONTINUOUS PATH/PROFILE

START/FINISH

Fig. 3.5.2 *Point-to-point and continuous control diagrams.*

Numerical Control

A numerically controlled machine controls motion using numerical data points. The machine is capable of automatically interpreting some of this data to provide several different types of motion control. The two most common types are point-to-point control and continuous, or contouring, control. The diagrams in Fig. 3.5.2 illustrate the differences in these two types of motion control.

Point-to-point control is most effective for cutting multiple holes in a plate or cutting in a straight line. Point-to-point control gives the operator limited control of the path that the system takes between the points that are specified in the program. **Continuous control** can direct motion to individual points and the path between the points using multiple axes simultaneously. In Fig. 3.5.3, continuous control must be used to cut the metal patterns efficiently.

A continuous control system is capable of following complex 3D morphology by controlling more than one axis simultaneously. The technique a machine tool uses to determine the most appropriate path between two points is called **interpolation**. Interpolation uses a mathematical model to determine the best-fitting path to approximate locations. Then it sets speeds that are appropriate for the application.

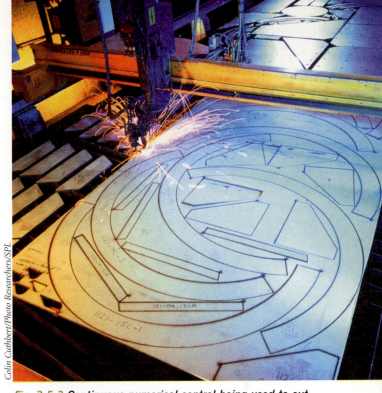

Colin Cuthbert/Photo Researchers/SPL

Fig. 3.5.3 *Continuous numerical control being used to cut sheet metal.*

Numerically controlled systems can be managed using two different control loop techniques: open-loop and closed-loop. The open-loop system uses impulses or signals to determine where the tool is. Open-loop systems "count" the number of impulses and associate a dimensional movement per pulse. This type of control can have problems if the system skips a pulse or loses track of the number of pulses.

A closed-loop system uses a sensor to determine the exact location of the tool, providing a feedback step. See Fig. 3.5.4. These sensors are called *encoders* and can be linear or rotary in design. The encoder gives positional signals that the system can monitor. Closed-loop control systems are more accurate than open-loop systems, but they can be more expensive to produce and require more software control.

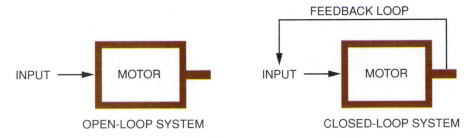

Fig. 3.5.4 *Open- and closed-loop systems.*

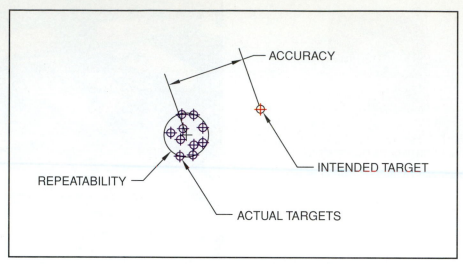

Fig. 3.5.5 Repeatability vs. accuracy.

Accuracy and Repeatability

Numerically controlled systems require two types of tolerances: accuracy and repeatability. **Accuracy** is the precision with which the machine tool can reach a specified location. **Repeatability** is the ability of the machine tool to hit the specified location reliably multiple times. Both of these tolerances are very important to the ability of the machine tool to make a component. In the example shown in Fig. 3.5.5, the machine tool has achieved good repeatability, but none of the hits is close to the intended target site. Either the tool is not sufficiently accurate for the application, or an adjustment needs to be made to the tool settings to increase its accuracy.

Machine Tool Selection

Care has to be taken when determining which machine tool to use to create each component in an assembly. As a rule, the more accurate the machine tool is, the longer it takes to remove material and the more expensive the machine tool is to buy and operate. The manufacturing facility must maintain a fine balance in order to stay profitable and produce the components that the customer can use. The manufacturer has to balance cost, schedule, and quality. Each factor weighs equally in the decision.

Projecting 2D Geometry from a 3D Model

In recent years, there has been more of an effort to go to a paperless office in which the designs are never detailed on paper. Some manufacturing and machine shops take the 3D solid model and produce a component using CNC equipment. This method is quite popular with some industries. Other industries are regulated by government agencies that require the components be fully documented with signed paper drawings.

While 3D models are effective representations of components during the design stage of development, they often are not adequate for manufacturing personnel to produce the components. As mentioned previously, the component needs to be represented using a uniform language that can be interpreted by the people producing the part to the specifications of the designer.

The transition of the design stage to the detail stage can be straightforward, depending on the software package being used. The AutoCAD software has features that translate a solid model into the projected multiview drawing formats that are needed to define the components. In Practice Exercise 3.5A, you will use the swivel bracket model to experiment with the creation of 2D multiview drawings from a 3D solid model.

Practice Exercise 3.5A

AutoCAD's SOLDRAW command allows you to create conventional 2D views from a 3D solid model. The software does not detail the views, but the ability to create the basic views reduces rework to almost nothing in cases in which both 3D models and detailed multiview drawings are required.

SOLDRAW works by creating visible and hidden lines to represent the silhouettes and edges of a solid model. It then projects these silhouettes and edges to a plane that is perpendicular to the current view.

1. Open the **3D BRACKET.dwg** file and change to the **SE Isometric** view.
2. Create a new UCS with the origin and axes oriented as shown in Fig. 3.5.6.
3. Change to the **Top** view, as shown in Fig. 3.5.7. From this point, you will need to view the model as a 3D wireframe.
4. Enter the **SOLDRAW** command. When AutoCAD prompts you to select objects, press **Enter**. AutoCAD automatically switches to Layout 1 and creates a new viewport for the preliminary top view.
5. Pick the **PAPER** button at the bottom of the screen to switch to model space.
6. Enter the **SOLPROF** command and select the bracket model. Accept the **Yes** defaults to display hidden lines on a separate layer, project profile lines onto a plane, and delete tangential edges.
7. Pick the **Model** tab at the bottom of the drawing area.
8. Change to the **SE Isometric** view.

Two new layers now exist in the drawing. The layer that begins with PH contains the projected hidden lines, and the layer that begins with PV contains the projected visible lines. In Fig. 3.5.8, the PH layer is shown in magenta and the PV layer is shown in blue to distinguish them. You can do this also by changing the layer colors in the Layer Properties Manager.

9. Freeze layer 0 (or whichever layer the model currently resides on) so that only the projected geometry is visible.

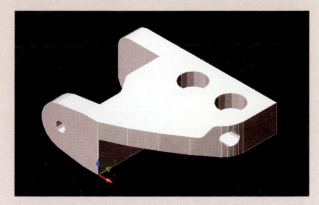

*Fig. 3.5.6 **Rotate the axes to create a new UCS.***

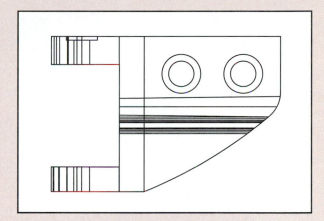

*Fig. 3.5.7 **The 3D wireframe of the top view.***

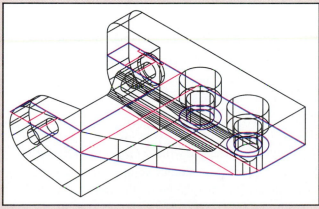

*Fig. 3.5.8 **The PH and PV layers from the SE Isometric view.***

10. Return to the **Top** view. The geometry should look similar to Fig. 3.5.9.

11. Use **SAVEAS** to save the drawing with a new name of **TOP VIEW.dwg**.

This completes the extraction of accurate 2D geometry for machining the outside profile with automated machining practices or to be detailed. To finish the view, you would need to add dimensions and notes and assign a plot style table to ensure that the lines are printed at their proper widths.

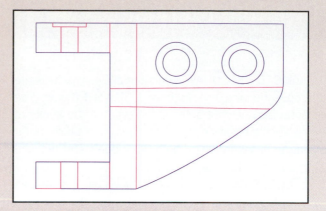

Fig. 3.5.9 *The extracted top view.*

CNC Code Generation

Many programs can be used to generate CNC code. Some CAD systems have embedded programs, and others rely on third party software developers. Code can be generated in many different ways. One of the most common ways to control CNC equipment is to use a combination of N-Code, G-Code, X-Code, Y-Code, Z-Code, A-Code, B-Code, C-Code, F-Code, S-Code, T-Code, R-Code, and M-Code. Each of these types of codes or words has a meaning that is briefly described in Fig. 3.5.10. Some are specific commands, and others are variables that can be set.

In the CNC language, these words are combined to produce a tool path. When the code words are used together, they provide a series of commands or instructions for the machine tool. Most CNC equipment will be able to understand these code words or something similar, although most machine tool manufacturers make subtle changes to the standard code to specialize it for their systems. Some of the more common control words are shown in Fig. 3.5.11.

Most programming is done automatically with the input of a 2D or 3D path. The program associates the desired profile with a cutter diameter and writes a program to create the profile. The program automatically compensates for cutter diameter and sharp corners to create the code to run the CNC machine tool.

Code Type	Meaning
N-Code	Indicates the sequence number of the step
G-Code	Indicates control functions
X-, Y-, Z-, A-, B-, and C-Codes	Indicate coordinate positions
F-Code	Indicates the feed speed of the machine tool
S-Code	Indicates the cutting speed of the machine tool
T-Code	Indicates the tool number that the system is using
R-Code	Indicates the clearance height of a drilling or plunge operation
M-Code	Is referred to as the "miscellaneous word"

Fig. 3.5.10 *Meanings of code types.*

Control Word	Meaning
G00	Rapid Traverse
G01	Linear Interpolation
G04	Dwell
M00	Program Stop
M06	Tool Change
M03	Spindle CW

Fig. 3.5.11 *CNC control words.*

Review Questions

1. What is a machine tool?
2. What is the relationship between machine tools and CNC machines?
3. Explain the difference between point-to-point control and continuous control.
4. In the context of machine tools, what is interpolation?
5. Which control technique is the most accurate: open-loop or closed-loop? Why?
6. What is an encoder and how does it work?
7. What is the difference between accuracy and repeatability? Why must a machine tool have both?
8. Name at least three factors that manufacturers must weigh when they determine which machine tool to use to create each component of an assembly.
9. Why is it sometimes necessary to generate 2D views from a 3D solid model?
10. Which two commands in AutoCAD can be used to project 2D views from 3D solid models?

Practice Problems

1. Open the TOP VIEW.dwg file you created in Practice Exercise 3.5A. Complete the drawing by adding the appropriate dimensions and notes. Be sure to place them on separate layers. Do not place any geometry or dimensions on the PH and PV layers. Information may be lost because these two layers are controlled by AutoCAD to store temporary information.
2. Create a 2D profile drawing of the front view of the swivel bracket. Save the file as FRONT VIEW.dwg. Add dimensions and notes as necessary.
3. Create a 2D profile drawing of the right-side view of the swivel bracket. Save the file as SIDE VIEW.dwg. Add dimensions and notes as necessary.
4. Create a new drawing file named MULTIVIEW.dwg. In this file, combine all of the views you created in Practice Problems 1 through 3 into a single multiview drawing. You may insert the geometry directly or use one of the views as the base drawing and incorporate the others by cross reference.

Portfolio Project

Bicycle Component (continued)

Use the 3D model you created to extract 2D profile views for a multiview drawing of the component.
1. Decide which and how many 2D views will be needed to describe the component completely.
2. Create each view independently and save it as a separate file.
3. In a new drawing file, combine all of the views into a single multiview drawing. You may insert the geometry directly or use one of the views as a base drawing and incorporate the others by cross reference.
4. Save and print the 2D multiview drawing.

Summary

- Solid models can be created in AutoCAD by using predefined primitives, by applying the processes of revolution and extrusion, and by using Boolean operations to refine a basic model.
- When designing a component or assembly using 3D modeling techniques, a drafter must pay careful attention to standard guidelines to ensure the manufacturability and feasibility of the product.

- An existing 3D solid model of a component can be used to create a solid model of an injection mold to produce the component.
- For government subcontractors and others who may require a full set of working drawings as well as a 3D model, the 2D profiles can be extracted directly from an existing 3D solid model.

Project 3 Analysis

1. Based on your work in Project 3, what qualifications do you think a drafter should have before working on solid models that will be used to drive CAD/CAM systems?
2. When a company requires both a 3D solid model and a set of 2D working drawings for a component or assembly, which should the drafter create first in most cases? Why?
3. Consider your work in 2D and 3D drafting so far in this textbook. Which do you consider the more difficult to achieve with accuracy? Explain.
4. A solid model of an injection mold cannot, in itself, be used to manufacture a component. What is the purpose of creating a 3D solid model of the injection mold?

Synthesis Projects

Apply the skills and techniques you have practiced in this project to these special follow-up projects.

1. In Project 3, you created a solid model of a single component. However, the importance of assembly modeling was also discussed. To experience assembly modeling, choose a relatively simple everyday object that consists of at least three separate components. Model all of the components, each in a separate drawing file. Then create an assembly file. Use one of the components as the base drawing and include the others as Xrefs. Practice manipulating and positioning the various components correctly.

2. Use the individual models you created in the above project to create a complete set of formal 2D working drawings for the entire assembly. Include ANSI borders and title blocks as needed. Generate all necessary parts lists.

Mechanical Shaft Assembly

When you have completed this project, you should be able to:

- Plan a concept drawing, given the required design parameters.
- Create a concept drawing using the power tools in the AutoCAD® Mechanical software.
- Create the same concept drawing using the basic AutoCAD software.
- Discuss the advantages and limitations of drawing mechanical parts using AutoCAD and AutoCAD Mechanical.

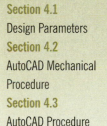

Section 4.1
Design Parameters
Section 4.2
AutoCAD Mechanical Procedure
Section 4.3
AutoCAD Procedure

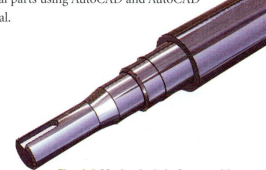

Fig. 4.1 Mechanical shaft assembly.

Introduction

AutoCAD is a very powerful CAD program that can do many different things. However, people who work in a specific area, such as architecture or mechanical design, often need or want design/drafting tools that are specifically targeted to their work.

Autodesk®, the company that developed the AutoCAD software, has therefore developed a series of specialty software programs that build on the basic AutoCAD command set. Examples of these programs include AutoCAD® Mechanical, Autodesk®

Mechanical Desktop, Autodesk® Civil Design, Autodesk® Architectural Desktop, Autodesk® Inventor, and so on.

In this project, you will create a concept drawing of a mechanical shaft. This project is perfectly suited to development using

AutoCAD Mechanical. For those who have access to AutoCAD Mechanical, the procedure is provided in Section 2. For those who have access only to AutoCAD, the information for creating the same shaft using AutoCAD only is presented in Section 3.

Design Parameters

Every product design has to begin with a concept. A **concept drawing** is one that presents a concept or idea for further review by engineers, administrators, and designer/drafters.

In this project, a concept drawing is developed for a mechanical shaft for use in an industrial tabletop mixer assembly. Only the shaft will be drawn in detail in the concept drawing. However, you need to understand how the mixer will work and the parts with which the shaft will interface so that you can design the shaft with the correct parameters and tolerances.

This section introduces the mixer concept as it was developed by the original developmental engineer. It identifies the components of the mixer and demonstrates how the various parts interface. Finally, it shows how the shaft to be developed in this project fits into the overall industrial tabletop mixer design.

Section Objectives
- Describe the use of concept drawings and concept models in industry.
- Develop a background knowledge for working with a concept drawing of a mechanical shaft assembly.

Key Terms
- concept drawing
- concept model
- NEMA frame size

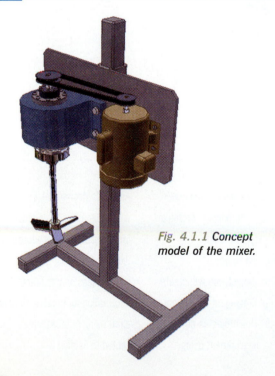

Fig. 4.1.1 Concept model of the mixer.

The Mixer Concept

Marketing has identified a market for an industrial table-top mixer. The mixer will be used for mixing paints and other chemicals. The 3D model in Fig. 4.1.1 shows the designer's preliminary idea for the mixer. This type of model is often called a **concept model**. Because photos cannot be used to show the concept, the designer or developmental engineer makes an accurate 3D model to show the major design features and how they interface.

The shaft will be one of the more expensive items in the mixer assembly to manufacture because it must be custom-machined. Most of the other components, such as the bearings, seals, motor, pulleys, and fasteners, can be purchased from a company that mass-produces them. The concept drawing to be created in this project will be used both for internal review and for gathering cost estimates from internal or third-party machine shops.

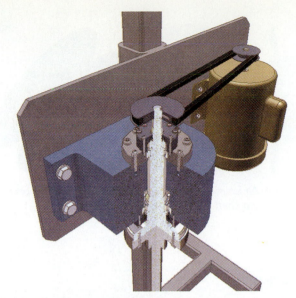

Fig. 4.1.2 *A half section taken through the shaft shows its position in the mixer assembly.*

It is important to understand that concept drawings and models are not intended to be final working drawings. These are simply the tools designers and engineers use to illustrate new product ideas. If the concept is approved, then formal engineering drawings are needed to specify the correct components, bearings, seals, and other parts.

These engineering details are not needed for the concept drawing. However, the drafter must choose the basic components for the assembly. In this case, the designer specified a typical motor, pulley, and drive belt for use in the mixer assembly and placed a housing around the shaft assembly.

Functional Design Concept

As you can see in Fig. 4.1.1, the mixer is designed to be mounted on an adjustable platform. The platform will adjust vertically to allow containers of paint or other liquid to be placed under the mixing blade. The blade can then be lowered into the liquid. When the mixing is finished, the blade will be raised so that the container can be removed. Note that the mechanism that allows the blade to be raised and lowered has not yet been designed into the concept. The current project concentrates on the central shaft for the mixer assembly. The mixer blade is attached to the lower end of the shaft. The upper end of the shaft connects to a pulley and drive belt. The drive belt attaches to a motor that turns the shaft, which turns the mixer blade.

The Mechanical Shaft Assembly

In Fig. 4.1.1, the mechanical shaft is mostly hidden by the housing. The half-section model of the shaft in Fig. 4.1.2 should give you a better idea of how the shaft fits into the mixer assembly.

The shaft assembly consists of the shaft and all of its supporting parts: bearings, seals, pulleys, and so on. Many of these items need to be included in the concept drawing of the shaft to show how the shaft interfaces with them. They will be described further as you create the actual concept drawing in later sections of this project.

Mixer Components

The following discussion concentrates on the major components of the mixer assembly. It is essential to understand the basic requirements for these items before you begin the shaft drawing. This understanding will allow you to create the shaft and its various parts to the correct tolerances.

Main Shaft

The main shaft is the part that will be developed in the concept drawing for this project. Its purpose is to drive the mixer blade. To do that, the shaft must interface with a pulley at one end and with the mixer blade at the other end.

The pulley end of the shaft has a nominal 16 mm locating diameter and a keyway designed to accept a $5 \times 5 \times 12$ mm key. See Fig. 4.1.3. Eventually, these features must be toleranced to create a precise fit with the pulley. For the concept drawing, however, you need only understand that this is a consideration for the final assembly.

Fig. 4.1.3 *The keyway at the pulley end of the shaft.*

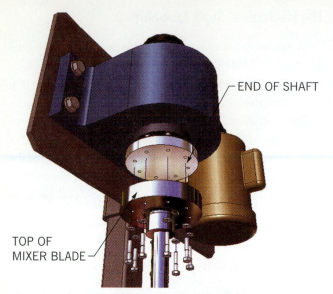

END OF SHAFT

TOP OF
MIXER BLADE

Fig. 4.1.4 *Assembly view of the shorter shaft.*

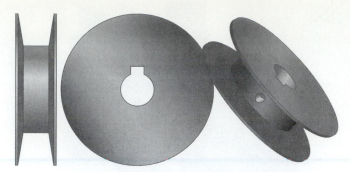

Fig. 4.1.5 *An example of the type of pulley needed to work with the main shaft.*

The other end of the shaft, on which the mixer blade will be mounted, has a nominal 120 mm OD locating diameter. This will need to be precisely toleranced to fit concentrically with the mixer blade ID locating diameter. The mixing blade is attached to this end of the flange with a circular bolt pattern of 90 mm. See Fig. 4.1.4. It is then bolted to the end of the shaft with M6 bolts and washers.

It is important to have a precise locating diameter at this connection. If the mixer blades were allowed to locate only with the bolt mounting pattern, then the mixing blades could locate off the centerline of the shaft. The blades would then spin eccentrically (off-balance).

Drive Pulleys

The pulleys the designer chose for this concept are V-belt pulleys, as shown in Fig. 4.1.5. The pulley that works with the main shaft needs to have a nominal bore size of 16 mm. It must have a nominal keyway to accept the key. On top of the keyway groove, there is a tapped hole to accept a cup-point setscrew. When the setscrew is tightened down onto the key, the cup point digs into the key to keep the key from sliding out of the keyway. This keeps the key pressed firmly against the shaft, which ensures that the pulley remains attached to the shaft. The key allows the pulley to drive the shaft.

Drive Belt

The drive belt chosen for this concept is a V-belt. The V-belt is designed to seat in the V-shaped pulley, giving more stability to the assembly when the motor is running. For this project, the designer specified a Browning Super Grip A-style belt. It is $1/2'' \times 5/16''$, with an outside length of 44.2″ and an inside length of 43.3″.

Tension must be applied to the drive belt to keep it in place and help prevent slippage. This tension is achieved by moving the motor in the direction opposite the shaft. The motor mount has slots that allow the motor to move in that direction. When the belt has been tightened, the motor is bolted into place to maintain the tension on the belt.

Electric Motor

The designer has suggested a general-purpose motor with a **NEMA frame size** of 48. NEMA (National Electrical Manufacturers Association) has assigned a series of numbers and letters to describe the dimensions and mounting type of various motors. If a designer specifies a NEMA 48, for example, any motor that meets the dimensional and mounting type specifications of NEMA 48 can be used, regardless of manufacturer. The number "48" in the NEMA 48 specification refers to the distance from the center of the drive shaft to the center bottom of the mount in sixteenths of an inch.

The specification of motors by NEMA frame size is good practice because it doesn't unnecessarily limit the proposal by naming a specific manufacturer. This is especially important at the design and concept stage of product development.

The designer also specified that the motor should be TEFC (totally enclosed, fan-cooled). TEFC motors are usually dust-tight, with a water-resistant seal. A TEFC motor is a good choice for the industrial mixer assembly. The enclosure will help keep any stray paint or chemical droplets from reaching and damaging the motor.

Shaft Assembly Components

The full-section model in Fig. 4.1.6 shows the mechanical shaft for this project in place with its various supporting components. You may need to refer to this drawing as you complete the concept drawing in the following sections.

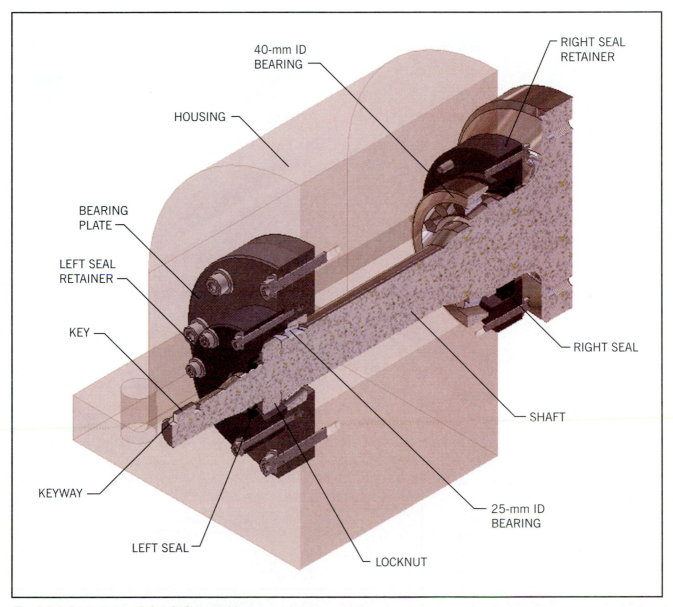

*Fig. 4.1.6 **Components of the shaft assembly.***

Review Questions

1. What is the purpose of a concept drawing?
2. Why are concept models often used in addition to concept drawings?
3. In a concept model of a mechanical shaft, why must all of the parts that make up the shaft assembly be shown?
4. Explain why it is important to understand the basic components for the entire mixer assembly before you begin working on a concept drawing for the main shaft.
5. What is the purpose of the main shaft in the mixer assembly?
6. On the mixer blade end of the shaft, why is it critical to specify an OD locating diameter that will fit at a close tolerance with the ID locating diameter of the mixer blade?
7. What is the purpose of the keyway on the pulley end of the shaft?
8. The pulley must operate under tension. What mechanism has the designer chosen to apply and maintain tension to the pulley?
9. What is the advantage of using standard NEMA frame sizes when specifying a motor for a concept project?
10. Why is a TEFC motor a good choice for an industrial tabletop mixer assembly?

Practice Problems

1. In some cases, a concept drawing for a new product is preceded by a simple, informal sketch. Sketching is a skill that drafters will always need. To practice your sketching technique, make a preliminary sketch of the mixer assembly. Use the model in Fig. 4.1.1 for ideas, but you may incorporate your own ideas as well.
2. Research general-purpose AC motors on the Internet. Aside from the characteristics described in this section, what other characteristics or parameters might be critical in choosing a proper motor for the mixer assembly?

Portfolio Project

Lever Arm Shaft

The lever arm shaft that will raise and lower the mixer blade into the paint or liquid in the mixer assembly has not yet been developed. For this portfolio project, you will design and create a concept drawing for the lever arm shaft.

1. Fig. 4.1.7 shows a designer's rough sketch of a possible design. Using the sketch as a basis, research the various parts that will be needed.
2. Create a more refined sketch of the lever arm shaft. Make design alterations if you wish using your own creative ideas. Remember that the shaft must work with the existing concept for the mixer assembly.
3. Present your refined sketch to your instructor for approval.

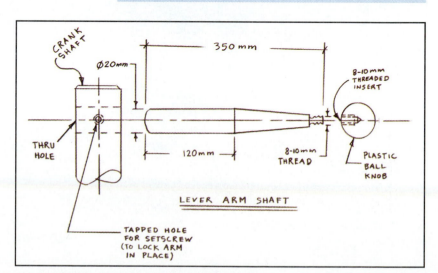

Fig. 4.1.7 *Designer's sketch of a lever arm shaft.*

AutoCAD Mechanical Procedure

AutoCAD Mechanical uses the same basic command set as AutoCAD, making it easy to use if you have a basic understanding of AutoCAD. However, to use the full power of AutoCAD Mechanical, you must be familiar with the additional tools it provides. Partly because of these additional tools, some of the procedures you have used in the other projects are slightly different in AutoCAD Mechanical.

At the beginning of this section, a brief overview of AutoCAD Mechanical is given to help you understand the major differences between AutoCAD and AutoCAD Mechanical. Then you will develop the mechanical shaft assembly for an industrial tabletop mixer using the tools available in AutoCAD Mechanical.

Reminder: If you do not have access to AutoCAD Mechanical, skip Section 4.2 and proceed directly to Section 4.3.

Section Objectives
- Navigate the AutoCAD Mechanical software easily.
- Set up a concept drawing using AutoCAD Mechanical parameters.
- Create a concept drawing using AutoCAD Mechanical power tools.
- Edit a concept drawing using AutoCAD Mechanical power tools.

Key Terms
- DIN
- power command
- power object
- power snap

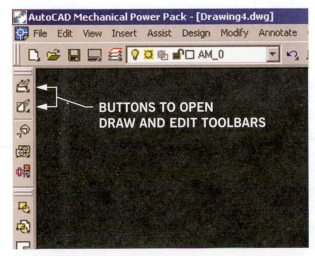

Fig. 4.2.1 *Buttons to display the Draw and Edit toolbars in AutoCAD Mechanical.*

Opening Screen

To students accustomed to working in AutoCAD, the opening screen in AutoCAD Mechanical may seem overwhelming at first. The default toolbars are arranged differently and contain different buttons than in AutoCAD. The familiar Draw and Edit toolbars are not visible. This does not mean that the Draw and Edit toolbars are not available, however. The purpose of two buttons at the top of the docked toolbar at the left side of the drawing area is to display these toolbars quickly and easily. See Fig. 4.2.1.

Aside from the toolbars, the basic screen layout should look very familiar to AutoCAD users. New drawings open automatically into model space, and two layout tabs are provided by default. The buttons at the bottom of the drawing area (SNAP, GRID, ORTHO, etc.) are in the same configuration and perform the same functions as in AutoCAD.

AutoCAD and AutoCAD Mechanical are both designed to allow many different forms of input. For example, some drafters prefer to enter commands by picking the buttons on toolbars. Others prefer to enter the commands or shortcuts at the keyboard. Still others use the pull-down menus at the top of the screen. Many people use a combination of these methods depending on which is easiest to use in a given situation. If you rely heavily on the pull-down menus in AutoCAD, be aware that the pull-down menus in AutoCAD Mechanical differ profoundly from those in AutoCAD. You should investigate them by using the cursor to open each menu and its submenus to see the options available.

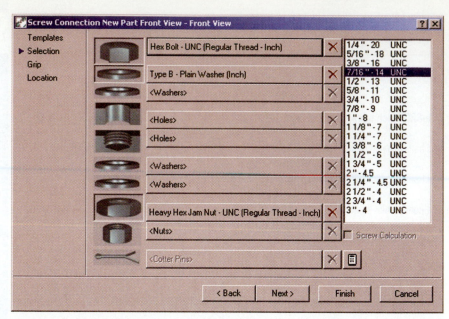

Fig. 4.2.3 *Edit window for the screw connection. Notice that each possible item for the assembly is listed in this box. From here, you could change the screw size, define additional washers or nuts, or remove items from the assembly.*

The AutoCAD Mechanical Environment

It is possible to work in AutoCAD Mechanical exactly as you would AutoCAD, using the basic drawing commands to create every object independently. However, AutoCAD Mechanical provides much more efficient ways to work with common mechanical objects such as screws, bearings, shafts, and structural steel.

Power Objects

In AutoCAD Mechanical, a **power object** is a predefined object such as a screw or shaft, or even a large object such as steel shapes (I-beams, etc.). These predefined objects have internal information and parameters attached to them that help define how they behave in the drawing.

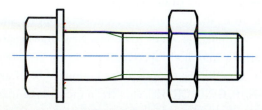

Fig. 4.2.2 *A screw connection consisting of a 7/16–14UNC hex bolt with a plain washer and a heavy hex jam nut.*

For example, one of the power objects is a screw connection, which consists of a screw and any washers or nuts associated with it, as shown in Fig. 4.2.2. In AutoCAD Mechanical, you can define the entire assembly in one operation. Later, if you change the screw size, the washers and nuts associated with it update automatically to accommodate this change. In a sense, power objects are intelligent objects.

Power Commands

To work with power objects, you need a different set of commands—the **power commands**. These commands update the entire object, including internal parameters. The power commands include Power Edit, Power Copy, Power Erase, Power Dimension, and Power View. You can access these commands by picking the buttons at the top of the screen, but the easiest way to use Power Edit is simply to doubleclick on the power object to be edited. The appropriate editing windows and options are displayed automatically. For example, doubleclicking the screw connection shown in Fig. 4.2.2 brings up the window shown in Fig. 4.2.3, from which you can change any part of the assembly.

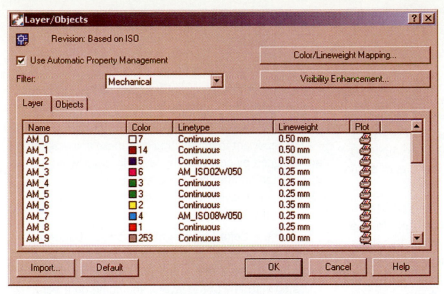

Fig. 4.2.4 *Partial list of AutoCAD Mechanical default layers.*

automatically places them on their own layers. By default, the layers are not intuitively named. For example, the layer for centerlines is named AM_7 for "AutoCAD Mechanical layer 7."

The software maintains a master set of layers that are present in all AutoCAD Mechanical drawings and sets the attributes of each layer as it is needed by the objects you insert. A partial list of the standard layers is shown in Fig. 4.2.4. To see the entire list, you can pick the Layers button to the left of the Layer Control dropdown box and scroll through the list of layers in the Layer Control window.

Drawing Setup

The process of setting up a new drawing in AutoCAD Mechanical is simpler than in AutoCAD. You do not have to set up drawing limits or layers, for example. Units and text style are largely determined by the template you use to begin the drawing.

Templates and Plot Styles

AutoCAD Mechanical does not have as many standard templates as AutoCAD has. One reason for this is that AutoCAD Mechanical does not use named plot styles. You may recall that in AutoCAD, you can specify an ANSI D-size sheet with either named plot styles or color-dependent plot styles. AutoCAD Mechanical uses only the color-dependent option.

Layers

In Projects 1, 2, and 3, you carefully constructed a set of drawing layers, giving each layer a specific color, name, linetype, and linewidth. You then proceeded to place various objects on their correct layers. Layers are handled much differently in AutoCAD Mechanical. You can still create and use your own layers, but if you use power objects, the software

Drawing Limits

Unlike in AutoCAD, in which you must set the limits of a drawing manually, AutoCAD Mechanical has an elastic set of limits. By default, the limits are literally defined by the size of the drawing. Therefore, the larger your drawing becomes, the larger the limits size will be.

Running Object Snaps

You can set running object snaps in AutoCAD Mechanical just as you can in AutoCAD using the Drafting Settings options. However, in Mechanical, you can also predefine up to four sets of **power snaps**. One way to do this is to enter AMPOWERSNAP at the keyboard. This presents the current running object snaps, as well as tabs that allow you to create four additional settings. Then simply enter AMPSNAP1, AMPSNAP2, AMPSNAP3, or AMPSNAP4 to load power snap sets 1, 2, 3, and 4, respectively.

Mechanical Shaft Drawing Setup

Practice Exercise 4.2A steps you through the process of setting up the drawing file needed for the concept drawing of the mechanical shaft in this project. If you have difficulty with any of the steps, refer to AutoCAD Mechanical's Help feature.

Practice Exercise

4.2A

Unlike the other drawings in this textbook, the mechanical shaft assembly drawing will be created using metric units. To set up the appropriate units and type style for a metric drawing in AutoCAD Mechanical, you need only select the proper template, as described below.

1. Open AutoCAD Mechanical. In the Today window, select the **Create Drawings** tab. In the dropdown box, select **Template**. If the templates are not displayed, pick the **A** row. Then select **am_iso.dwt** as the template to use.

2. If you are new to AutoCAD Mechanical, take a minute to explore the user interface. Move the cursor slowly over the buttons on the toolbars to see tooltips that give the names of the buttons. Also explore the pull-down menus at the top of the screen.

3. Enter the **AMOPTIONS** command at the keyboard to display the Mechanical Options window, as shown in Fig. 4.2.5. From this window, you can control most of the default values AutoCAD Mechanical uses within the drawing.

4. On the **Standards** tab (shown in Fig. 4.2.5), pick the **Layers/Objects** item in the box on the right to display the Layer/Objects window. Scroll down through the layers, noting their names, colors, and linetypes. Also note that each layer has been assigned a lineweight (linewidth).

It is important to understand that these lineweights are assigned by color, not by layer. In other words, if you created another layer and assigned the color red (1) to it, that layer would have a linewidth of .25 mm, just like layer AM_8.

5. Pick the **Color/Lineweight Mapping...** button at the top right corner of the Layer/Objects window. If you wanted to change the lineweight associated with any of the layers, you could do it here. Be aware, however, that AutoCAD Mechanical uses these layers to create power objects with the correct line thicknesses, so it is better not to change them unless absolutely necessary. For this project, we will use the defaults. Pick **Cancel** to return to the Mechanical Options window.

6. Doubleclick each of the items in the box at the right of the Mechanical Options window and note the options that are available. All of these options default to the ISO standard because you have chosen to use an ISO template (am_iso.dwt). Cancel out of each window without changing the defaults.

As you work your way down through the list, notice that AutoCAD Mechanical supports automatic balloons for part identification numbers, as well as a full complement of welding symbols. One of the most helpful tools is the parts list and BOM (bill of materials) support. Because the power objects in AutoCAD Mechanical have intrinsic properties, the software can track the specified parts and produce a parts list automatically. You will work with this feature later in this section.

7. Save the new drawing file using a filename of **PROJECT 4 - MECH.dwg**.

Fig. 4.2.5 The Mechanical Options window.

Bearing Geometry

The bearings for a shaft assembly are usually chosen early in the design process. The type of bearing chosen depends on many things, including bearing directional loads and lubrication methods, for example.

For this project, the designer has specified that the shaft will be supported on the mixer stand by two ISO 355 tapered roller bearings. A 3D rendering of the bearings is shown in Fig. 4.2.6. These bearings can be created quickly and easily in AutoCAD Mechanical using power objects. Follow the procedure in Practice Exercise 4.2B to create the tapered roller bearings.

Fig. 4.2.6 The tapered roller bearings that will support the mechanical shaft.

Practice Exercise

Before you can draw the bearings, you must create a centerline for the shaft that they will support. In AutoCAD Mechanical, you can specify a centerline and the attributes are assigned automatically.

1. Open the **PROJECT 4.dwg** file you created in Practice Exercise 4.2A.
2. To give yourself a feel for the current drawing size, use the **F7** key to display the AutoCAD grid. **ZOOM All**.
3. Instead of using the LINE command, go to the **Design** pulldown menu and select **Centerlines** and **Center Line** from the menu. Specify a start-ing point at the left side of the drawing area, about halfway down from the top. For now, make the centerline 160 mm long by specifying a second point using the polar coordinate **@160<0**.

Notice that even though you did not specify any particular layer, the centerline appears with the correct linetype. If you click on the line and look at the display in the Layer Control dropdown box, you will see that AutoCAD Mechanical automatically put the centerline on layer AM_7—the layer the software has reserved for centerlines.

4. The next task is to create working geometry to locate the bearings. Using the dimensions shown in Fig. 4.2.7, create reference lines for the inside diameter of both bearings. Space the reference lines 142.5 mm apart, as shown.

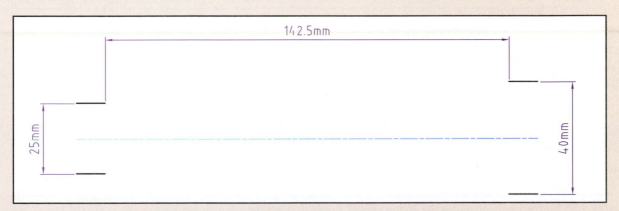

Fig. 4.2.7 Reference lines to locate the inside diameters of the two tapered roller bearings.

Fig. 4.2.8 *Location of the Roller Bearings button.*

When you create reference geometry like that in step 4, you can use the ordinary AutoCAD drawing and editing commands. To make this easier, you may want to display the Draw and Edit toolbars. If so, pick the first two buttons on the top toolbar docked on the left side of the drawing area. Refer again to Fig. 4.2.1 for the location of these buttons.

5. To create the left bearing, look once more at the toolbars that are docked by default at the left side of the AutoCAD Mechanical screen. The lowest one contains the buttons used to create power objects. Use tooltips as necessary to find the **Shaft Generator** button. Press and hold the left mouse button over this button to see the fly-out menu. Select **Roller Bearings**, as shown in Fig. 4.2.8. *Note:* If this menu is not present at the left side of the drawing area, move up to the top of the screen and right-click near the toolbars. Select **AMPP** and **ACAD/M_PP Content** to display the toolbar. You may want to dock it on the left side of the screen by dragging it to the left edge until it docks.

6. The Select a Roller Bearing window appears, giving you pictorial representations and text descriptions of the types of roller bearings available. Click once on **Radial**. New pictorial representations, called thumbnails, appear and display the options for radial roller bearings. *Note:* When the entire text below a thumbnail is too long to display in the allotted space, you can rest the cursor over that thumbnail to see a tooltip that gives the full name of the option. Choose **ISO 355**. In the next window, select **Front View**.

7. For the insertion point, snap to the inner endpoint of the upper line that represents the 25-mm inside diameter of the bearing, as shown in Fig. 4.2.9. Then select a point on the centerline when prompted.

8. The ISO 355 bearing window appears. Note that the inner diameter defaults to 25 mm because you selected the predefined 25-mm reference line as the insertion point. Without changing any of the defaults, pick the **Next** button to display the Calculation portion of the window. Again without changing any defaults, pick the **Next** button to display the Result portion of the window. From the list of bearing results, pick **2CC-25x50x17**. Then pick **Finish**.

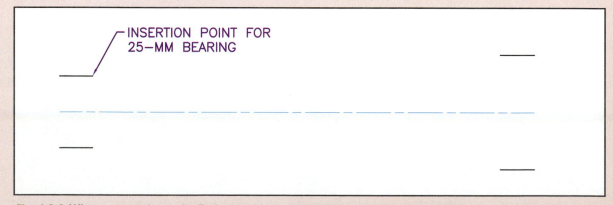

Fig. 4.2.9 *When prompted, use the Endpoint object snap to select the point shown here.*

9. With Ortho off, drag the resulting bearing until its associated tool tip shows a bearing size of **ISO 355-2CC-25x50x17**. *Hint:* You may need to let the cursor rest in place for a few seconds before the tooltip appears.

10. In the Select Part Size window, select the second option to finalize the selection of **ISO 355-2CC-25x50x17**. Pick **OK**. The bearing representation should appear both above and below the center-line, as shown in Fig. 4.2.10.

11. Repeat steps 5 through 8 to create the 40-mm bearing. However, for this bearing, choose bearing size **ISO 355-4CB-40x75x19**. *Note:* The Select a Roller Bearing window defaults to the view selection for the same bearing you created the last time you used this option. Therefore, steps 5 and most of 6 have already been done for you. You can simply pick **Front View** and pick up the sequence from there. However, if you choose to begin again at step 5 in this procedure, you can doubleclick the **Radial** folder at the top of the box on the left side of the window to display the bearing thumbnails.

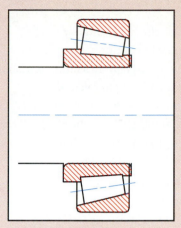

Fig. 4.2.10 Result of placing the 25-mm bearing.

12. Locate and face the bearing as shown in Fig. 4.2.11, with the small end of the taper facing inboard. Drag the bearing result as you did in step 9. You may not be able to achieve the exact width for the bearing. Drag to a size of **ISO 355-4CB-40x75x26**. In the Select Part Size window, select **ISO 355-4CB-40x75x19**. Pick **OK**.

13. Erase the four lines of the bearing reference geometry to avoid confusion later.

Fig. 4.2.11 Locate the bearing as shown here.

Shaft Generation

Shafts are typically made of alloy steel and are heat-treated to improve strength and durability. They are usually machined on lathes and precision-ground to tight tolerances.

Like the bearings, the shaft itself can be generated as a power object in AutoCAD Mechanical. The 3D model in Fig. 4.2.12 will give you a better idea of how the finished shaft will look. In Practice Exercise 4.2C, you will create the basic shaft for the mixer assembly. Refer to Figs. 4.1.6 and 4.2.12 as necessary while you work.

Fig. 4.2.12 **A 3D model of the finished mechanical shaft.**

Practice Exercise

1. From the same toolbar you used to create the bearings, pick the **Shaft Generator** button. (This is the same button you picked and held to display the flyout toolbar for the roller bearings.)

2. For the starting point, use object snap tracking to pick the point on the main centerline that is perpendicular to the right edge of the 40-mm bearing, as shown in Fig. 4.2.13.

3. When the Shaft Generator window appears, pick the **Config...** button at the bottom of the window.

4. In the Shaft Generator - Configuration window, change the settings as follows:
   • Under Stationary Shaft, check **Prompt**.
   • Under Front View, check **Radius Reflection Line** and **Check Contour**.
   • Under Side and Sectional Views, check **Radius Reflection Line** and **Always Update**.
   • Pick **OK**.

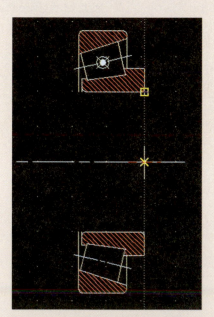

Fig. 4.2.13 **Use object snap tracking as shown here to find the point on the main centerline that is perpendicular to the right edge of the 40-mm bearing.**

The Outer Contour tab of the Shaft Generator window is shown in Fig. 4.2.14. Notice that there are two buttons for generating cylindrical shaft segments. Both are labeled "Cylinder," but the bottom one shows length and width dimensions in green. In the following steps, it is important to choose the Cylinder button that includes the length and width dimensions. Choosing the other Cylinder button will cause the procedure not to work properly.

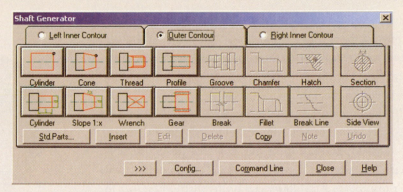

Fig. 4.2.14 *The Outer Contour tab of the Shaft Generator, showing the two Cylinder buttons on the left side.*

5. To create the first segment of the cylinder, pick **Cylinder** from the Outer Contour menu. Specify a length of **–25** and a diameter of **40**. The negative value for length tells AutoCAD Mechanical that you want to extend the shaft to the left of the starting point. By default, positive numbers create the geometry to the right.

6. To create the second segment, pick the **Slope 1:x** button (just to the right of the Cylinder button). Specify a length of **–10**, a diameter starting point of **40**, and a diameter endpoint of **33**.

7. For the third cylinder segment, pick **Cylinder**. Specify a length of **–90** and a diameter of **33**.

8. For the fourth cylinder segment, pick **Cylinder**. Specify a length of **–18** and a diameter of **25**.

9. To create the fifth segment of the cylinder, pick **Thread** from the top row of buttons. In the window that appears, choose **ISO 261 External Threads (Regular Thread)**. In the thread window, choose the thread size **M24 x 2**. Specify a length of **–12**. Next to **Start from**, make sure that the **Left** radio button is selected. Pick **OK**. *Note:* It doesn't matter if the shaft begins to run off the screen to the left. Work through the remaining steps without closing the shaft generator.

10. To create the sixth segment, pick **Cylinder**. Specify a length of **–15** and a diameter of **20**.

11. For the seventh segment, pick **Slope 1:x**. Specify a length of **–10**, a diameter starting point of **20**, and a diameter endpoint of **16**.

12. For the last segment, pick **Cylinder**. Specify a length of **–35** and a diameter of **16**. Pick **Close** to end the shaft generation procedure. The shaft should now look like Fig. 4.2.15.

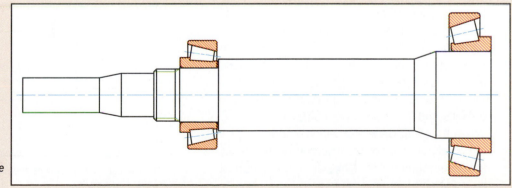

Fig. 4.2.15 *The basic shaft.*

Shaft Modification

It is possible to edit the shaft as an entity, as opposed to simply editing the lines that define it. You can even add segments to the shaft. To do these things without violating the integrity of the internal shaft parameters, you must use the power edit commands.

If you compare the model in Fig. 4.2.12 with the geometry in your drawing file, you will see that more segments need to be added to the shaft to fit the industrial mixer assembly. Specifically, we need to add the segments that interface with the mixer blade at the lower end of the shaft. In Fig. 4.2.16, the lower end of the shaft is shown fastened to the upper flange of the mixer blade.

In Practice Exercise 4.2D, you will add the upper segments using the power edit commands. Refer to Fig. 4.2.16 as necessary to understand the orientation and location of the additional segments.

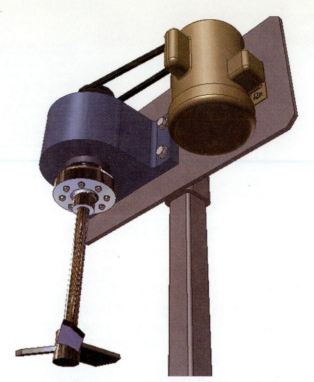

Fig. 4.2.16 *The 120-mm segment on the lower end of the shaft fastens to the mixer blade, as shown here.*

Practice Exercise

The first editing task will be to add segments to the bottom (right) end of the shaft. Before you begin, adjust the zoom level so that the right end of the shaft is located on the left side of the screen, with room on the right for additional segments.

1. Pick the **Power Edit** button from the toolbar at the top of the screen. At the prompt, select the rightmost shaft segment (Ø40 × 25 long). A series of prompts will appear at the Command line to allow you to change the size of the segment. Press **Enter** at each prompt to accept the defaults and keep the current size.

2. When the Shaft Generator window appears, pick the **Insert** button. For the insertion point, specify the intersection of the right side of the rightmost segment and the main centerline.

3. Pick **Cylinder** (with width and diameter inputs). Specify a length of **30** and a diameter of **50**.

4. Pick **Cylinder** again to create another segment. Specify a length of **20** and a diameter of **120**. Pick **Close**. The right end of the shaft should now resemble Fig. 4.2.17.

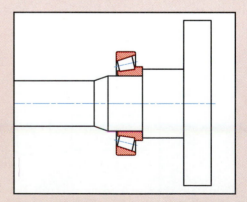

Fig. 4.2.17 *Right end of the shaft showing added segments.*

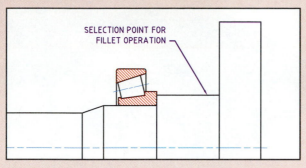

Fig. 4.2.18 *Select the top outer diameter line of the second shaft segment, being sure to select a point near the right end of the line.*

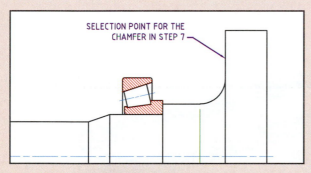

Fig. 4.2.19 *Select the left side of the first shaft segment for the first chamfer operation.*

Notice the sharp corners on the upper segments of the shaft. The next task will be to chamfer and fillet the edges to refine the shaft.

5. Pick the **Power Edit** button again. Pick the rightmost shaft segment. Press **Enter** to move through the default size information until the Shaft Generator window appears.

6. Pick the **Fillet** button. Select the second shaft segment from the right. Pick the upper right outside diameter line, as shown in Fig. 4.2.18. Reference circles appear to indicate the proposed fillet location. Enter a radius of **12**.

7. In the Shaft Generator, pick the **Chamfer** button and select the left face of the far right segment, as shown in Fig. 4.2.19. Enter a chamfer length of **3** and an angle of **45**.

8. Pick the **Chamfer** button again. This time, pick the upper outside diameter line of the second segment from the right, being sure to pick close to the left end of the line. Enter a chamfer length of **2** and an angle of **15**. Pick **Close**.

9. Change the zoom window so that you can see the left end of the shaft. Pick the **Power Edit** button again and press **Enter** through the defaults to see the Shaft Generator window. Pick **Chamfer** and select the left face of the leftmost shaft segment. Enter a chamfer length of **1** and an angle of **45**. Pick **Close**. The shaft should now look similar to the one in Fig. 4.2.20.

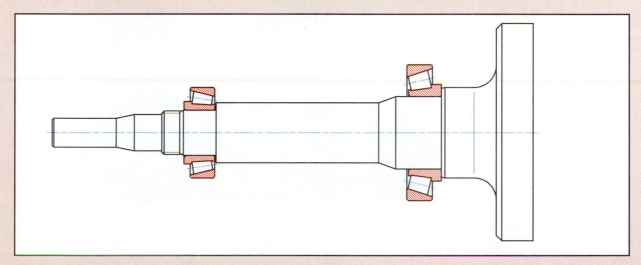

Fig. 4.2.20 *The shaft after the filleting and chamfering operations.*

Key and Keyway

At the top of the shaft, the key and keyway must be added at the interface with the drive pulley. The purpose of the key is to keep the drive pulley from slipping. Fig. 4.2.21 shows a closeup view of the key and keyway extending from the top of the housing.

Fig. 4.2.22 is an exploded view that shows the relationship of the pulley and key. Notice the cup-point setscrew that threads into the pulley to tighten down onto the key.

In Practice Exercise 4.2E, you will use power commands to create the key and keyway for the mixer. Refer to Figs. 4.2.21 and 4.2.22 as you complete the exercise.

Fig. 4.2.21 *Closeup showing the key and keyway in position at the top of the shaft.*

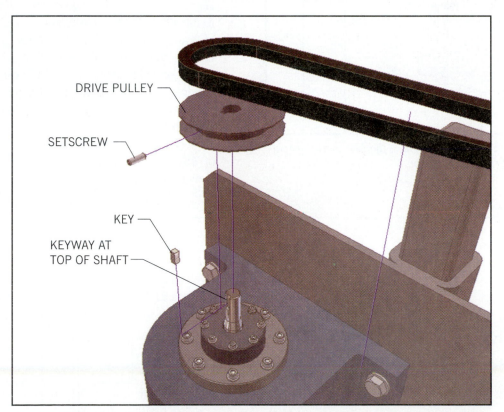

Fig. 4.2.22 *Exploded view showing how the key and keyway at the top of the shaft are used in the mixer assembly.*

Practice Exercise 4.2E

Like the other parts of the shaft, the key and keyway can be added using power commands. The key specified in the concept drawing is a standard DIN 6885B parallel key, and it is included in the parts database maintained by AutoCAD Mechanical. **DIN** stands for "Deutsche Industrie Norm" (German Industry Standard). You will encounter DIN standards often when you work on metric or ISO projects. Follow the steps below to add the key and keyway to the shaft drawing.

1. Create a reference line for inserting the keyway by offsetting the leftmost vertical line of the shaft to the right by **18.5**. Use grips to extend the line upward to cross the upper shaft diameter line .

2. Pick the **Power Edit** button and select the leftmost segment to edit. Press **Enter** to accept the default sizes. When the Shaft Generator window appears, pick the **Std. Parts...** button.

3. In the pictorial menu that appears, pick **Parallel Keys - On End of Shaft**. (You will need to use the tooltips to find the correct parallel key item.)

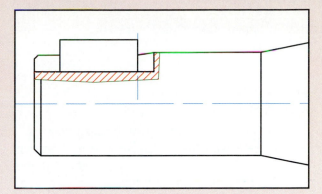

Fig. 4.2.23 *The key and keyway.*

4. Choose **Front View**.

5. From the types of parallel keys displayed, choose **DIN 6885 B**. For the insertion point, select the intersection of the top diameter line and the reference line you offset in step 1. Specify the direction as **Left** and press **Finish**. Drag the size to the left until the size (shown at the left end of the status bar) is **B 5 x 5 x 12**. Close the Shaft Generator window and erase the reference line from step 1. The keyway and key should appear as shown in Fig. 4.2.23.

Seals, Housing, and Fasteners

Several items besides the shaft itself make up the shaft sub-assembly for the industrial tabletop mixer. These include the right and left shaft seals, the seal retainers, the housing, and various fasteners. These need to be shown in enough detail to show how the shaft fits into the overall assembly.

The housing and seals help keep dirt and debris out of the bearings and moving parts, extending the life and reliability of the assembly. See Fig. 4.2.24. The type of seal needed depends on the maximum speed of the shaft and the kind of lubrication needed for the chosen bearings. For example, different types of seals are needed for bearings lubricated with oil and grease. The seals chosen for this project are radial lip seals that are compliant with DIN 3760 and are made of NBR (nitrile-butadine rubber). In Practice Exercise 4.2F, you will add the seals, seal retainers, housing, bearing plates, and fasteners to the concept drawing.

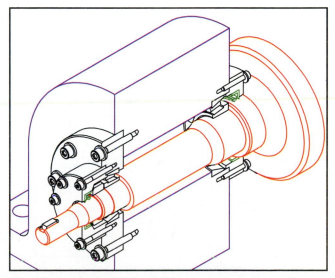

Fig. 4.2.24 *The seals (green) fit against the shaft to help keep out dirt and debris.*

Practice Exercise 4.2F

The shaft assembly requires two lip seals: one above the upper bearing, and the other below the lower bearing. Start with the lower seal.

1. Zoom in on the right side of the shaft drawing.

2. Create a temporary reference line by offsetting the left face of the second shaft segment to the right by **4 mm**. (Be sure to select the left face, not the chamfer line.) With Ortho on, use grips to extend the line to cross the top diameter line of the segment. See Fig. 4.2.25.

3. Pick the **Power Edit** button and select the second shaft segment from the right. Press **Enter** to accept the default size and features of the segment. Choose **Std. Parts...** and pick **Shaft Seals** from the pictorial menu that appears. On the next menu, select **DIN 3760 AS (Left)**. For the insertion point, select the intersection of the top diameter line of the second shaft segment and the temporary reference line you created in step 2. Specify the direction as **Right**. Drag to the right until the tooltip displays **DIN 3760 AS 50 x 68 x 8 NBR**. Close the Shaft Generator window and delete the temporary reference line. The seal should appear as shown in Fig. 4.2.26.

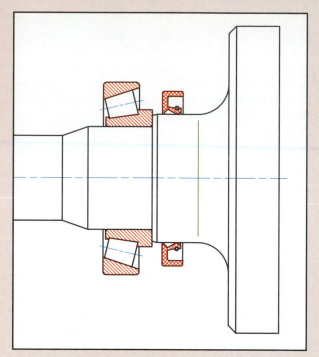

Fig. 4.2.26 *The upper (right) seal.*

4. Zoom in on the left side of the shaft drawing and prepare to add the upper (left) seal.

5. Create a temporary reference line by offsetting the left face of the third shaft segment from the left to the right by **5 mm**. With Ortho on, use grips to extend the line so that it crosses the top diameter line of the segment. See Fig. 4.2.27.

6. Pick the **Power Edit** button and select the third shaft segment from the left. Press **Enter** to

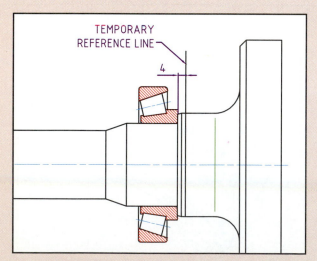

Fig. 4.2.25 *Place the temporary reference line 4 mm from the left edge of the second shaft segment, as shown here.*

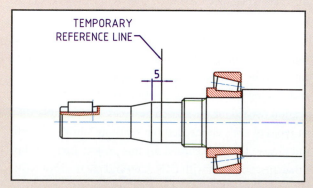

Fig. 4.2.27 *Place the temporary reference line 5 mm from the left edge of the third shaft segment.*

accept the defaults. Choose **Std. Parts...** in the Shaft Generator window. Notice that the window displays shaft seals by default. Select **DIN 3760 AS (Left)**. For the insertion point, select the intersection of the temporary reference line you created in step 5 and the top diameter line of the shaft segment. Specify the direction as **Right**. Drag the size to **DIN 3760 AS 20 x 35 x 7 NBR**. Close the Shaft Generator window and delete the temporary reference line. The seal should appear as shown in Fig. 4.2.28.

7. A lock nut is needed to retain the left bearing. Use basic AutoCAD commands to draw a basic representation of the lock nut using the dimensions shown in Fig. 4.2.29. Place the lines on the **AM_0** layer.

8. Add the main housing and bearing plate using basic AutoCAD commands. Use the dimensions shown in Fig. 4.2.30.

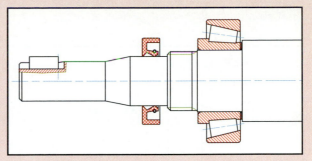

Fig. 4.2.28 The upper (left) seal.

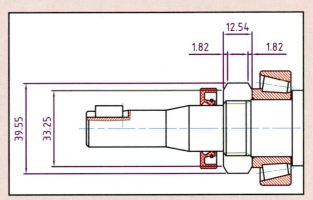

Fig. 4.2.29 Dimensions for the lock nut.

Fig. 4.2.30 Dimensions for the housing and bearing plate.

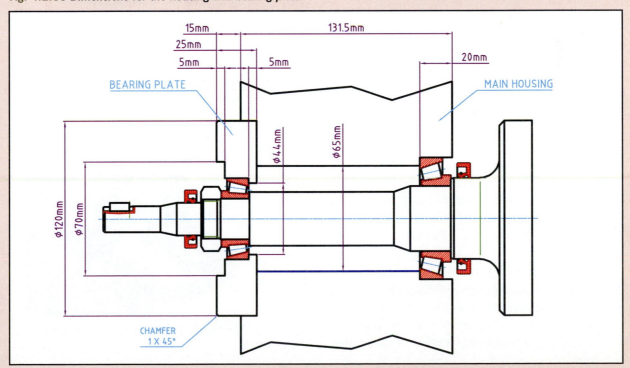

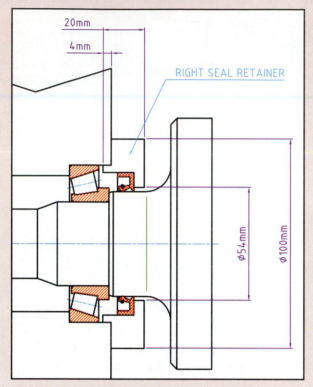

Fig. 4.2.31 *Dimensions for the right seal retainer.*

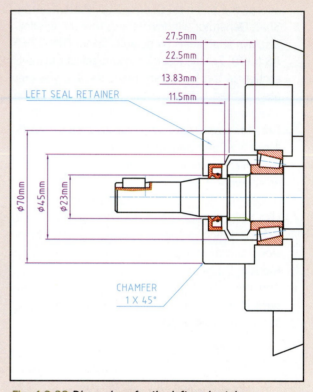

Fig. 4.2.32 *Dimensions for the left seal retainer.*

9. Use basic AutoCAD commands to create the right seal retainer. Use the dimensions shown in Fig. 4.2.31.

10. Add the left seal retainer using the dimensions shown in Fig. 4.2.32.

The next step is to add the mounting screws for the bearing plate and the left seal plate. See Fig. 4.2.33. You will create the bolt circle diameter (B.C.D.) for each set of screws using basic AutoCAD commands. However, you can use power objects for the screws themselves. Refer to Fig. 4.2.34 as you create them.

11. Create two centerlines to represent the upper and lower limits of a 100-mm B.C.D. around the main horizontal centerline of the shaft.

12. Pick the **Screws** button at the top of the lowest menu docked on the left side of the AutoCAD Mechanical screen to view the Screw Connection window.

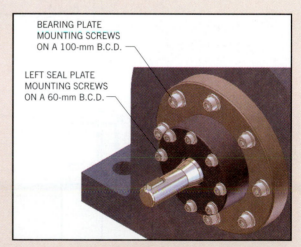

Fig. 4.2.33 *The bearing plate and left seal plate mounting screws.*

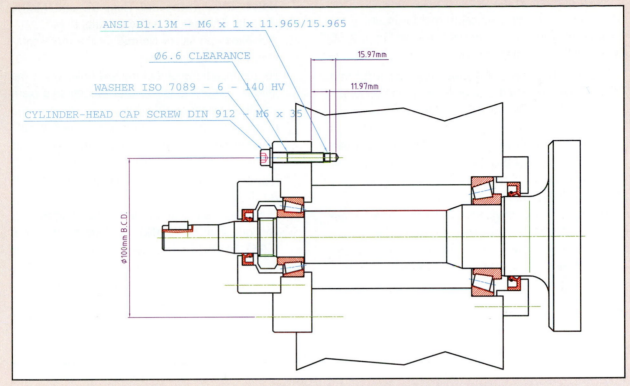

Fig. 4.2.34 Placement and specifications of the bearing plate mounting screws.

13. Pick the **<Screws>** button at the top of the window. From the resulting pictorial menu, pick **Socket Head Types**. Scroll down through the pictorial menu to find and select **DIN EN ISO 4762 (Regular Thread)**. Pick **Front View**. On the right side of the Screw Connection – Front View window, select **M6** from the list of sizes shown.

14. Pick the first **<Washers>** button on the Screw Connection window. From the following pictorial menus, pick **Plain** and **ISO 7089**. Notice that the size defaults to M6 to match the screw size.

15. Pick the first **<Holes>** button on the Screw Connection window. From the pictorial menus, pick **Through Cylindrical** and **ISO 273 Normal**. Again, the size defaults to 6 mm.

16. Pick the second **<Holes>** button on the Screw Connection window. From the pictorial menus, pick **Tapped Holes**, **Blind (Standard Runout)**, and **Metric (Regular Thread)**. The size defaults to 6 mm.

17. Pick the **Next** button at the bottom of the Screw Connection window. Pick **Normal** for the representation. Pick **Next** again.

18. For the insertion point of the first hole, select the intersection of the left face of the bearing plate and the line representing the upper limit of the 100-mm B.C.D. For the endpoint of the first hole (gap between the holes) specify the intersection of the right face of the bearing plate and the line representing the upper limit of the 100-mm B.C.D. Pick **Finish**.

19. Drag the screw size to a length of **M6 x 35**. Drag the tap and drill lengths to **ANSI B1.13M-M6 x 1 x 11.965/15.965** deep.

Now proceed to the left seal plate fasteners. The procedure is almost identical to that for inserting the bearing plate mounting screws. Refer to Fig. 4.2.35 as you follow the steps below.

20. Create the upper and lower lines for a 60-mm B.C.D. for the left seal plate mounting screws.

21. Pick the **Screws** button at the left side of the drawing area to begin another screw connection. Pick the **<Screws>** button. Select **Socket Head Types**, **DIN EN ISO 4762 (Regular Thread)**, and **Front View**. In the Screw Connection window, select a size of **M4**.

22. Pick the first **<Washers>** button. Select **Plain** and **ISO 7089**. The size defaults to M4.

23. Pick the first **<Holes>** button. Select **Through Cylindrical** and **ISO 273 Normal**. The size defaults to M4.

24. Pick the second **<Holes>** button. Select **Tapped Holes**, **Blind (Standard Runout)**, and **Metric (Regular Thread)**. The size defaults to M4.

25. Pick **Next** and select **Normal** for the representation. Pick **Next** again.

26. For the insertion point of the first hole, select the intersection of the left face of the left seal plate and the line representing the upper limit of the 60-mm B.C.D. For the endpoint of the first hole (gap between holes), select the intersection of the right face of the left seal plate and the line representing the upper limit of the 60-mm B.C.D. Pick **Finish**.

27. Drag the screw length to **M4 x 35**. Drag the tap and drill lengths to **ANSI B1.13M-M4 x 0.7 x 9.051/11.851** deep. When you finish, your drawing should look like the one in Fig. 4.2.35, without the dimensions and specifications.

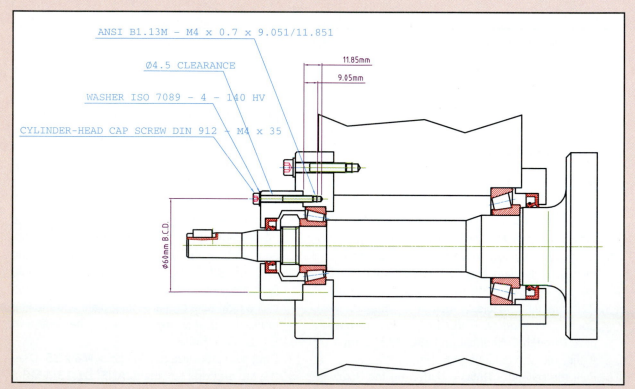

Fig. 4.2.35 Placement and specifications of the left seal plate mounting screws.

Finally, create the mounting screws for the right seal plate. Consult the dimensions, specifications, and placement shown in Fig. 4.2.36 as you follow the steps below.

28. Create the upper and lower lines for an 88-mm B.C.D. for the right seal plate mounting screws.

29. Pick the **Screws** button at the left side of the drawing area to begin another screw connection. Pick the **<Screws>** button. Select **Socket Head Types**, **DIN EN ISO 4762 (Regular Thread)**, and **Front View**. In the Screw Connection window, select a size of **M4**.

30. Pick the first **<Washers>** button. Select **Plain** and **ISO 7089**. The size defaults to M4.

31. Pick the first **<Holes>** button. Select **Through Cylindrical** and **ISO 273 Normal**. The size defaults to M4.

32. Pick the second **<Holes>** button. Select **Tapped Holes**, **Blind (Standard Runout)**, and **Metric (Regular Thread)**. The size defaults to M4.

33. Pick **Next** and select **Normal** for the representation. Pick **Next** again.

34. For the insertion point of the first hole, select the intersection of the right face of the right seal plate and the line representing the upper limit of the 88-mm B.C.D. For the endpoint of the first hole (gap between holes), select the intersection of the left face of the right seal plate and the line representing the upper limit of the 88-mm B.C.D. Pick **Finish**.

35. Drag the screw length to **M4 x 25**. Drag the tap and drill lengths to **ANSI B1.13M-M4 x 0.7 x 11.023/13.823** deep. When you finish, your drawing should look like the one in Fig. 4.2.36, without the dimensions and specifications.

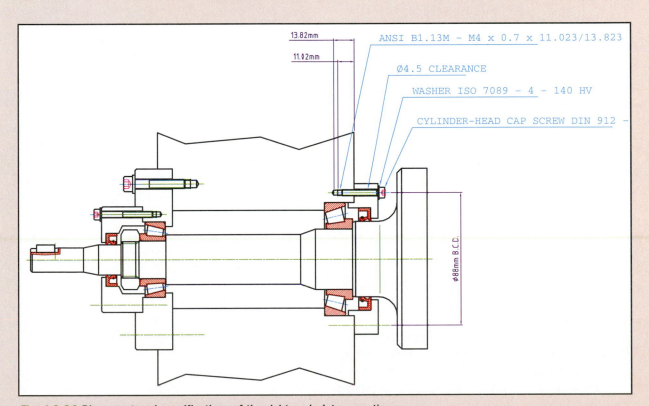

13.82mm
11.02mm
ANSI B1.13M – M4 x 0.7 x 11.023/13.823
Ø4.5 CLEARANCE
WASHER ISO 7089 – 4 – 140 HV
CYLINDER-HEAD CAP SCREW DIN 912 –
Ø88mm B.C.D.

Fig. 4.2.36 Placement and specifications of the right seal plate mounting screws.

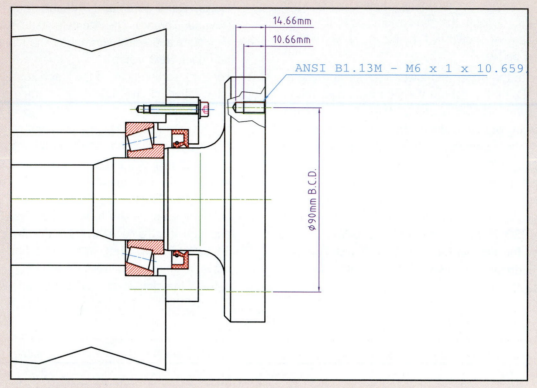

Fig. 4.2.37 *Dimensions and specifications for the tapped holes in the lower end of the shaft.*

The 120-mm segment at the bottom (right) of the shaft must be prepared to receive the mounting screws of the mixing blade. This process is similar to that of adding the actual fasteners, as you did in previous steps. Like the fasteners, the holes can be created using power objects.

For the concept drawing, you will break a small piece from the end of the shaft to show the internal hole clearly in cross section. Refer to Fig. 4.2.37 as you complete the following steps.

36. Create the upper and lower centerlines to describe the diameter of a **90-mm** B.C.D. on the right face of the shaft.

37. From the Content menu at the lower left edge of the drawing area, hold down the **Holes** button, (the next button after the Screws button) to display a flyout menu. From the flyout, pick **Tapped Blind Holes**.

38. Pick **Metric (Regular Thread)** and **Front View**. For the insertion point, specify the intersection of the right face of the shaft and the line representing the 90-mm B.C.D. Specify a rotation angle of **0** and select a size of **M6**. Drag the tap and drill to a length of **ANSI B1.13M-M6 x 1 x 10.659/ 14.659** deep.

39. To finish the geometry, create the hatching to distinguish the individual components in the cross-section, as shown in Fig. 4.2.38. Recall that, by drafting convention, fasteners shown in section are not hatched.

40. Add the lines to indicate the pulley and mixer blade and label them as shown in Fig. 4.2.38.

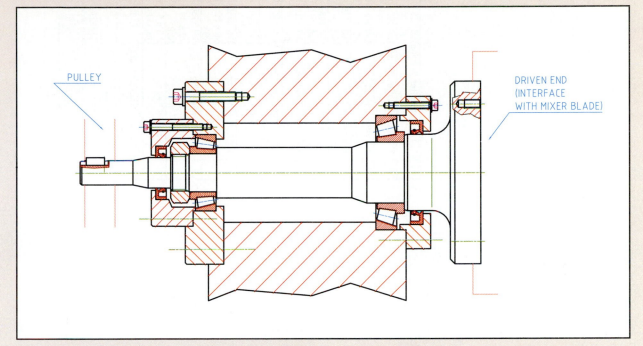

PULLEY

DRIVEN END
(INTERFACE
WITH MIXER BLADE)

Fig. 4.2.38 *Add hatches and labels to finish the geometry.*

Finish the Drawing

To finish the concept drawing, you need only place it on a layout, add the border and title block, and create the parts list. These tasks will be slightly different for this project than in previous projects.

Drawing Sheet

Because this is a metric drawing, you will place it on an ISO size A2 sheet and use an ISO-standard title block. This is approximately the size of an ANSI size C drawing sheet. Recall that the drawing sheet size was not specified by the ISO template that you used to create the drawing, so you will need to specify it before you can begin your layout.

Parts List

Parts lists are much easier to create in AutoCAD Mechanical than in AutoCAD because AutoCAD Mechanical has an automatic parts list generator. You can create the parts list easily without using an external spreadsheet.

In Practice Exercise 4.2G, you will add a layout with an appropriate border and title block. You will then complete the concept drawing by adding an automatically generated parts list. The finished concept drawing is shown in Fig. 4.2.39. Refer to this figure as necessary as you complete the drawing exercise.

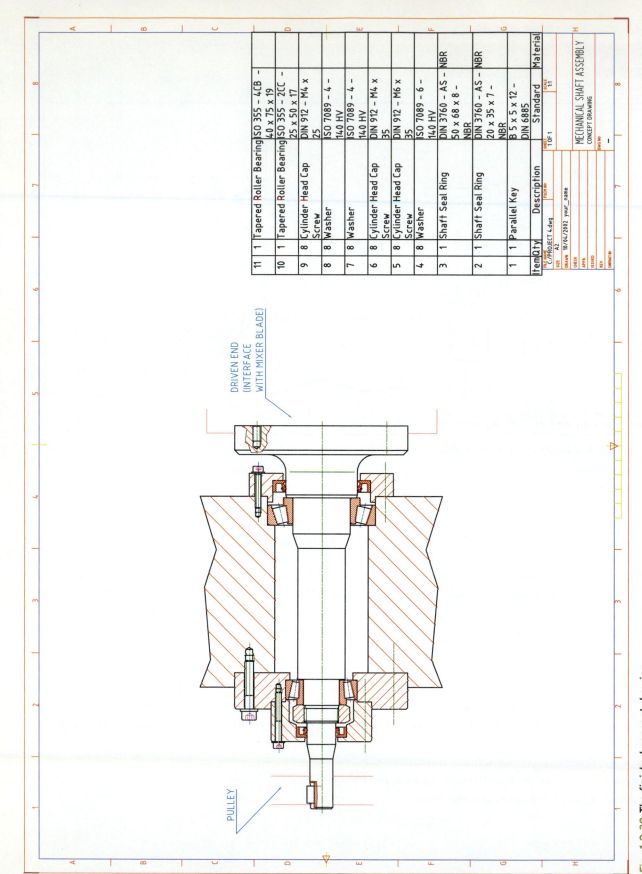

Item	Qty	Description		FSCM NO		Material
11	1	Tapered Roller Bearing	ISO 355 – 4CB – 40 x 75 x 19			
10	1	Tapered Roller Bearing	ISO 355 – 2CC – 25 x 50 x 17			
9	8	Cylinder Head Cap Screw	DIN 912 – M4 x 25			
8	8	Washer	ISO 7089 – 4 – 140 HV			
7	8	Washer	ISO 7089 – 4 – 140 HV			
6	8	Cylinder Head Cap Screw	DIN 912 – M4 x 35			
5	8	Cylinder Head Cap Screw	DIN 912 – M6 x 35			
4	8	Washer	ISO 7089 – 6 – 140 HV			
3	1	Shaft Seal Ring	DIN 3760 – AS – 50 x 68 x 8 – NBR			
2	1	Shaft Seal Ring	DIN 3760 – AS – 20 x 35 x 7 – NBR			
1	1	Parallel Key	B 5 x 5 x 12 – DIN 6885			
			Standard			

FILE NAME C:/PROJECT 4.dwg
SIZE A2
DRAWN 10/04/2002 your_name
CHECK
APPR.
ISSUED
REV
CONTRACT NO

SCALE 1:1

MECHANICAL SHAFT ASSEMBLY
CONCEPT DRAWING

SHEET 1 OF 1
DWG NO —

DRIVEN END
(INTERFACE WITH MIXER BLADE)

PULLEY

Fig. 4.2.39 The finished concept drawing.

Practice Exercise 4.2G

The template you used to create the concept drawing did not include a border and title block. Notice at the bottom of the drawing area that the layout tabs are labeled Layout1 and Layout2. The first thing you must do is specify the size of the drawing sheet. Then you can add an ISO-standard border and title block.

1. Click the **Layout1** tab. The Page Setup - Layout1 window appears, as shown in Fig. 4.2.40. At the top of the window, change the layout name to **Concept Drawing - Mechanical Shaft**. *Do not press Enter yet.* Click in the box next to Paper size and scroll down to select a paper size of **ISO A2 (420.00 x 594.00 MM)**. Pick **OK**.

The layout is now sized correctly for an A2 sheet. Notice that the tab at the bottom of the drawing area has changed to the name you selected for this layout. Now you can add an ISO-standard border and title block.

2. From the **Annotate** menu at the top of the screen, select **Drawing Title/Revisions** and then **Drawing Title/Borders...** to display the Drawing Borders with Title Block window. Select the paper format **A2 (420.00 x 594.00mm)**. (This should be the default.) For Title Block, choose ISO Title Block A. Leave the scale at 1:1 and pick **OK**. Pick **OK** again in the Page Setup window without changing the defaults. Pick a point at the lower left corner of the drawing sheet to insert the border into its proper position.

3. The Change Title Block Entry window appears automatically. Your name (or the default name on the computer you are using), the date, and the file name and path appear automatically also. If your name does not appear in the Drawn by box, doubleclick the text and change it to your name. For the drawing title, enter **MECHANICAL SHAFT ASSEMBLY**. For the drawing subtitle, enter **CONCEPT DRAWING**. Pick **OK**. The border and title block appear as shown in Fig. 4.2.40.

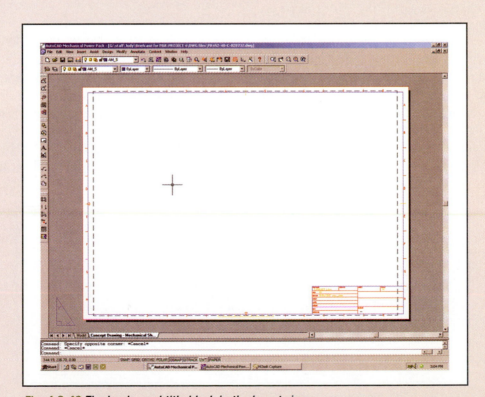

Fig. 4.2.40 The border and title block in the layout view.

Notice that the text in the title block is yellow. This is because the title block was designed primarily for use in model space. For this project, in which the drawing will be displayed and printed in paper space, you should change the color of the text.

4. The title block exists as a block with attributes. Unlike AutoCAD, the AutoCAD Mechanical software can easily change the color of the text without exploding the block. At the keyboard, enter the following text exactly as it appears:

_.EATTEDIT

This displays the Enhanced Attribute Editor. *Note:* If you simply doubleclick the block, AutoCAD's ordinary Attribute Editor appears, allowing you to change only the values of the attributes. To change other characteristics, such as color, you must use the Enhanced Attribute Editor.

5. The Enhanced Attribute Editor has three tabs. The first one, Attribute, contains a scrolling list of the attributes defined in the block. Since you must change the color for each attribute individually, you can change only those for which you have specified values. The first of these is **File Name**. Click to select this attribute. Then select the third tab, **Properties**, to display the properties for that attribute. Pick the box next to **Color** and change it to **Black**. Do not pick OK yet.

6. Repeat step 5 for each of the attributes to which you have assigned a value, switching back and forth between the Attribute and Properties tabs as necessary. When you are finished, pick **OK** to close the window. The text should appear as shown in Fig. 4.2.41.

7. Make **AM_VIEWS** the current layer. Add the viewport. From the **View** menu, select **Viewports** and **1 Viewport**. Use the mouse to specify a viewport to fill the drawing area.

8. Pick and hold the **Power Edit** button to display the flyout menu and pick **Object Properties** from the flyout. Select the viewport you just created. Click the box next to **Standard scale** and change the scale to **1:1**. Remove the Properties window from the screen.

9. Pick the **PAPER** button at the bottom of the drawing area to switch to model space. Use the **PAN** command to move the concept drawing toward the left side of the drawing sheet to make room for the parts list. Pick the **MODEL** button to return to paper space.

Now you can begin working with the parts list. AutoCAD Mechanical keeps track of the parts you create as power objects and includes them in its BOM (bill of materials) database. You should view and edit this database before you ask the software to create the parts list.

10. Use tooltips to find the **BOM Database** button docked to the left side of the drawing area and pick it. (By default, it is located just above the Screws button.) If you are prompted for the BOM table to use, press **Enter** to accept the Main default. When the BOM window appears, you

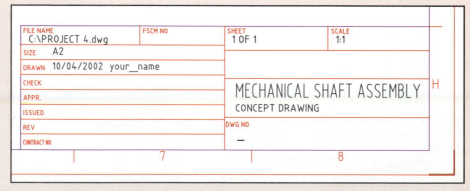

Fig. 4.2.41 Title block after editing.

should lengthen the window so that you can see all of the columns through "Material." To do this, position the cursor over the right edge of the window. When a double arrow appears, drag to the right to lengthen the window. See Fig. 4.2.42.

Notice that all of the parts you entered as power objects appear in the list of materials. The software keeps track of how many times you insert each power object and enters this number in the BOM database automatically. Notice that the quantity of each part, including the fasteners, is "1." That is because, in the concept drawing, you created only one of each set of fasteners. In a finished working drawing, all of the fasteners would be shown in one view or another, so the count would be more accurate. Fortunately, you can edit the quantity to correct the numbers for this concept drawing.

11. Look at each entry in the BOM database carefully to see if its quantity is correct. For example, the design calls for only one key, so item #3 is correct.

The first item that is not correct is #6. If you look back at your drawing (or at the steps you followed to create the drawing), you will see that the M6 washers are used with the bearing plate mounting screws. The bearing plate requires 8 screws on a 100-mm B.C.D., so the quantity needs to be changed to 8.

12. In the BOM window, *rightclick* the **1** in the Qty column for item #6. In the shortcut menu that appears, click **Do not calculate number of items**. Then doubleclick the **1** and change it to **8**.

13. Item #5 is the cap screws associated with the washers in item #6, so there should be 8 of them also. Repeat the procedure outlined in step 12 to change the quantity to **8**.

Item #8 and item #11 may appear to be exactly the same at first glance. However, if you expand the column (position the cursor over the right end of the top of the Name column and drag it to the right), you will see that the screws are actually two different lengths. This justifies having individual entries for them. Items #9 and #10, which are the washers for the screws in #8 and #11, are identical.

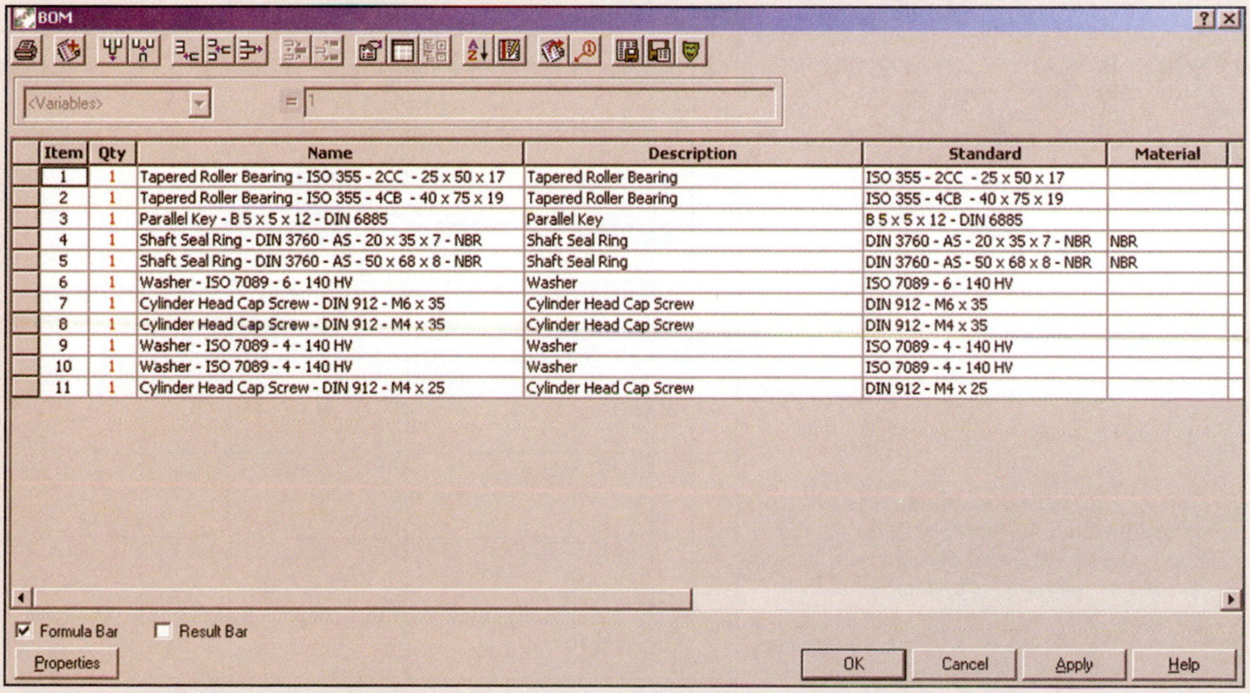

Item	Qty	Name	Description	Standard	Material
1	1	Tapered Roller Bearing - ISO 355 - 2CC - 25 x 50 x 17	Tapered Roller Bearing	ISO 355 - 2CC - 25 x 50 x 17	
2	1	Tapered Roller Bearing - ISO 355 - 4CB - 40 x 75 x 19	Tapered Roller Bearing	ISO 355 - 4CB - 40 x 75 x 19	
3	1	Parallel Key - B 5 x 5 x 12 - DIN 6885	Parallel Key	B 5 x 5 x 12 - DIN 6885	
4	1	Shaft Seal Ring - DIN 3760 - AS - 20 x 35 x 7 - NBR	Shaft Seal Ring	DIN 3760 - AS - 20 x 35 x 7 - NBR	NBR
5	1	Shaft Seal Ring - DIN 3760 - AS - 50 x 68 x 8 - NBR	Shaft Seal Ring	DIN 3760 - AS - 50 x 68 x 8 - NBR	NBR
6	1	Washer - ISO 7089 - 6 - 140 HV	Washer	ISO 7089 - 6 - 140 HV	
7	1	Cylinder Head Cap Screw - DIN 912 - M6 x 35	Cylinder Head Cap Screw	DIN 912 - M6 x 35	
8	1	Cylinder Head Cap Screw - DIN 912 - M4 x 35	Cylinder Head Cap Screw	DIN 912 - M4 x 35	
9	1	Washer - ISO 7089 - 4 - 140 HV	Washer	ISO 7089 - 4 - 140 HV	
10	1	Washer - ISO 7089 - 4 - 140 HV	Washer	ISO 7089 - 4 - 140 HV	
11	1	Cylinder Head Cap Screw - DIN 912 - M4 x 25	Cylinder Head Cap Screw	DIN 912 - M4 x 25	

Fig. 4.2.42 Lengthened view of the BOM window.

14. Item #8 refers to the mounting screws for the left seal plate, and the concept requires 8 of them. Change the quantity for this item to **8**.

Even though items #9 and #10 are identical, you should leave them as two separate entries because of AutoCAD Mechanical's automatic balloon/part number system. For a set of working drawings, you could mess up the part identification process if you delete a part number. We will not be adding part identification numbers and balloons to this concept drawing, but we should follow best practices nevertheless.

15. Eight washers are needed for the left seal plate (item #9) and 8 more are needed for the right seal plate (item #10). Change the quantity for both of these items to **8**.

16. Item #11 refers to the mounting screws for the right seal plate, and 8 screws are required. Change the quantity to **8**.

17. Items #1 and #2 are correct, because you only need one of each type of bearing. Pick **OK** to close the BOM database.

18. Make **AM_1** the current layer.

19. To insert the parts list into the drawing, pick and hold the **BOM Database** button to display a fly-out menu. From the flyout, pick **Parts List**. A window appears showing the items in the BOM database in the default ISO parts list format. Pick **OK**. The parts list appears and is attached to the cursor. Place the parts list just above the title block, as in previous projects.

20. To change the yellow grid lines to black, select the parts list, which exists as a block, right-click, and pick **Object Properties**. Select the box next to **Color** and change the color to **White**. The lines will now show up on the white paper as black. The parts list should now look like the one in Fig. 4.2.43. Note that the actual part numbers on the parts list may be different from those in the BOM database, as shown in Fig. 4.2.43.

21. Freeze the **AM_VIEWS** layer.

22. Pick **PAPER** to switch to model space. Use **PAN** to position the concept drawing so that it does not overlap the parts list. The completed drawing should look like the one in Fig. 4.2.39.

11	1	Tapered Roller Bearing	ISO 355 – 4CB – 40 x 75 x 19	
10	1	Tapered Roller Bearing	ISO 355 – 2CC – 25 x 50 x 17	
9	8	Cylinder Head Cap Screw	DIN 912 – M4 x 25	
8	8	Washer	ISO 7089 – 4 – 140 HV	
7	8	Washer	ISO 7089 – 4 – 140 HV	
6	8	Cylinder Head Cap Screw	DIN 912 – M4 x 35	
5	8	Cylinder Head Cap Screw	DIN 912 – M6 x 35	
4	8	Washer	ISO 7089 – 6 – 140 HV	
3	1	Shaft Seal Ring	DIN 3760 – AS – 50 x 68 x 8 – NBR	NBR
2	1	Shaft Seal Ring	DIN 3760 – AS – 20 x 35 x 7 – NBR	NBR
1	1	Parallel Key	B 5 x 5 x 12 – DIN 6885	
Item	Qty	Description	Standard	Material

Fig. 4.2.43 The ISO standard parts list generated by AutoCAD Mechanical for the concept drawing. Note that the item numbers in this list do not match the item numbers in the BOM database. The items are in the same order, but the BOM database numbers the items from the top down, and the parts list numbers them from the bottom up.

Review Questions

1. How can drafters who wish to use basic AutoCAD drawing and editing commands in AutoCAD Mechanical display the standard Draw and Edit toolbars?
2. How is a power object different from an object created using the standard AutoCAD commands?
3. Why is it necessary to use power commands to edit power objects?
4. Explain layer setup in AutoCAD Mechanical. Why should you allow the software to use its default layers?
5. How many sets of object snaps can you predefine in AutoCAD Mechanical? What are they called?
6. What command in AutoCAD Mechanical allows you to display and change the defaults for layers, drawing sheets, dimensions, among other settings?
7. How do you add segments to an existing shaft using the power commands?
8. What command is used to modify shaft segments to include fillets or chamfers?
9. Explain the purpose of the key and keyway.
10. Name at least two things that should be taken into consideration when specifying seals for an assembly.
11. How do you enter title block information in AutoCAD Mechanical?
12. What is the difference between the standard ATTEDIT command and the _.EATTEDIT command?
13. Why is it sometimes necessary to edit the items in the automatically generated parts list?
14. Explain how to change the quantity of an item in the BOM database for a drawing.

Practice Problems

1. Create a new drawing in AutoCAD Mechanical. Name the drawing Section 4-2 problem 1.dwg. Use the power commands to create a 1.00-8UNC hex-head bolt with a plain washer (type A), a through hole with a free fit, and an ANSI B18.22.1-1 UNC wide type A hex nut.

Portfolio Project

Lever Arm Shaft (continued)

Begin working on the concept drawing for the lever arm shaft you have designed.

1. Use AutoCAD Mechanical power objects as much as possible in the drawing.
2. Include at least one of each type of fastener to show their location and purpose in the drawing. Use power objects for the fasteners.
3. Create a layout for the drawing and place a border and title block on it using the AutoCAD Mechanical functions.
4. Edit the drawing's BOM database for accuracy and place the parts list in the drawing.
5. Save your work.

2. Create a new drawing in AutoCAD Mechanical. Name the drawing Section 4-2 problem 2.dwg. Use power commands to create a shaft that incorporates the following segments (left to right):
   - Ø40 cylindrical segment, 50 mm long
   - Sloped segment, 30 mm long, with a starting diameter of 40 and an ending diameter of 25
   - Ø25 cylindrical segment, 110 mm long
   - Sloped segment, 15 mm long, with a starting diameter of 25 and an ending diameter of 62
   - Ø62 cylindrical segment, 76 mm long
   - Ø105 cylindrical segment, 22 mm long

3. In a drawing named Section 4-2 problem 3.dwg, use the power commands to create a shaft with the dimensions and segments shown in Fig. 4.2.44.

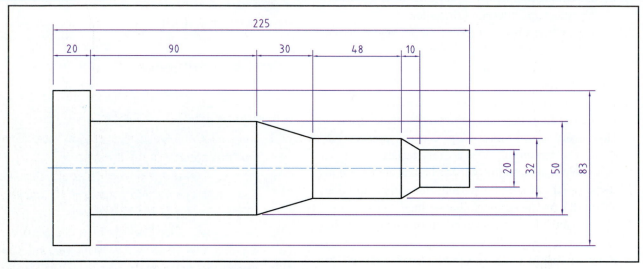

Fig. 4.2.44

4. After saving your work from Practice Problem 3, use the SAVEAS command to save the drawing with a new name of Section 4-2 problem 4.dwg. In this drawing, use the power commands to add the fillets and chamfer shown in Fig. 4.2.45.

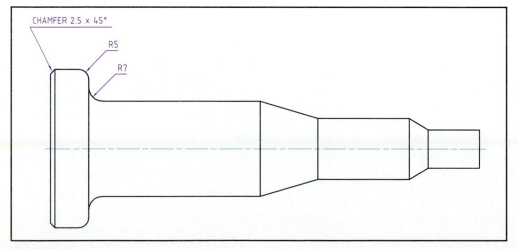

Fig. 4.2.45

AutoCAD Procedure

AutoCAD can perform all of the same tasks that AutoCAD Mechanical performs. Because AutoCAD is intended for a broader variety of uses, however, it lacks some of the "bells and whistles" that are available in AutoCAD Mechanical for creating mechanical parts. Even if you do not have access to AutoCAD Mechanical, you may want to read through Section 4.2 to give yourself an idea of the differences between the two programs.

This section is designed for students who have access to the AutoCAD software but do not have access to the AutoCAD Mechanical software. It does not teach any new AutoCAD commands or procedures. Instead, it provides the necessary steps and dimensions for building the mechanical shaft assembly using only the AutoCAD software. If you have both AutoCAD and AutoCAD Mechanical, you may want to perform the steps in both sections and compare both the procedures and the results.

Section Objectives
- Create a concept drawing using the AutoCAD software.
- Edit a concept drawing using the AutoCAD software.

Key Terms
- DIN
- NBR

Drawing Setup

Unlike the drawings in previous projects, the concept drawing of a mechanical shaft will be created in metric units. All of the parts will conform to ISO standards and/or **DIN** ("Deutsche Industrie Norme," or German Industry Standard). You will therefore need to use a different drawing template. AutoCAD supplies standard ISO templates as well as ANSI templates.

Other than the use of metric units, drawing setup will be much the same as it was for the previous three projects. You will use the same set of layers. Certain aspects of the drawing may look different, however, because using an ISO template automatically changes other settings to the ISO defaults. For example, dimensions will automatically appear in the preferred ISO style, as shown in Fig. 4.3.1.

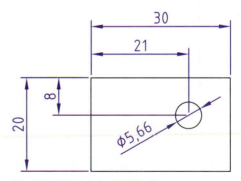

*Fig. 4.3.1 **Example of ISO dimensions.***

In Practice Exercise 4.3A, you will create and set up the drawing file for the mechanical shaft drawing. Even though this is a metric drawing, you will be able to reuse some elements from earlier projects in this textbook.

Practice Exercise

Many of the preparations you made to the ANSI drawing files in earlier projects are made automatically when you use the ISO template. For example, the drawing limits are already set to the correct size for an A2 drawing, so you don't have to set them manually. The units are assumed to be millimeters. Use the following steps to create and prepare the drawing file.

1. Open AutoCAD and create a new drawing using the **ISO A2 - Named Plot Styles.dwt** drawing template. The new file opens into paper space, showing the border and title block. Pick the **PAPER** button at the bottom of the drawing area to switch to model space. Save the drawing file as **PROJECT 4 - AutoCAD.dwg**.

2. Pick the **Layers** button to display the Layer Properties Manager. Notice that several layers have already been set up. These layers are used in paper space in connection with the border and title block. You will not use these layers for the geometry, but do not delete them. They are needed for the layout. Pick **OK** to close the Layer Properties Manager.

3. Turn on the grid (**F7**) and **ZOOM All**.

4. Pick the **DesignCenter** button to display the AutoCAD DesignCenter. Navigate to the folder where your Project 2 files are stored. Pick any of the final drawing files from Project 2. In the box on the right side of DesignCenter, doubleclick the **Layers** icon to display the layers in Project 2. Then highlight all of the custom layers and drag them into the drawing area to add the layers from Project 2 to the current drawing file. Click the **Up** button at the top of DesignCenter (the file folder icon with the up arrow) to return to the list of attributes in Project 2. Doubleclick **Linetypes** and highlight and drag the linetypes into the current drawing. Close DesignCenter.

ISO uses a different text style than ANSI. For the concept drawing in this project, we will change the text to a commonly accepted standard ISO style.

5. Enter the **STYLE** command. Pick the arrow in the **Style Name** box and change the style name to **ISO Proportional**. Notice in the Font Name box that the default changes to isocp.shx, the standard AutoCAD font for ISO drawings.

6. Because this is a metric drawing, the linetype scale needs to be changed. Enter the **LTSCALE** command and specify a new linetype of **24.3**.

Shaft Generation

Shafts are typically made of alloy steel and are heat-treated to improve strength and durability. They are usually machined on lathes and precision-ground to tight tolerances.

In AutoCAD Mechanical, the shaft is constructed segment by segment. In AutoCAD, however, the best way to construct the shaft is to create its outline using the OFFSET, LINE, EXTEND, and TRIM commands. In Practice Exercise 4.3B, you will create the basic shaft.

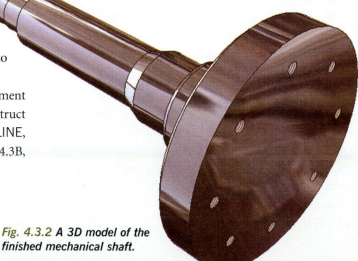

Fig. 4.3.2 A 3D model of the finished mechanical shaft.

Practice Exercise 4.3B

1. Make **Center** the current layer. Then create a horizontal centerline that extends across the middle of the screen, almost from edge to edge of the drawing area. This will be the main shaft centerline. Because the shaft is symmetrical about this centerline, you can use it as a basis to create the shaft.

2. Draw the shaft segments using the dimensions shown in Fig. 4.3.3. You can offset the upper and lower lines from the centerline, but you might want to work on one segment at a time to avoid confusion. Change the lines that represent the shaft to the **Object** layer. Remember that the basic unit is the millimeter in this drawing, so if you enter a value of 18, for example, AutoCAD interprets it as 18 mm.

3. Zoom in on the lower (right) end of the shaft. The left edge of the rightmost segment will have a 3-mm × 45° chamfer, as shown in Fig. 4.3.4. Use the **CHAMFER** command to add the chamfer. Remember that the shaft is a round object, so you will need to show the chamfer at both the top and the bottom of the shaft segment.

4. The edge of the chamfer shows as a line in the front view, so add a vertical line between the top and bottom of the chamfer, as shown in Fig. 4.3.4.

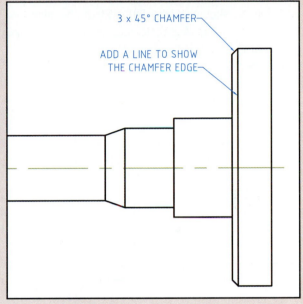

Fig. 4.3.4 *Chamfer for the rightmost shaft segment.*

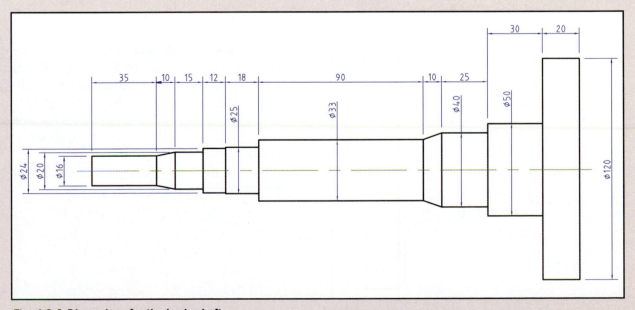

Fig. 4.3.3 *Dimensions for the basic shaft.*

5. Use the **FILLET** command to add the R12 fillet between the rightmost shaft segment and the next segment, as shown in Fig. 4.3.5. Remember to create the fillet at both the top and the bottom of the cylindrical segment. Add a vertical line as shown at the left end of the fillet, as shown in Fig. 4.3.5, to suggest the cylindrical characteristic of the fillet. Place the line on the **Dimensions** layer.

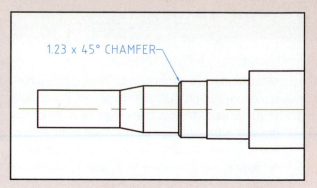

1.23 x 45° CHAMFER

Fig. 4.3.7 **Chamfer the fourth segment from the left.**

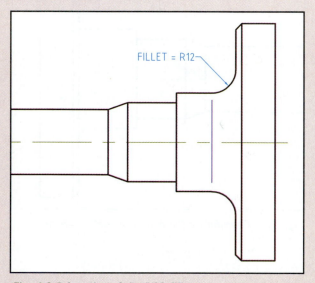

FILLET = R12

Fig. 4.3.5 **Location of the R12 fillet.**

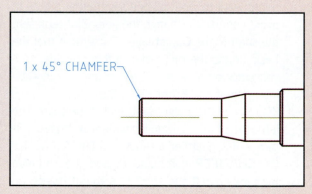

1 x 45° CHAMFER

Fig. 4.3.8 **Chamfer the leftmost segment.**

6. At the left edge of the second segment from the right, create a chamfer of **2 mm × 15°**, as shown in Fig. 4.3.6. Add the vertical line to connect the top and bottom edges of the chamfer.
7. Add a **1.23 mm × 45°** chamfer to the left face of the fourth segment from the left, as shown in Fig. 4.5.7.
8. Add a **1 mm × 45°** chamfer to the left face of the leftmost segment, as shown in Fig. 4.5.8.

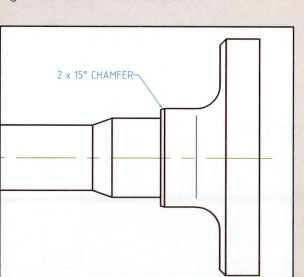

2 x 15° CHAMFER

Fig. 4.3.6 **Chamfer the left face of the second segment.**

Key and Keyway

At the top of the shaft, the key and keyway must be added at the interface with the drive pulley. The purpose of the key is to keep the drive pulley from slipping. Fig. 4.3.9 shows a closeup view of the key and keyway extending from the top of the housing.

Fig. 4.3.10 is an exploded view that shows the relationship of the pulley and key. Notice the cup-point setscrew that threads into the pulley to tighten down onto the key.

In Practice Exercise 4.2C, you will use power commands to create the key and keyway for the mixer. Refer to Figs. 4.3.9 and 4.3.10 as you complete the exercise.

Fig. 4.3.9 Closeup showing the key and keyway in position at the top of the shaft.

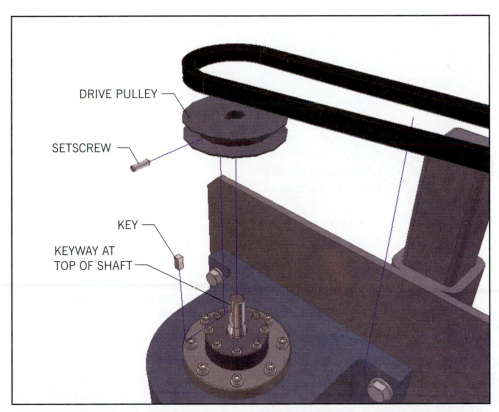

Fig. 4.3.10 Exploded view showing how the key and keyway at the top of the shaft are used in the mixer assembly.

Practice Exercise

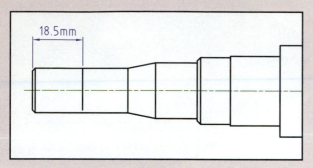

Fig. 4.3.11 *Location of the right edge of the keyway.*

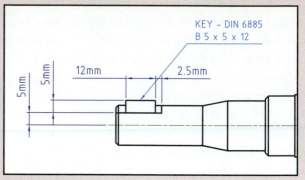

Fig. 4.3.12 *Dimensions and specifications for the key and keyway.*

The key specified for the concept drawing is a standard DIN 6885B parallel key. Follow the steps below to add the key and keyway to the shaft drawing.

1. Offset the left face of the leftmost shaft segment by **18.5** to locate the right edge of the keyway. Use grips to extend the line upward so that it touches the upper diameter line of the shaft, as shown in Fig. 4.3.11.

2. Establish the depth of the keyway by offsetting the centerline up by **5** mm. Change the offset line to the **Object** layer and trim the lines as shown in Fig. 4.3.12.

3. Add the **5 × 5 × 12 DIN 6885B** parallel key according to the location dimensions shown in Fig. 4.3.12.

4. The outer surface of the keyway needs to be lowered slightly. Change the upper diameter line as shown in Fig. 4.3.13. Be sure to change the line on both sides of the key, as shown. Trim the bevel line from the chamfer away from the keyway, as shown.

5. Finally, the keyway is being shown in section to show how the key seats in it. Create a broken-out section around the keyway. Use **BHATCH** to hatch the main surface of the shaft to show that it is being shown in section. (Keys and keyways are not hatched in section.) Use the **ANSI31** hatch pattern at a scale of **.3**. See Fig. 4.3.14.

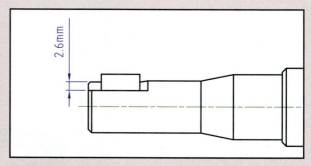

Fig. 4.3.13 *Lower the top edge of the keyway.*

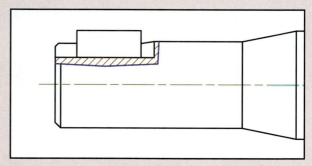

Fig. 4.3.14 *The finished representation of the keyway and key, shown in broken-out section.*

6. Trim the bevel edge of the chamfer to the lower edge of the section. When you finish, the keyway and key should look like the illustration in Fig. 4.3.14.

Bearing Geometry

The bearings for a shaft assembly are usually chosen early in the design process. The type of bearing chosen depends on many things, including bearing directional loads and lubrication methods, for example.

For this project, the designer has specified that the shaft will be supported on the mixer stand by two ISO 355 tapered roller bearings. A 3D rendering of the bearings is shown in Fig. 4.3.15. Follow the procedure in Practice Exercise 4.3D to create the tapered roller bearings.

Fig. 4.3.15 *The tapered roller bearings that will support the mechanical shaft.*

Practice Exercise 4.3D

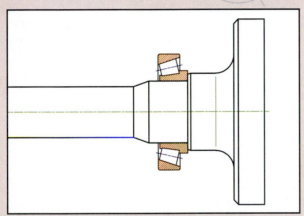

Fig. 4.3.16 *The lower bearing.*

The ISO 355 tapered roller bearings are standard parts, so they do not need to be represented in great detail. However, they do need enough accurate detail to show how they fit into the assembly. Like the shaft, they will be shown in section.

The specification for the lower bearing is ISO 355-4CB 40 × 75 × 19 (metric). It fits against the chamfer on the left face of the sec-

ond shaft segment from the right. You will need to show the bearing as it appears in cross section at both the upper and lower diameters of the shaft. See Fig. 4.3.16. Start with the upper representation of the bearing, as described in the following steps.

1. Zoom in on the lower (right) end of the shaft.
2. Use the dimensions shown in Fig. 4.3.17 to create the upper representation of the lower (right) bearing. *Hint:* For ease of maneuvering, you may wish to create the bearing representation at a clear space in the drawing area somewhere near the lower end of the shaft. When the bearing is complete, then you can move it into place.
3. Hatch the sectioned surfaces using **ANSI31** and a scale of **.3**.

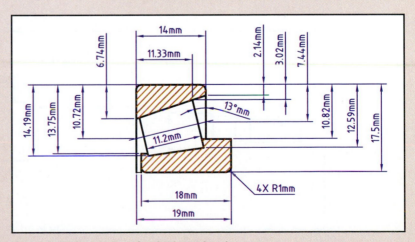

Fig. 4.3.17 *Dimensions for the lower bearing.*

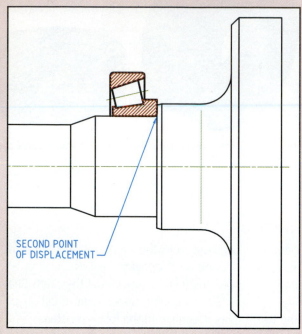

Fig. 4.3.18 *Placement of the upper bearing representation on the shaft.*

SECOND POINT OF DISPLACEMENT

4. Enter the **MOVE** command and select the entire upper bearing representation. For the base point, use object snap tracking to acquire the imaginary intersection of the lower right corner of the bearing (the intersection before the corner was filleted). Move the representation into place on the shaft, as shown in Fig. 4.3.18, by picking the intersection of the chamfer on the left face of the second segment from the right with the upper diameter line of the third segment.

5. To place the lower representation of the bearing, enter the **MIRROR** command and select all of the objects that make up the upper representation. Use the main shaft centerline as the mirror line and do not delete source objects. When you finish this step, your drawing should look like the one in Fig. 4.3.16.

6. Before you begin drawing the left bearing, offset the rightmost face of the right bearing to the left by **142.5 mm**. With Ortho on, use grips to extend this line up and down so that it intersects the top and bottom lines of the shaft. See Fig. 4.3.19. Note that this vertical line lies almost, but not quite, on top of the left face of one of the shaft segments. In Fig. 4.3.19, the line color of this temporary line has been changed to red to avoid confusion with permanent lines in the drawing. The point at which this line crosses the upper shaft will be the bearing insertion point.

The specification for the lower bearing is ISO 355-2CC 25 × 50 × 17 (metric). It fits against the fifth shaft segment from the left. You will need to show the bearing as it appears in cross section at both the upper and lower diameters of the shaft. Start with the upper representation of the bearing, as described in the following steps.

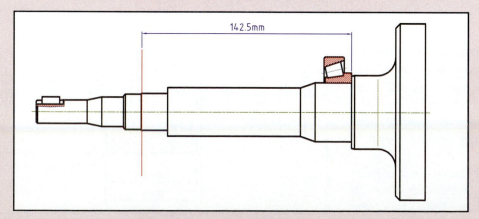

142.5mm

Fig. 4.3.19 *Temporary placement line for the left bearing.*

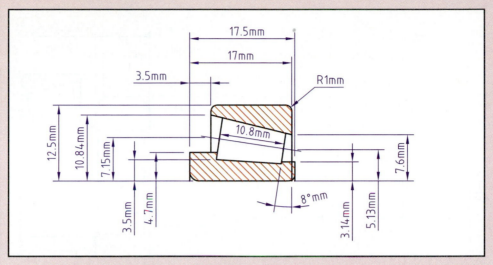

Fig. 4.3.20 Dimensions for the left bearing.

7. In a clear space on the drawing, create the upper representation of the left bearing. Use the dimensions and specifications in Fig. 4.3.20.
8. Move the bearing into place, as shown in Fig. 4.3.21.
9. Erase the temporary placement line.
10. Enter the **MIRROR** command and mirror the upper representation about the main horizontal centerline to place the lower representation of the bearing accurately. When you finish, the bearing should look like the one in Fig. 4.3.22.

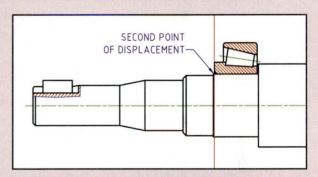

Fig. 4.3.21 Placement of the upper representation of the left bearing.

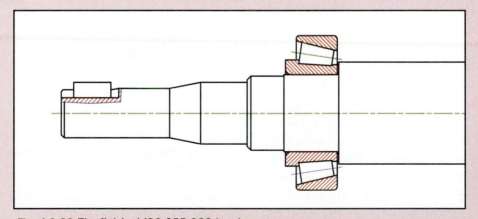

Fig. 4.3.22 The finished ISO 355-2CC bearing.

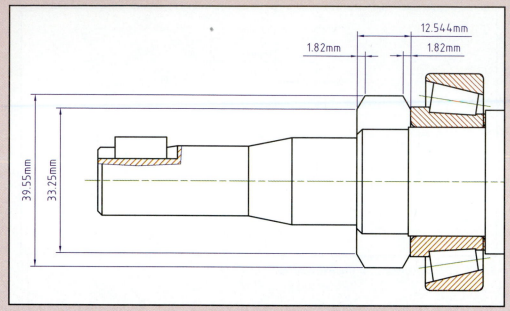

Fig. 4.3.23 *Dimensions and specifications for the lock nut.*

11. A lock nut is needed to retain the left bearing. Add the top representation of the lock nut using the dimensions shown in Fig. 4.3.23.

12. Mirror the top representation to create the bottom representation of the lock nut, as shown in Fig. 4.3.23.

Seals, Housing, and Fasteners

Several items besides the shaft itself make up the shaft subassembly for the industrial tabletop mixer. These include the right and left shaft seals, the seal retainers, the housing, and various fasteners. See Fig. 4.3.24. These need to be shown in enough detail to show how the shaft fits into the overall assembly.

The housing and seals in a shaft assembly help keep dirt and debris out of the bearings and moving parts, extending the life and reliability of the assembly. The type of seal needed for a shaft assembly depends on the maximum speed of the shaft and the kind of lubrication needed for the bearings that were chosen. For example, different types of seals are needed for bearings lubricated with oil and grease. The seals chosen for this project are radial lip seals that are compliant with DIN 3760 and are made of **NBR** (nitrile-butadine rubber).

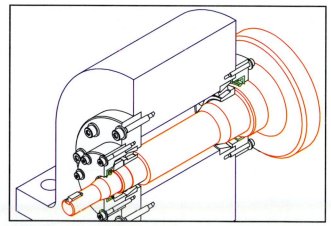

Fig. 4.3.24 *The seals (green) fit against the shaft to help keep out dirt and debris.*

Now proceed to Practice Exercise 4.2E. In this exercise, you will add the seals, seal retainers, housing, bearing plates, and fasteners to the concept drawing.

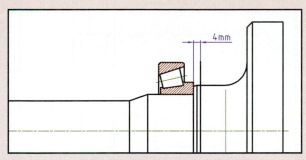

Fig. 4.3.25 *Location of the left face of the lower lip seal.*

The shaft assembly requires two lip seals: one above the upper bearing, and the other below the lower bearing. Start with the lower seal, which is a DIN 3760 50 × 68 × 8 NBR radial lip seal.

1. Zoom in on the right side of the shaft drawing.
2. Locate the left face of the seal by offsetting the left face of the second shaft segment to the right by **4 mm**. (Be sure to select the left face of the shaft, not the chamfer line.) With Ortho on, use grips to extend the line to cross the top diameter line of the segment. See Fig. 4.3.25.
3. Add the upper representation of the lip seal using the dimensions and specifications shown in Fig. 4.3.26 and using the reference line you created in the previous step for placement.
4. Mirror the upper representation about the main horizontal centerline of the shaft to place the lower representation of the seal accurately.
5. Erase or trim away any part of the reference line from step 2 that was not incorporated into the upper representation of the lip seal. The seal should resemble the one shown in Fig. 4.3.27.
6. The upper seal is a DIN 3760 20 × 35 × 7 NBR radial lip seal. Create the upper representation of the seal according to the dimensions and specifications shown in Fig. 4.3.28. Place the left face of the seal **5 mm** from the left face of the third segment from the left end of the shaft.

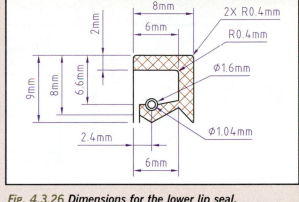

Fig. 4.3.26 *Dimensions for the lower lip seal.*

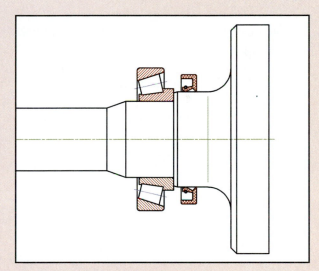

Fig. 4.3.27 *The finished lower lip seal.*

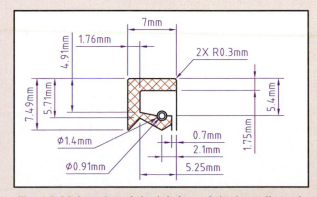

Fig. 4.3.28 *Location of the left face of the lower lip seal.*

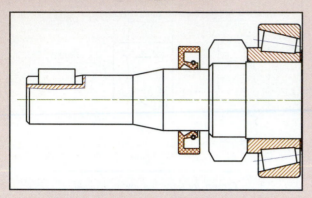

Fig. 4.3.29 The finished left (upper) lip seal.

7. Mirror the upper representation of the seal to create the lower representation. The finished upper seal should look like the one in Fig. 4.3.29.
8. Add the bearing plate for the upper bearing. Use the dimensions and specifications shown in Fig. 4.3.30.
9. Add the housing representation, as shown in Fig. 4.3.31. You need only draw enough of the housing to show how the shaft assembly fits into it.

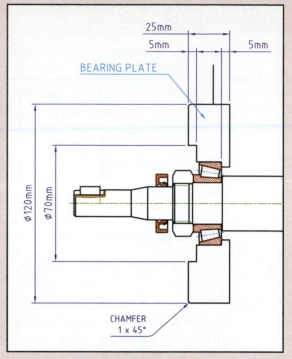

Fig. 4.3.30 Dimensions for the bearing plate.

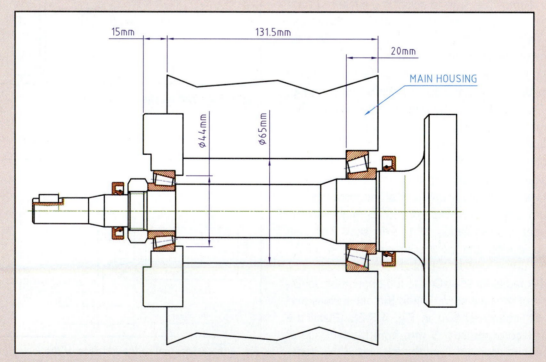

Fig. 4.3.31 Placement of the housing in relation to the bearing plate and bearings.

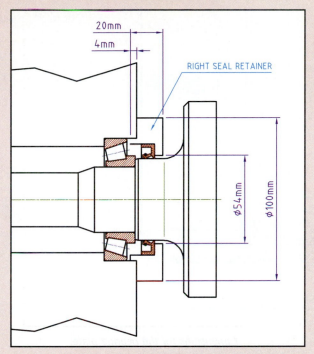

Fig. 4.3.32 *Placement of the right (lower) seal retainer.*

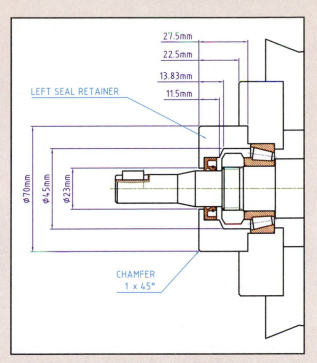

Fig. 4.3.33 *Placement of the left (upper) seal retainer.*

10. Add the right (lower) seal retainer. Use the dimensions shown in Fig. 4.3.32.

11. Add the left (upper) seal retainer. Use the dimensions shown in Fig. 4.3.33.

The next step is to add the mounting screws for the bearing plate and the left seal plate. See Fig. 4.3.34. All of the screws used in this assembly are standard metric sizes and conform to ANSI B1.13M. Because these are standard parts, you can create basic representations of them. If you need dimensions other than those given in the illustrations, you can either look them up in the appropriate ANSI table or estimate them for this concept drawing.

12. Offset the main horizontal centerline to create two centerlines to represent the upper and lower diameters of a 100-mm B.C.D. around the main horizontal centerline of the shaft.

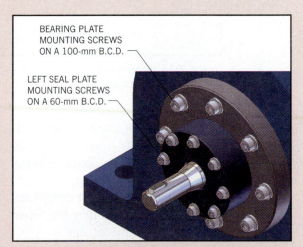

Fig. 4.3.34 *The bearing plate and left seal plate mounting screws.*

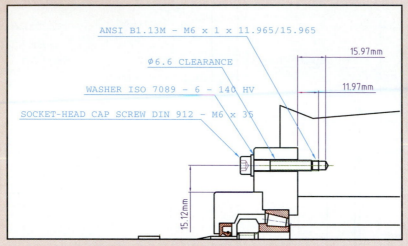

*Fig. 4.3.35 **Placement of the representative screw connection for the bearing plate mounting screws.***

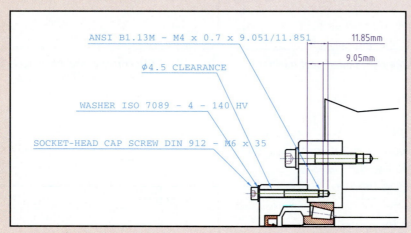

*Fig. 4.3.36 **Placement of the representative screw connection for the left seal plate mounting screws.***

13. On the top B.C.D. centerline, create the **M6 × 1 × 11.965/15.965** socket-head cap screw with an **M6** tapped hole and an accompanying washer, as shown in Fig. 4.3.35.

14. Trim the top B.C.D. centerline so that it only runs through the screw connection. Trim the lower B.C.D. centerline to match the top one.

Now proceed to the left seal plate fasteners. The procedure is identical to that for inserting the bearing plate mounting screws. Refer to Fig. 4.3.36.

15. Create the upper and lower lines for a 60-mm B.C.D. for the left seal plate mounting screws.

16. On the top B.C.D. centerline, create the **M4 × 0.7 × 9.051/11.851** socket-head cap screw with an **M4** tapped hole and an accompanying washer, as shown in Fig. 4.3.36.

17. Trim the top B.C.D. centerline so that it only runs through the screw connection. Trim the lower B.C.D. centerline to match the top one.

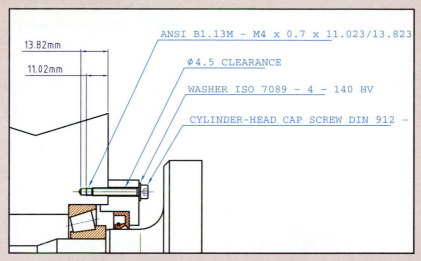

13.82mm

11.02mm

ANSI B1.13M – M4 x 0.7 x 11.023/13.823

Ø4.5 CLEARANCE

WASHER ISO 7089 – 4 – 140 HV

CYLINDER-HEAD CAP SCREW DIN 912 –

Fig. 4.3.37 Placement of the representative screw connection for the right seal plate mounting screws.

Create the representative mounting screw for the right seal plate. Consult the dimensions, specifications, and placement shown in Fig. 4.3.37 as you follow the steps below.

18. Create the upper and lower lines for a 88-mm B.C.D. for the right seal plate mounting screws.

19. On the top B.C.D. centerline, create the **M4 × 0.7 × 11.023/13.823** socket-head cap screw with an **M4** tapped hole and an accompanying washer, as shown in Fig. 4.3.37.

20. Trim the top B.C.D. centerline so that it only runs through the screw connection. Trim the lower B.C.D. centerline to match the top one.

The 120-mm segment at the bottom (right) of the shaft must be prepared to receive the mounting screws of the mixing blade. For the concept drawing, you will break a small piece from the end of the shaft to show a representative internal hole clearly. Refer to Fig. 4.3.38 as you complete the following steps.

21. Create the upper and lower centerlines to describe the diameter of a **90-mm** B.C.D. on the right face of the shaft.

22. On the top B.C.D. centerline, create the tapped, blind hole, creating the tap and drill lengths to ANSI B1.13M-**M6 × 1 × 10.659/14.659** deep, as shown in Fig. 4.3.38.

23. Trim the top B.C.D. centerline so that it only runs through the screw connection. Trim the lower B.C.D. centerline to match the top one.

24. Use the **PLINE** or **LINE** command to create a broken-out section around the internal tapped hole, as shown in Fig. 4.3.38. Trim the bevel line of the chamfer away from the section.

25. Add the lines to suggest the pulley on the left end and the mixer blade on the other, as shown in Fig. 4.3.39.

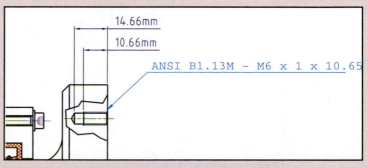

14.66mm

10.66mm

ANSI B1.13M – M6 x 1 x 10.65

Fig. 4.3.38 Placement of the representative tapped hole.

26. Add the labels for the pulley and driven end, as shown in Fig. 4.3.39.

27. To finish the concept drawing geometry, add the hatching as shown in Fig. 4.3.39. Use the **ANSI31** hatch pattern. Be sure to hatch adjoining parts with lines at different angles.

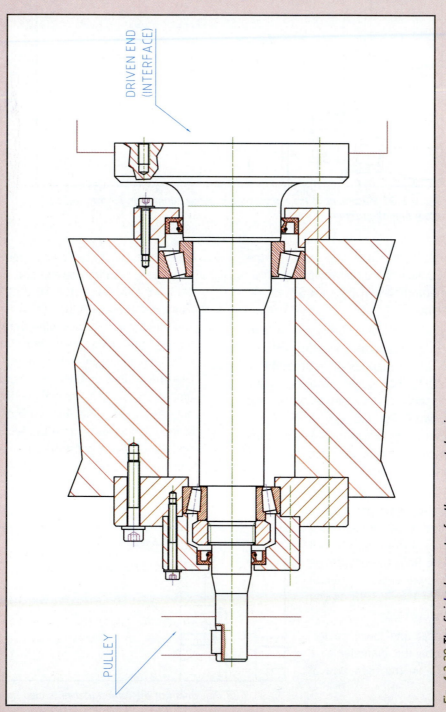

DRIVEN END (INTERFACE)

PULLEY

Fig. 4.3.39 The finished geometry for the concept drawing.

Finish the Drawing

To finish the concept drawing, you need only place it on a layout, add the title block, and create the parts list. The ISO size A2 drawing sheet was created automatically when you used AutoCAD's A2 template to create the drawing. An ISO title block was also inserted at that time. In fact, even an appropriate viewport exists on the ISO A2 Title Block layout in this drawing, so you have little work left except for the parts list generation.

In Practice Exercise 4.3F, you will place the concept drawing on the ISO A2 Title Block layout. You will then complete the concept drawing by adding the parts list.

Practice Exercise 4.3F

1. Pick the **ISO A2 Title Block** tab at the bottom of the drawing area to see the A2 layout. Notice that the drawing appears automatically.
2. Rightclick the tab name, choose **Rename**, and rename it **CONCEPT DRAWING**.
3. Pick the **PAPER** button at the bottom of the screen to enter model space. Use the **PAN** command to center the drawing between the left and right edges of the layout. Place it high enough on the layout to allow space for the parts list.

4. The parts list for the concept drawing is shown in Fig. 4.3.40. Use any of the methods described in earlier projects to create the parts list.
5. Place the parts list just above the title block on the layout.
6. Zoom in on the title block and enter the correct information for this drawing. The finished drawing is shown in Fig. 4.3.41.

11	1	TAPERED ROLLER BEARING	ISO 355 - 4CB - 40 x 75 x 19	
10	1	TAPERED ROLLER BEARING	ISO 355 - 2CC - 25 x 50 x 17	
9	8	CYLINDER-HEAD CAP SCREW	DIN 912 - M4 x 25	
8	8	WASHER	ISO 7089 - 4 - 140 HV	
7	8	WASHER	ISO 7089 - 4 - 140 HV	
6	8	CYLINDER-HEAD CAP SCREW	DIN 912 - M4 x 35	
5	8	CYLINDER-HEAD CAP SCREW	DIN 912 - M6 x 35	
4	8	WASHER	ISO 7089 - 6 - 140 HV	
3	1	SHAFT SEAL RING	DIN 3760 - AS - 50 x 68 x 8 - NBR	NBR
2	1	SHAFT SEAL RING	DIN 3760 - AS - 20 x 35 x 7 - NBR	NBR
1	1	PARALLEL KEY	8.5 x 5 x 12 - DIN 6885	
ITEM	QTY.	DESCRIPTION	STANDARD	MATL.

Fig. 4.3.40 *Parts list for the concept drawing.*

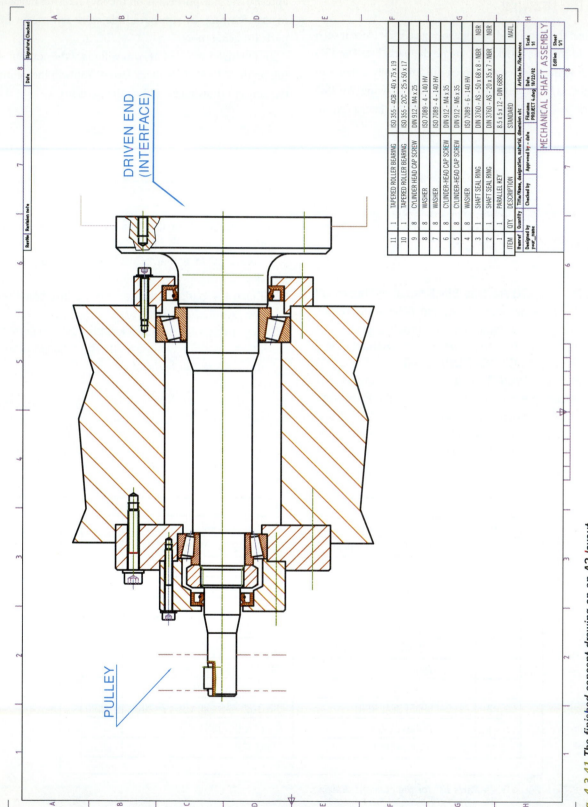

DRIVEN END
(INTERFACE)

PULLEY

ITEM	QTY.	DESCRIPTION	Article No./Reference
11	1	TAPERED ROLLER BEARING	ISO 355 - 4CB - 40 x 75 x 19
10	1	TAPERED ROLLER BEARING	ISO 355 - 2CC - 25 x 50 x 17
9	8	CYLINDER HEAD CAP SCREW	DIN 912 - M4 x 25
8	8	WASHER	ISO 7089 - 4 - 140 HV
7	8	WASHER	ISO 7089 - 4 - 140 HV
6	8	CYLINDER-HEAD CAP SCREW	DIN 912 - M4 x 35
5	8	CYLINDER-HEAD CAP SCREW	DIN 912 - M6 x 35
4	8	WASHER	ISO 7089 - 6 - 140 HV
3	1	SHAFT SEAL RING	DIN 3760 - AS - 50 x 68 x 8 - NBR
2	1	SHAFT SEAL RING	DIN 3760 - AS - 20 x 35 x 7 - NBR
1	1	PARALLEL KEY	8.5 x 5 x 12 - DIN 6885
Item-ref	Quantity	Title/Name, designation, material, dimension etc.	STANDARD	MATL

Designed by your_name	Checked by	Approved by - date	Date 10/22/02	Filename PROJECT 4.dwg

MECHANICAL SHAFT ASSEMBLY

Edition	Scale 1:1
	Sheet 1/1

Fig. 4.3.41 *The finished concept drawing on an A2 layout.*

Review Questions

1. What two sets of standards are used in creating metric drawings?
2. What drawing settings are specified automatically when you use an ISO template?
3. What is the preferred text style for metric drawings created in AutoCAD?
4. What layers does AutoCAD create automatically when you use the ISO A2 template? What is the purpose of these layers?
5. Explain the purpose of the key and keyway.
6. Why is it not necessary to include every tiny detail in the bearings, seals, and fasteners on a concept drawing?
7. Name at least two things that should be taken into consideration when specifying seals for an assembly.
8. What factors influence the decision to use a certain type of bearings in an assembly?
9. What is the purpose of the seals and housing in the shaft assembly?
10. For what two parts are broken-out sections needed in the concept drawing? Why?

Portfolio Project

Lever Arm Shaft (continued)

Begin working on the concept drawing for the lever arm shaft you have designed.

1. Include at least one of each type of fastener to show their location and purpose in the drawing. Use power objects for the fasteners.
2. Create a layout for the drawing, place the drawing on it, and edit the title block as necessary.
3. Create a parts list using any of the methods described in this textbook and add it to the concept drawing.
4. Save your work.

Practice Problems

1. Create a new drawing in AutoCAD. Name the drawing Section 4-3 problem 1.dwg. Create a shaft that incorporates the following segments (left to right):
   - Ø62 cylindrical segment, 35 mm long
   - Sloped segment, 20 mm long, with a starting diameter of 62 and an ending diameter of 45
   - Ø45 cylindrical segment, 35 mm long
   - Sloped segment, 60 mm long, with a starting diameter of 45 and an ending diameter of 20
   - Ø20 cylindrical segment, 50 mm long
   - Ø50 cylindrical segment, 25 mm long
   - Ø86 cylindrical segment, 86 mm long

2. Create a new drawing in AutoCAD. Name the drawing Section 4-3 problem 2.dwg. Create a shaft that incorporates the following segments (left to right):
   - Ø40 cylindrical segment, 50 mm long
   - Sloped segment, 30 mm long, with a starting diameter of 40 and an ending diameter of 25
   - Ø25 cylindrical segment, 110 mm long
   - Sloped segment, 15 mm long, with a starting diameter of 25 and an ending diameter of 62
   - Ø62 cylindrical segment, 76 mm long
   - Ø105 cylindrical segment, 22 mm long

3. In a drawing named Section 4-3 problem 3.dwg, create a shaft with the dimensions and segments shown in Fig. 4.3.42.

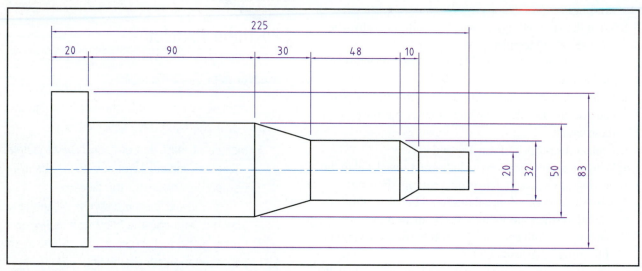

Fig. 4.3.42

4. After saving your work from Practice Problem 3, use the SAVEAS command to save the drawing with a new name of Section 4-3 problem 4.dwg. In this drawing, add the fillets and chamfer shown in Fig. 4.3.43.

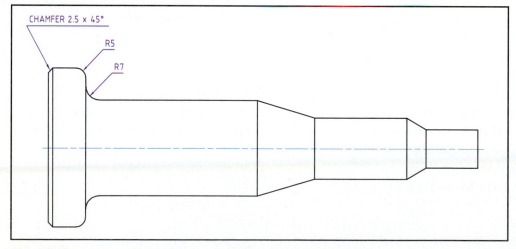

Fig. 4.3.43

Review

Summary

- A concept drawing presents ideas for new products to engineers and administrators for approval.
- In a concept drawing, the part being considered must be shown in assembly with any mating parts and other parts that are required for its proper use.
- The AutoCAD Mechanical software is suited specifically for creating mechanical assemblies.
- The power tools in AutoCAD Mechanical allow drafters to create mechanical assemblies, such as the concept drawing of the mechanical shaft assembly, that have built-in specifications and parameters.
- The basic AutoCAD software can be used to create all types of drawings for mechanical, architectural, civil, and other uses, so it is also appropriate software to use in developing the concept drawing of the mechanical shaft.

Project 4 Analysis

1. Explain the philosophical difference between a concept drawing such as the one in Project 4 and the set of working drawings developed in Projects 1 and 2.
2. What items should you consider when creating a concept drawing for a product?
3. Name at least two advantages of using power objects in AutoCAD Mechanical, as opposed to the basic AutoCAD commands.
4. What disadvantages might there be to using a specialized program such as AutoCAD Mechanical to create parts or assemblies?
5. The order in which the parts are added to the assembly in the concept drawing is different in the AutoCAD and AutoCAD Mechanical procedures. Why might a drafter choose to use a different drawing order depending on the software being used to create the drawing?
6. The concept drawing in this project is specifically for a shaft to be used in a mixer assembly. In addition to the shaft assembly drawing, what other concept drawings would be needed for the entire mixer assembly?

Synthesis Projects

Apply the skills and techniques you have practiced in this project to these special follow-up projects.

1. It is common practice in industry for manufacturers to reuse components and assemblies for more than one product whenever possible. This practice reduces design time and allows companies to market more products in a given amount of time. The mechanical shaft that provides the basis of Project 4 is meant for use in a tabletop mixer assembly. However, with minor modifications, it might be reused for other types of products. Determine at least one other product for which the mechanical shaft could be modified and used. Create a concept drawing of the product assembly.

2. On review, one of the engineers suggested an improvement to the mixer assembly. She suggested the addition of an idler pulley assembly to apply tension to the drive belt. The idler pulley would have an internal bearing that would allow it to spin freely as the belt runs. This would slide onto one end of the shaft and be retained with a socket-head cap screw

Continued on next page

Review

(Continued from previous page)

and washer that screws into the end of the shaft. The other end of the shaft would locate in a slot in the shaft mount. An additional socket-head cap screw and washer would thread into this end of the shaft. The shaft would move up and down in the slot, then lock into place with the cap screw after the correct tension is applied to the belt. The engineer's original sketches are shown below. Using the sketches as a guide, create a concept drawing of the idler pulley assembly.

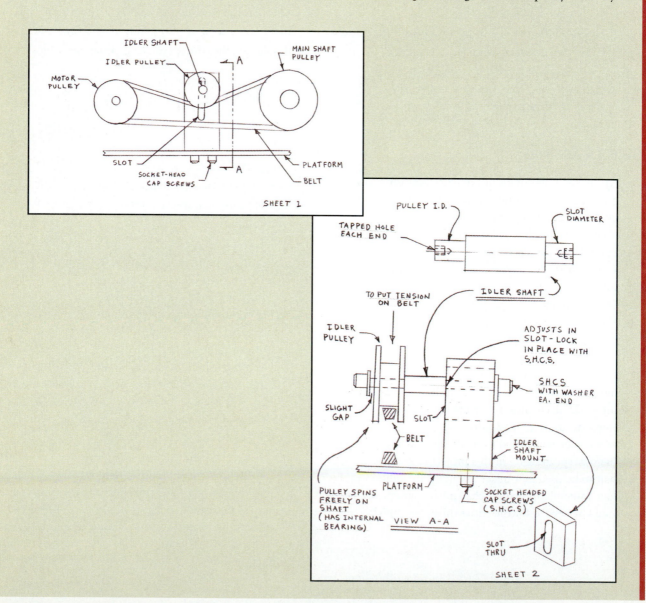

CAD
REFERENCE

CAD
REFERENCE CONTENTS

Adding a Printer or Plotter

Note: There are fine differences between printers and plotters, but those differences are not important in this discussion. Therefore, we will use the word *printer* to mean either a printer or a plotter.

AutoCAD automatically detects any printers that you have set up as system printers; that is, printers that you have set up through the Windows operating system for use with your computer. However, if you want to print to a printer that you do not ordinarily use for other applications—one that will print large-format drawings, for example—you must define it in AutoCAD before you can use it.

AutoCAD manages all nonsystem printers through its Plotter Manager. When you open the Plotter Manager (on the File menu), AutoCAD displays a window with all of the printers you have defined, as well as the Add-A-Plotter wizard. See Fig. CR-1. This wizard presents a series of windows that step you through the process of defining additional printers.

If more than one Autodesk program is loaded on your computer, the first screen asks you to identify the program with which you want to use the printer. Next, you can choose whether to configure the plotter from your computer or from a network server. The next window presents a list of manufacturers. When you choose a manufacturer, a list of models appears in the window on the right. Select the model you wish to set up. If you have old pcp or pc2 files that were set up for a previous version of AutoCAD, you can import them in the next window.

Next, select the method of printing the drawing. If you choose Plot to a port, you must select the port on your computer to which the printer is attached. If you select Plot to File or AutoSpool, you do not need to select a port.

Note: If you want to print large drawings but do not have a printer capable of printing the size you need, set up a printer description for a printer that is available somewhere else, such as at your school or at a local printing service. You will need to know the exact model of the printer, but you do not have to have the printer connected to your computer to set up the printer description. Select

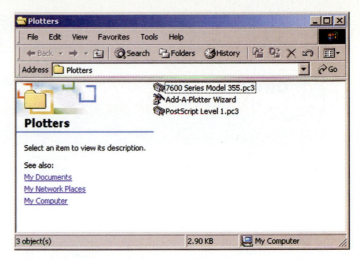

Fig. CR-1 *Selecting the Plotter Manager option from the File menu produces a window that shows all of the printers set up for use with AutoCAD, as well as a wizard that allows you to add other printers.*

Plot to File to create a plot file. Then take the file to the location where the printer is and print it from there.

Next, choose a name for the printer. It is usually best to name the printer according to its model name or number, unless you have two printers of the same model. This makes it easier to select the correct printer later when you want to plot the drawing.

Finally, you can configure or calibrate the printer if necessary. Consult the documentation that came with your printer for more information about configuration and calibration. Press the Finish button to complete the printer setup.

Autodesk Point A

Beginning with AutoCAD 2000i, Autodesk has provided an Internet connectivity system called Autodesk Point A. See Fig. CR-2. This system comprises the lower half of the AutoCAD Today window. It links AutoCAD users to support materials, references, catalogs, and other information. It also provides a secure means for companies to share drawing information with other company sites or with clients.

If you have never used Autodesk Point A before, take a few minutes to pick each tab and explore the many resources available through this system. Notice that you

Fig. CR-2 *Autodesk Point A serves as a portal to many useful goods and services, as well as information regarding various drafting standards.*

can update your version of AutoCAD, read about new developments in the CAD world, and link to CAD-related journals and magazines, among many other activities. It is worthwhile to explore this feature thoroughly even if you won't be using it every day.

BHATCH Command

AutoCAD automates the process of hatching (adding section lines to) a drawing. Either the HATCH command or the BHATCH command can be used, but BHATCH is much more efficient. The older HATCH command employs options that must be entered at the command line. BHATCH presents the Boundary Hatch window shown in Fig. CR-3. This window has two tabs: Quick and Advanced. The options on the Quick tab are sufficient for most applications.

AutoCAD provides several predefined ANSI and ISO hatch patterns, as well as other common hatches. The only one currently used in technical drawings is ANSI31, as shown in Fig. CR-3.

To choose the area to be hatched, you can use the Pick Points or the Select Objects method. The Pick Points method is more convenient when you want to hatch an area that has one or more interior objects that should not be hatched. When picking points, you simply pick a point in the area you want to hatch. AutoCAD determines the

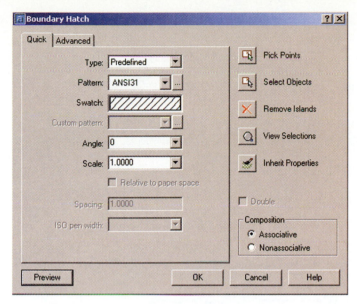

Fig. CR-3 *The Boundary Hatch window.*

AutoCAD provides two commands for blocking sets of objects: BLOCK and WBLOCK. The BLOCK command defines a block within the current drawing file. The WBLOCK command defines a block as a separate drawing file so that it can be inserted more easily into other drawings. In other words, a block created using WBLOCK is defined as an independent file rather than in the current drawing's database.

The BLOCK command opens the Block Definition window shown in Fig. CR-4. At a minimum, you must provide a block name, select objects to include in the block, and select a base point for the block. By default, AutoCAD converts the selected objects to a block when it creates the block definition in the file's database. However, you can also choose to retain the independent items or to delete the objects altogether. These choices do not affect the definition of the block. Again by default, the command creates an icon for quick identification in DesignCenter, but you can choose not to create the icon if file size is a consideration.

most logical hatch boundaries based on your selection and highlights them in dashed lines. It automatically excludes any interior objects or "islands." To include one of the objects in the hatch, pick the Remove Islands button and select the interior of the object to be removed. Note that you can pick as many points as necessary to define the entire hatch area. The hatch area does not need to be contiguous; hatches in more than one area can be part of the same hatch object.

Using the Select Objects method, you pick specific objects to be hatched. AutoCAD ignores any interior objects, so the entire object receives the hatch.

You may also set the angle and scale at which the hatch is formed. Different angles should be used for hatched surfaces that share a boundary. You should set a scale that is appropriate for the current drawing.

BLOCK and WBLOCK Commands

A block is a set of one or more objects in AutoCAD that have been named and defined as a single object in the drawing database for an individual drawing file. The advantage of blocking a set of objects is that you can then insert the block any number of times without redrawing the geometry.

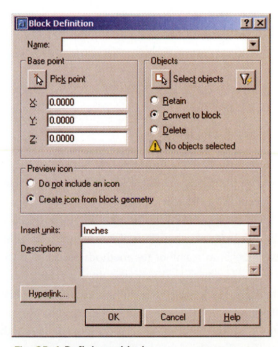

Fig. CR-4 *Defining a block.*

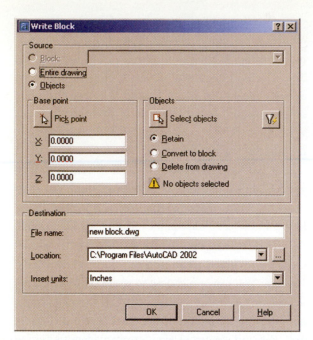

Fig. CR-5 *Writing a block as a separate file.*

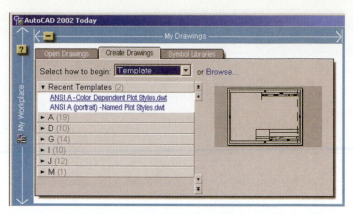

Fig. CR-6 *Using a template to create a new drawing.*

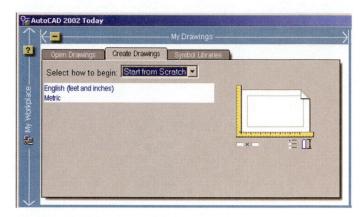

Fig. CR-7 *Creating a new drawing from scratch.*

The WBLOCK command presents the Write Block window shown in Fig. CR-5. As you can see, the options are similar to those produced by the BLOCK command. At the top of the window, however, you are given the option of creating the block from a previously defined block, from the entire drawing, or from objects that you will select. Instead of a block name, you provide a file name. AutoCAD also allows you to specify where to store the new file.

File Creation

To set up a new file in AutoCAD, work from the My Drawings portion of the AutoCAD Today window. Select the Create Drawings tab.

There are several ways to set up a file from within the AutoCAD Today window. Pick the arrow next to Select how to begin to see all of the methods available to you. For example, you can create a new drawing using one of AutoCAD's predefined templates or a custom template that your company has created. Notice in Fig. CR-6 that AutoCAD displays the most recently used templates at the top of the list. Other predefined templates are listed in alphabetical order. When you single-click the name of a template, a thumbnail view appears.

You can also create drawings without using a template, known as creating a file "from scratch." When you create a file from scratch, you have a choice of using English (U.S. Customary) or Metric units as a basis for the drawing, as shown in Fig. CR-7. Files created from scratch use Auto-CAD's default acad.dwt template.

If you don't want to use a template but need help setting up a drawing file, you can use a wizard to set up the drawing. See Fig. CR-8. AutoCAD provides two file setup wizards. The Basic wizard guides you through the process of setting up the units and drawing area (limits) for a new drawing. The rest of the settings are taken from the acad.dwt default template. The Advanced wizard is similar to the Basic wizard, but it helps you set up the units, angles, angle measures and direction, as well as the drawing limits.

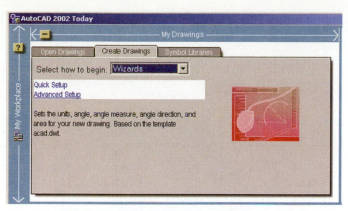

Fig. CR-8 *AutoCAD provides two wizards to help you set up the basic parameters of a file.*

Layer Setup

For most applications, you will need to work on more than one layer. Layers in AutoCAD are similar to the overlays of traditional drawings. They allow you to control what information is visible at any time. For example, by placing dimensions on a separate layer, you can print a drawing with or without the dimensions—without doing anything more than freezing the layer that contains the dimensions.

To control layers in AutoCAD, use the Layer Properties Manager. You can display this dialog box by entering the LAYER command or by picking the Layers button on the Object Properties toolbar. Fig. CR-9 shows an example of layers set up for an architectural drawing.

LIMITS Command

The drawing area in AutoCAD is set and controlled using the LIMITS command. LIMITS defines the lower left corner and the upper right corner of the drawing boundary. The drawing limits are important because all objects are created at their full size in AutoCAD. If necessary, they can later be scaled to fit on printed sheets. By default, the lower left corner is set at coordinate 0,0. The default upper right corner is set at 9,12. This size works well with ANSI sheet size A.

Before you change the limits of a drawing, you should set up the type of units you will use in the file. For English (U.S. Customary) drawings, AutoCAD assumes all quan-

tities to be in inches unless you specify otherwise. So, for example, if you were working on an architectural drawing and wanted to create a drawing area of $300' \times 200'$, you would either need to specify feet using the foot symbol or calculate how many inches are in $300'$ and $200'$. It is obviously easier to use the foot mark, but if the drawing units are set to decimal inches, AutoCAD will not accept the foot mark in limits specifications.

To change the size of the drawing area, you should generally change only the limit that defines the upper right corner. Leave the lower left corner at the default origin of 0,0,0. Follow these steps:

1. Enter the **LIMITS** command at the keyboard and press **Enter** to accept the <0.000,0.000> default.
2. Set the upper right limit to a coordinate value large enough to hold the drawing you intend to create.
3. **ZOOM All** and press the **F7** key to show the grid. If it is too dense to display, enter the **GRID** command and set the grid to a larger interval.

Regardless of whether you change the limits of a drawing, you should always turn on the grid (F7) and enter ZOOM All to see the extents of the drawing area. This puts the drawing into perspective for you. You can then zoom

Fig. CR-9 *Adding layers and layer attributes in the Layer Properties Manager.*

in to see a smaller area if necessary. If AutoCAD displays a message that the grid is too dense to display, enter the GRID command and set the grid spacing to a more appropriate scale for the drawing. An appropriate grid spacing for a 300″ × 200″ drawing is 10″.

AutoCAD can also do a certain amount of limits checking. By entering the **LIMITS** command and then entering **ON**, you can set AutoCAD to reject any points you specify that fall outside the drawing area. Circles and other geometry that calculate their own points may still fall outside the drawing area, but you will not be able to specify endpoints for objects outside of the defined drawing area. If you turn limits checking off, AutoCAD stores the values for the drawing limits but does not enforce them unless you enter the LIMITS command and once again turn limit checking on.

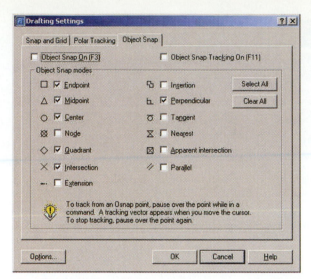

Fig. CR-10 *Setting running object snaps from the Drafting Settings window.*

Object Snaps

To create a drawing accurately, it is often necessary to pick precise points such as the endpoint or midpoint of a line. These and many other "special" points can be picked easily using AutoCAD's object snap feature. The object snap functions like a magnet. When the the object snap is active and the cursor moves near one of the defined types of points, a symbol appears on the screen at that point to let you know that if you click the line or other object, the cursor will snap to that point.

There are two ways to use object snaps in AutoCAD. First, if you find you need to use an object snap while a command is active, you can type the first three letters of the desired object snap's name to activate it for that single instance. If you will need to use one or more of the object snaps several times, it is more efficient to activate them in advance. Object snaps that are activated in this way are called *running object snaps* because they run in the background. You do not have to specify them every time you need to use them. To set running object snaps, enter the **DSETTINGS** command to display the Drafting Settings window and pick the **Object Snap** tab, as shown in Fig. CR-10. This window lists all of the available object snaps along with their associated symbols. Pick the check box next to each object snap you want to run in the background. Pick the **OK** button to return to the drawing.

Polar Coordinates

AutoCAD uses the Cartesian coordinate system to allow the CAD operator to specify any point in two- or three-dimensional space by entering its XY or XYZ coordinates. For example, on a 2D drawing that uses U.S. Customary units, you can place the endpoint of a line at exactly 3 inches to the right of the origin (point 0,0) and exactly 4.7 inches above the origin by entering the coordinates 3,47.

However, in many cases, you won't know the exact point on the coordinate grid where you want to place a point. Instead, you may need to place the point so that a line is exactly 2 inches long and extends at a 30° angle from an existing line. To achieve this, you can use polar coordinates.

Suppose that you have entered the LINE command and placed a first point at absolute coordinate 2,1. The "sentence structure" of the polar coordinate to place a second point is as follows:

@3.75<90

The @ symbol tells AutoCAD that this will be a polar coordinate. The total length of the line will be 3.75″. The <90 signifies a 90° angle. Therefore, in this case, the line would extend 3.75″ straight up from its first point. See Fig. CR-11.

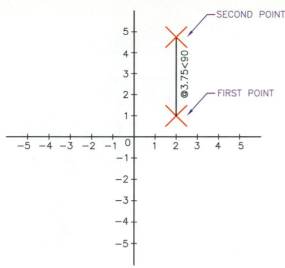

Fig. CR-11 *The effect of polar coordinate entry.*

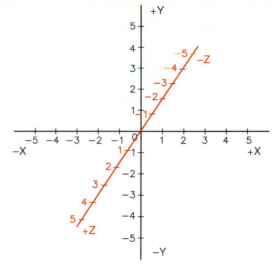

Fig. CR-12 *The Z axis.*

UCS Manipulation

In 2D drafting, drafters rarely have to change the current UCS. In 3D modeling, however, understanding how to work with UCSs is critical to obtain accurate results.

When you work in 3D, a third axis, the Z axis, is added to the coordinate system, as shown in Fig. CR-12. This axis runs perpendicular to both the X and the Y axis. In the positive direction, it runs out of the screen toward the drafter. In the negative direction, it runs back from the screen away from the drafter.

When you create a new UCS, you have the opportunity not only to place the origin, but also to rotate the three axes in any direction. There are many ways to create a new UCS. You will probably use all of them at one time or another, depending on the requirements for specific drawings. The six options to create a new UCS are:

- **ZAxis**—creates a new UCS with the origin and positive Z axis you specify.
- **3point**—creates and orients the UCS based on three points that you specify. The first point locates the origin, the second specifies the direction of the positive X axis, and the third specifies the direction of the positive Y axis. See Fig. CR-13.
- **Object**—creates and orients the UCS based on an existing 3D object in the drawing. The options for creating a UCS from an object are complex and depend on the object selected.

- **Face**—creates and orients the UCS based on one face of a 3D solid in the drawing. The drafter can cycle through the various faces using the Next option until the face to be used is highlighted. It is also possible to rotate the axes after the face has been selected.
- **View**—creates and orients the UCS so that the X and Y axes are flat to the screen.
- **X/Y/Z**—rotates the axes around the specified axis to create a new UCS.

In addition, you can change the location of the origin without changing the orientation of the axes by entering **O** (for Origin) and then selecting the new origin for the UCS.

Units

By default, AutoCAD uses a nonspecific unit that you can interpret to be anything you want. For example, a unit might equal an inch, a foot, a meter, or even a kilometer or a lightyear. For convenience, however, AutoCAD allows you to set up a drawing using various measurement systems.

When you use a wizard to set up a drawing or set it up from scratch, AutoCAD allows you to specify the basic measurement system when you create the file. For more precise control, use the DDUNITS command after the file has been created to display the Drawing Units window, as

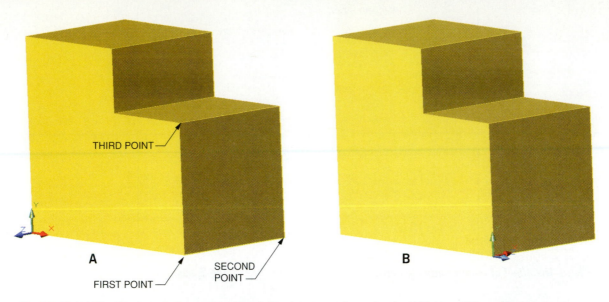

THIRD POINT

A

FIRST POINT

SECOND POINT

B

Fig. CR-13 *(A) The XY plane is flat to the side of the object, as shown by the UCS icon. When the UCS 3point option is used with the three points shown, the UCS moves to the front face of the object, as shown in (B).*

shown in Fig. CR-14. You can also define the precision to which units are displayed, as well as how AutoCAD handles angles and undimensioned units.

You can also set up the way AutoCAD calculates and displays angles. In addition to decimal degrees, for example, you can choose to display angle measurements as radians; grads; surveyor's units; or degrees, minutes, and seconds.

By default, AutoCAD measures and displays all angles in the counterclockwise direction. For certain applications, however, this may be bulky or nonintuitive. For these circumstances, the Drawing Units window allows you to set the software to measure in the clockwise direction instead. When the box next to Clockwise is checked, all angles are measured and displayed in the clockwise direction.

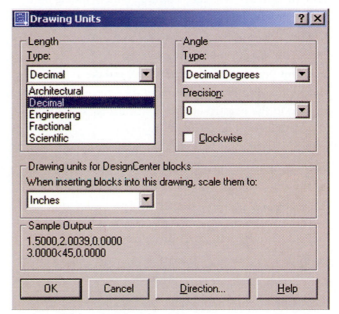

Fig. CR-14 *The Drawing Units window gives you precise control over the units a drawing will use.*

APPENDIXES

Appendix A

Appendix B

Appendix C

Appendix D

ASME B16.5-1996 Class 150
Plate Flange Reference

Pipe Size	Outside Diameter	Inside Diameter	Number of Bolt Holes	Bolt-Hole Diameter	Bolt Circle
$1/2$"	$3\,1/2$	0.88	4	$5/8$	$2\,3/8$
$3/4$"	$3\,7/8$	1.09	4	$5/8$	$2\,3/4$
1"	$4\,1/4$	1.36	4	$5/8$	$3\,1/8$
$1\,1/4$"	$4\,5/8$	1.70	4	$5/8$	$3\,1/2$
$1\,1/2$"	5	1.95	4	$5/8$	$3\,7/8$
2"	6	2.44	4	$3/4$	$4\,3/4$
$2\,1/2$"	7	2.94	4	$3/4$	$5\,1/2$
3"	$7\,1/2$	3.57	4	$3/4$	6
$3\,1/2$	$8\,1/2$	4.07	8	$3/4$	7
4"	9	4.60	8	$3/4$	$7\,1/2$
5"	10	5.66	8	$7/8$	$8\,1/2$
6"	11	6.72	8	$7/8$	$9\,1/2$
8"	$13\,1/2$	8.72	8	$7/8$	$11\,3/4$
10"	16	10.88	12	1	$14\,1/4$
12"	19	12.88	12	1	17
14"	21	14.14	12	$1\,1/8$	$18\,3/4$
16"	$23\,1/2$	16.16	16	$1\,1/8$	$21\,1/4$
18"	25	18.18	16	$1\,1/4$	$22\,3/4$
20"	$27\,1/2$	20.20	20	$1\,1/4$	25
24"	32	24.25	20	$1\,3/8$	$29\,1/2$

Adapted from ASME B16.5-1996.

ASME B16.5-1996 Class 300
Plate Flange Reference

Pipe Size	Outside Diameter	Inside Diameter	Number of Bolt Holes	Bolt-Hole Diameter	Bolt Circle
$1/2$"	$3\ 3/4$	0.88	4	$5/8$	$2\ 5/8$
$3/4$"	$4\ 5/8$	1.09	4	$3/4$	$3\ 1/4$
1"	$4\ 7/8$	1.36	4	$3/4$	$3\ 1/2$
$1\ 1/4$"	$5\ 1/4$	1.70	4	$3/4$	$3\ 7/8$
$1\ 1/2$"	$6\ 1/8$	1.95	4	$7/8$	$4\ 1/2$
2"	$6\ 1/2$	2.44	8	$3/4$	5
$2\ 1/2$"	$7\ 1/2$	2.94	8	$7/8$	$5\ 7/8$
3"	$8\ 1/4$	3.57	8	$7/8$	$6\ 5/8$
$3\ 1/2$"	9	4.07	8	$7/8$	$7\ 1/4$
4"	10	4.57	8	$7/8$	$7\ 7/8$
5"	11	5.66	8	$7/8$	$9\ 1/4$
6"	$12\ 1/2$	6.72	12	$7/8$	$10\ 5/8$
8"	15	8.72	12	1	13
10"	$17\ 1/2$	10.88	16	$1\ 1/8$	$15\ 1/4$
12"	$20\ 1/2$	12.88	16	$1\ 1/4$	$17\ 3/4$
14"	23	14.14	20	$1\ 1/4$	$20\ 1/4$
16"	$25\ 1/2$	16.16	20	$1\ 3/8$	$22\ 1/2$
18"	28	18.18	24	$1\ 3/8$	$24\ 3/4$
20"	$30\ 1/2$	20.20	24	$1\ 3/8$	27
24"	36	24.25	24	$1\ 5/8$	32

Adapted from ASME B16.5-1996.

ASME B16.5-1996 Class 600
Plate Flange Reference

Pipe Size	Outside Diameter	Inside Diameter	Number of Bolt Holes	Bolt-Hole Diameter	Bolt Circle
$1/2$"	$3\ 3/4$	0.88	4	$5/8$	$2\ 5/8$
$3/4$"	$4\ 5/8$	1.09	4	$3/4$	$3\ 1/4$
1"	$4\ 7/8$	1.36	4	$3/4$	$3\ 1/2$
$1\ 1/4$"	$5\ 1/4$	1.70	4	$3/4$	$3\ 7/8$
$1\ 1/2$"	$6\ 1/8$	1.95	4	$7/8$	$4\ 1/2$
2"	$6\ 1/2$	2.44	8	$3/4$	5
$2\ 1/2$"	$7\ 1/2$	2.94	8	$7/8$	$5\ 7/8$
3"	$8\ 1/4$	3.57	8	$7/8$	$6\ 5/8$
$3\ 1/2$"	9	4.07	8	1	$7\ 1/4$
4"	$10\ 3/4$	4.57	8	1	$8\ 1/2$
5"	13	5.66	8	$1\ 1/2$	$10\ 1/2$
6"	14	6.72	12	$1\ 1/2$	$11\ 1/2$
8"	$16\ 1/2$	8.72	12	$1\ 1/4$	$13\ 3/4$
10"	20	10.88	16	$1\ 3/8$	17
12"	22	12.88	20	$1\ 3/8$	$19\ 1/4$
14"	$23\ 3/4$	14.14	20	$1\ 1/2$	$20\ 3/4$
16"	27	16.16	20	$1\ 5/8$	$23\ 3/4$
18"	$29\ 1/4$	18.18	20	$1\ 3/4$	$25\ 3/4$
20"	32	20.20	24	$1\ 3/4$	$28\ 1/2$
24"	37	24.25	24	2	33

Adapted from ASME B16.5-1996.

Welding Processes: Standard Letter Designations

Abbreviation or Acronym	Welding Process
AAC	air carbon arc cutting
AAW	air acetylene welding
AB	arc brazing
ABD	adhesive bonding
AC	arc cutting
AHW	atomic hydrogen welding
AOC	oxygen arc cutting
ASP	arc spraying
AW	carbon arc welding
B	brazing
BB	block brazing
BMAW	bare metal arc welding
CAB	carbon arc brazing
CAC	carbon arc cutting
CAW	carbon arc welding
CAW-G	gas carbon arc welding
CAW-S	shielded carbon arc welding
CAW-T	twin carbon arc welding
CEW	coextrusion welding
CW	cold welding
DB	dip brazing
DFB	diffusion brazing
DFW	diffusion welding
DS	dip soldering
EBC	electron beam cutting

Adapted from ANSI/AWS A2.4-91.

Welding Processes:
Standard Letter Designations *(continued)*

Abbreviation or Acronym	Welding Process
EBW	electron beam welding
EBW-HV	electron beam welding – high vacuum
EBW-MV	electron beam welding – medium vacuum
EBW-NV	electron beam welding – nonvacuum
EGW	electrogas welding
ESW	electroslag welding
EXW	explosion welding
FB	furnace brazing
FCAW	flux-cored arc welding
FLB	flow brazing
FLOW	flow welding
FLSP	flame spraying
FOC	chemical flux cutting
FOW	forge welding
FRW	friction welding
FS	furnace soldering
FW	flash welding
GMAC	gas metal arc cutting
GMAW	gas metal arc welding
GMAW-P	gas metal arc welding – pulsed arc
GMAW-S	gas metal arc welding – short-circuiting arc
GTAC	gas tunsten arc cutting
GTAW	gas tungsten arc welding
GTAW-P	gas tunsten arc welding – pulsed arc
HPW	hot pressure welding

Welding Processes:
Standard Letter Designations *(continued)*

Abbreviation or Acronym	Welding Process
IB	induction brazing
INS	iron soldering
IRB	infrared brazing
IRS	infrared soldering
IS	induction soldering
IW	induction welding
LBC	laser beam cutting
LBC-A	laser beam cutting – air
LBC-EV	laser beam cutting – evaporative
LBC-IG	laser beam cutting – inert gas
LBC-O	laser beam cutting – oxygen
LBW	laser beam welding
LOC	oxygen lance cutting
MAC	metal arc cutting
OAW	oxyacetylene welding
OC	oxygen cutting
OFC	oxyfuel gas cutting
OFC-A	oxyacetylene cutting
OFC-H	oxyhydrogen cutting
OFC-N	oxynatural gas culling
OFC-P	oxypropane cutting
OFW	oxyfuel gas welding
OHW	oxyhydrogen welding
PAC	plasma arc cutting
PAW	plasma arc welding

Welding Processes:
Standard Letter Designations *(continued)*

Abbreviation or Acronym	Welding Process
PEW	percussion welding
PGW	pressure gas welding
POC	metal powder cutting
PSP	plasma spraying
PW	projection welding
RB	resistance brazing
RS	resistance soldering
RSEW	resistance seam welding
RSEW-HF	resistance seam welding – high frequency
RSEW-1	resistance seam welding – induction
RSW	resistance spot welding
ROW	roll welding
RW	resistance welding
S	soldering
SAW	submerged arc welding
SAW-S	series submerged arc welding
SMAC	shielded metal arc cutting
SMAW	shielded metal arc welding
SSW	solid state welding
SW	stud arc welding

Roller Chain

Single Strand—Riveted

Single Strand—Cottered

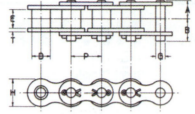

Roller Chain

Specifications — Single Strand Roller Chain

Table No. 1

Chain Pitch P	Chain No.	Chain		Dimensions						
		Average Tensile Strength Lbs.	Average Weight per Ft. Lbs.	Connecting Links		Rollers		Pins	Side Plates	
				A	B	D	E	G	H	T
				STANDARD						
1/4"	25	875	.09	.156"	.188"	.130"	1/8"	.0905"	.234"	.030"
3/8	35	2,100	.21	.233	.267	.200	3/16	.141	.350	.050
1/2	41	2,000	.25	.256	.322	.306	1/4	.141	.383	.050
1/2	40	3,700	.42	.315	.380	.312	5/16	.156	.466	.060
5/8	50	6,100	.69	.395	.460	.400	3/8	.200	.584	.080
3/4	60	8,500	1.00	.495	.586	.468	1/2	.234	.700	.094
1	80	14,500	1.71	.637	.741	.625	5/8	.312	.934	.125
1 1/4	100	24,000	2.58	.778	.923	.750	3/4	.375	1.166	.156
1 1/2	120	34,000	3.87	.980	1.150	.875	1	.437	1.400	.187
1 3/4	140	46,000	4.95	1.059	1.215	1.000	1	.500	1.634	.219
2	160	58,000	6.61	1.261	1.451	1.125	1 1/4	.562	1.866	.250
2 1/2	200	95,000	10.96	1.560	1.777	1.562	1 1/2	.781	2.250	.312

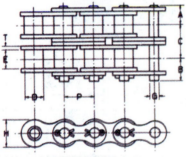

Roller Chain

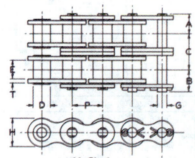

LL Chain

Double Strand—Riveted

Double Strand—Cottered

Specifications — Double Strand Roller Chain

Table No. 1

Chain Pitch P	Chain No.	Chain		Dimensions							
		Average Tensile Strength Lbs.	Average Weight per Ft. Lbs.	Connecting Links		Spacing	Rollers		Pins	Side Plates	
				A	B	C	D	E	G	H	T
				STANDARD							
3/8"	35-2	4,200	.40	.233"	.267"	.400"	.200"	3/16"	.141"	.350"	.050"
1/2	40-2	7,400	.82	.315	.380	.564	.312	5/16	.156	.466	.060
5/8	50-2	12,200	1.36	.395	.460	.730	.400	3/8	.200	.584	.080
3/4	60-2	17,000	1.99	.495	.586	.904	.468	1/2	.234	.700	.094
1	80-2	29,000	3.40	.637	.741	1.158	.625	5/8	.312	.934	.125
1 1/4	100-2	48,000	5.10	.778	.923	1.406	.750	3/4	.375	1.166	.156
1 1/2	120-2	68,000	7.65	.980	1.150	1.791	.875	1	.437	1.400	.187
1 3/4	140-2	92,000	9.80	1.059	1.215	1.933	1.000	1	.500	1.634	.219
2	160-2	116,000	13.10	1.261	1.451	2.327	1.125	1 1/4	.562	1.866	.250
2 1/2	200-2	190,000	21.50	1.560	1.777	2.912	1.562	1 1/2	.781	2.250	.312
3	240-2	260,000	33.20	1.895	2.187	3.458	1.875	1 7/8	.937	2.800	.375

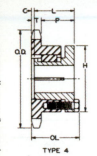

Browning

Roller Chain Sprockets **120**

Sprockets for **No. 120**, 1½″ Pitch ANSI Chain

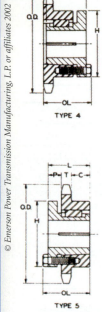

TYPE 4

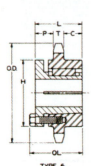

TYPE 5

TYPE 6

Table No. 1   **Steel Single Sprockets with Split Taper Bushings**

Part No.	Bushing	Bore Range	DIAMETERS Outside	DIAMETERS Pitch	No. Teeth	Type	T Nom.	OL	L	P	C	H	Wt. Less Bushing
H120Q11	Q1	3/4 - 2 11/16"	6.01"	5.324"	11	4	.924"	2 25/32"	2 1/2"	1 9/16"	0	4 1/8"	4.8
H120Q12	Q1	3/4 - 2 11/16"	6.50	5.796	12	4	.924	2 25/32	2 1/2	1 9/16	0	4 1/8	6.3
H120Q13	Q1	3/4 - 2 11/16"	6.99	6.268	13	4	.924	2 25/32	2 1/2	1 9/16	0	4 1/8	7.9
H120Q14	Q1	3/4 - 2 11/16"	7.47	6.741	14	4	.924	2 25/32	2 1/2	1 9/16	0	4 1/8	9.1
H120Q15	Q1	3/4 - 2 11/16"	7.96	7.215	15	4	.924	2 25/32	2 1/2	1 9/16	0	4 1/8	10.4
H120Q16	Q1	3/4 - 2 11/16"	8.39	7.689	16	4	.924	2 25/32	2 1/2	1 9/16	0	4 1/8	11.8
H120R16	R1	1 1/8 - 3 3/4	8.39	7.689	16	4	.924	3 5/32	2 7/8	1 15/16	0	5 3/8	12.3
H120Q17	Q1	3/4 - 2 11/16	8.88	8.163	17	4	.924	2 25/32	2 1/2	1 9/16	0	4 1/8	13.4
H120R17	R1	1 1/8 - 3 3/4	8.88	8.163	17	4	.924	3 5/32	2 7/8	1 15/16	0	5 3/8	13.6
H120Q18	Q1	3/4 - 2 11/16	9.41	8.638	18	4	.924	2 25/32	2 1/2	1 9/16	0	4 1/8	15.6
H120R18	R1	1 1/8 - 3 3/4	9.41	8.638	18	4	.924	3 5/32	3 5/32	1 9/16	0	5 3/8	15.9
H120R19	R1	1 1/8 - 3 3/4	9.89	9.113	19	4	.924	3 5/32	3 5/32	1 9/16	0	5 3/8	16.8
H120R20	R1	1 1/8 - 3 3/4	10.37	9.589	20	4	.924	3 5/32	2 7/8	1 15/16	0	5 3/8	18.8
H120R21	R1	1 1/8 - 3 3/4	10.85	10.064	21	4	.924	3 5/32	2 7/8	1 15/16	0	5 3/8	21.0
H120R22	R1	1 1/8 - 3 3/4	11.33	10.540	22	4	.924	3 5/32	2 7/8	1 15/16	0	5 3/8	22.5
H120R23	R1	1 1/8 - 3 3/4	11.81	11.016	23	4	.924	3 5/32	2 7/8	1 15/16	0	5 3/8	24.8
H120R24	R1	1 1/8 - 3 3/4	12.29	11.492	24	4	.924	3 5/32	2 7/8	1 15/16	0	5 3/8	26.9
H120R25	R1	1 1/8 - 3 3/4	12.77	11.968	25	4	.924	3 5/32	2 7/8	1 15/16	0	5 3/8	29.8
H120R26	R1	1 1/8 - 3 3/4	13.25	12.444	26	5	.924	3 5/32	2 7/8	7/8	1 1/16	5 3/8	32.9
H120R28	R1	1 1/8 - 3 3/4	14.21	13.397	28	5	.924	3 5/32	2 7/8	7/8	1 1/16	5 3/8	38.3
H120R30	R1	1 1/8 - 3 3/4	15.17	14.350	30	5	.924	3 5/32	2 7/8	7/8	1 1/16	5 3/8	43.4
120R32	R1	1 1/8 - 3 3/4	16.13	15.303	32	5	.924	3 5/32	2 7/8	7/8	1 1/16	5 3/8	49.4
120R35	R2	1 3/8 - 3 5/8	17.57	16.734	35	6	.924	5 5/32	4 7/8	1 15/16	2	5 3/8	68.0
120R36	R2	1 3/8 - 3 5/8	18.05	17.211	36	6	.924	5 5/32	4 7/8	1 15/16	2	5 3/8	72.0
120R40	R2	1 3/8 - 3 5/8	19.96	19.118	40	6	.924	5 5/32	4 7/8	1 15/16	2	5 3/8	82.0
120S40	S1	1 11/16 - 4 1/4	19.96	19.118	40	5	.924	4 3/4	4 3/8	1 1/16	2 3/8	6 3/8	83..0
120S42	S1	1 11/16 - 4 1/4	20.92	20.072	42	5	.924	4 3/4	4 3/8	1 1/16	2 3/8	6 3/8	90.0
120R45	R2	1 3/8 - 3 5/8	22.35	21.503	45	6	.924	5 5/32	4 7/8	1 15/16	2	5 3/8	102
120S45	S1	1 11/16 - 4 1/4	22.35	21.503	45	5	.924	4 3/4	4 3/8	1 1/16	2 3/8	6 3/8	100
120S48	S1	1 11/16 - 4 1/4	23.79	22.935	48	5	.924	4 3/4	4 3/8	1 1/16	2 3/8	6 3/8	111
120S54	S1	1 11/16 - 4 1/4	26.65	25.798	54	5	.924	4 3/4	4 3/8	1 1/16	2 3/8	6 3/8	138
120R60	R2	1 3/8 - 3 5/8	29.52	28.661	60	6	.924	5 5/32	4 7/8	1 15/16	2	5 3/8	179
120S60	S1	1 11/16 - 4 1/4	29.52	28.661	60	5	.924	4 3/4	4 3/8	1 1/16	2 3/8	6 3/8	180
120R70	R2	1 3/8 - 3 5/8	34.30	33.434	70	6	.924	5 5/32	4 7/8	1 15/16	2	5 3/8	148
120S70	S2	1 11/16 - 4 1/4	34.30	33.434	70	5	.924	7 1/8	6 3/4	2 15/16	2 7/8	6 3/8	167
120R80	R2	1 3/8 - 3 5/8	39.08	38.207	80	6	.924	5 5/32	4 7/8	1 15/16	2	5 3/8	291
120S80	S2	1 7/8 - 4 3/16	39.08	38.207	80	6	.924	7 1/8	6 3/4	2 15/16	2 7/8	6 3/8	305

HARDENED TEETH

Where two sprockets with the same number of teeth but different bushings are offered, we suggest using the one with the larger bushing for heavier service drives.

VTWS-200

Browning Ball Bearings

V	VALUE & QUALITY
T	TAKE-UP
W	WIDE SLOT
S	SET SCREW

2	200
0	NORMAL
0	DUTY

Lock: Set Screw
Seal: Contact
Housing: Cast Iron
Temperature: -20°F to 200°F
Self Alignment: ±1.5°
Inserts: VS-200
Dimensions pg 16
Ratings pg 17

Bore Size	Fitting
1/2" - 1 1/4" S	1/4" - 28NF
1 1/4" - 2 15/16"	1/8" NPT

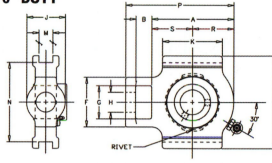

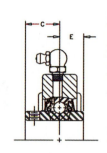

SHAFT DIA. IN	UNIT NO.	BRG. NO.	Dimensions In Inches															UNIT WT.
			A	B	C	E	F	G	H CORE	J HUB	K	M	N	O	P	R	S	
1/2	VTWS-208	VS-208	2 11/16	5/8	5/8	13/32	1 7/8	1 1/4	3/4	1 3/8	2	17/32	3	3 1/2	3 11/16	1 5/16	1 3/8	1.5
5/8	VTWS-210	VS-210																
3/4	VTWS-212	VS-212	3	5/8	23/32	1/2	1 7/8	1 1/4	3/4	1 3/8	2 1/4	17/32	3	3 1/2	4	1 7/16	1 9/16	1.7
7/8	VTWS-214	VS-214																
15/16	VTWS-215	VS-215	3 3/32	5/8	13/16	9/16	1 7/8	1 1/4	3/4	1 1/2	2 1/4	17/32	3	3 1/2	4 3/32	1 9/16	1 17/32	1.6
1	VTWS-216	VS-216																
1 1/8	VTWS-218	VS-218	3 3/8	3/4	7/8	5/8	2 1/8	1 7/16	7/8	1 3/4	2 1/2	17/32	3 1/2	4	4 5/8	1 25/32	1 19/32	2.4
1 3/16	VTWS-219	VS-219																
1 1/4	VTWS-220	VS-220	3 3/4	3/4	1	11/16	2 1/8	1 7/16	7/8	1 3/4	2 3/4	17/32	3 1/2	4	5	2	1 3/4	3.3
1 3/8	VTWS-222	VS-222																
1 7/16	VTWS-223	VS-223																
1 1/2	VTWS-224	VS-224	4 7/32	3/4	1 3/16	3/4	2 3/4	1 15/16	1 1/8	2 1/8	3 1/4	11/16	4	4 1/2	5 19/32	2 3/16	2 1/32	5.0
1 5/8	VTWS-226	VS-226																
1 11/16	VTWS-227	VS-227	4 3/8	3/4	1 3/16	3/4	2 3/4	1 15/16	1 1/8	2 1/8	3 1/4	11/16	4	4 1/2	5 3/4	2 1/4	2 1/8	5.6
1 3/4	VTWS-228	VS-228																
1 15/16	VTWS-231	VS-231	4 17/32	3/4	1 9/32	3/4	2 3/4	1 15/16	1 1/8	2 1/8	3 3/8	11/16	4	4 1/2	5 29/32	2 5/16	2 7/32	5.8
2	VTWS-232	VS-232	5 3/32	1	1 5/16	7/8	3 5/8	2 1/2	1 3/8	2 1/2	3 3/4	1 1/16	5 1/8	5 3/4	6 27/32	2 9/16	2 17/32	8.4
2 3/16	VTWS-235	VS-235																
2 7/16	VTWS-239	VS-239	5 1/2	1	1 9/16	1	3 5/8	2 1/2	1 3/8	2 1/2	4	1 1/16	5 1/8	5 3/4	7 1/4	2 27/32	2 21/32	9.3

NK Nickel Plated (NK) To order specify NK suffix, i.e. VTWS-216 NK

REV 0

VTWE-200

V	VALUE & QUALITY
T	TAKE-UP
W	WIDE SLOT
E	ECCENTRIC

2	200
0	NORMAL
0	DUTY

Browning®
Ball Bearings

Lock: Eccentric
Seal: Contact
Housing: Cast Iron
Temperature: -20°F to 200°F
Self Alignment: ±1.5°
Inserts: VE-200
Dimensions pg 74
Ratings pg 75

Bore Size	Fitting
1/2" - 1 1/4" S	1/4" - 28NF
1 1/4" - 2 15/16"	1/8" NPT

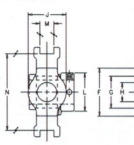
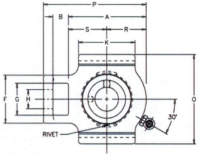
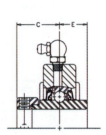

SHAFT DIA. IN	UNIT NO.	BRG. NO.	Dimensions In Inches																UNIT WT
			A	B	C	E	F	G	H CORE	J HUB	K	L	M	N	O	P	R	S	
3/4	VTWE-212	VE-212	3	5/8	1 3/64	43/64	1 7/8	1 1/4	3/4	1 3/8	2 1/4	1 5/16	17/32	3	3 1/2	4	1 7/16	1 9/16	1.8
7/8	VTWE-214	VE-214																	
15/16	VTWE-215	VE-215	3 3/32	5/8	1 1/16	11/16	1 7/8	1 1/4	3/4	1 1/2	2 1/4	1 1/2	17/32	3	3 1/2	4 3/32	1 9/16	1 17/32	1.9
1	VTWE-216	VE-216																	
1 1/8	VTWE-218	VE-218	3 3/8	3/4	1 3/16	23/32	2 1/8	1 7/16	7/8	1 3/4	2 1/2	1 3/4	17/32	3 1/2	4	4 5/8	1 25/32	1 19/32	2.8
1 3/16	VTWE-219	VE-219																	
1 1/4	VTWE-220	VE-220																	
1 3/8	VTWE-222	VE-222	3 3/4	3/4	1 9/32		2 1/8	1 7/16	7/8	1 3/4	2 3/4	2 3/16	17/32	3 1/2	4	5	2	1 3/4	3.5
1 7/16	VTWE-223	VE-223																	
1 1/2	VTWE-224	VE-224	4 7/32	3/4	1 3/8	27/32	2 3/4	1 15/16	1 1/8	2 1/8	3 1/4	2 3/8	11/16	4	4 1/2	5 19/32	2 3/16	2 1/32	5.2
1 5/8	VTWE-226	VE-226																	
1 11/16	VTWE-227	VE-227	4 3/8	3/4	1 3/8	27/32	2 3/4	1 15/16	1 1/8	2 1/8	3 1/4	2 1/2	11/16	4	4 1/2	5 3/4	2 1/4	2 1/8	5.8
1 3/4	VTWE-228	VE-228																	
1 15/16	VTWE-231	VE-231	4 17/32	3/4	1 1/2	31/32	2 3/4	1 15/16	1 1/8	2 1/8	3 3/8	2 3/4	11/16	4	4 1/2	5 29/32	2 5/16	2 7/32	6.1
2	VTWE-232	VE-232	5 3/32	1	1 23/32	1 3/32	3 5/8	2 1/2	1 3/8	2 1/2	3 3/4	3	1 1/16	5 1/8	5 3/4	6 27/32	2 9/16	2 17/32	8.7
2 3/16	VTWE-235	VE-235																	
2 7/16	VTWE-239	VE-239	5 1/2	1	1 27/32	1 7/32	3 5/8	2 1/2	1 3/8	2 1/2	4	3 3/8	1 1/16	5 1/8	5 3/4	7 1/4	2 27/32	2 21/32	9.6

REV 0

PBE920

Lock: Set Screw
Seal: Contact
Housing: Cast Iron
Temperatures: -20° to 200°F
Fitting: 1/8" NPT
Bearing Insert: Tapered Roller

Load Rating Tables: Page 111
Interchange Tables: Page 162

Browning
Tapered Roller Bearings

P PILLOW
B BLOCK
E TYPE 'E'

9 920
2 SET SCREW LOCK
0 TAPERED ROLLER BEARING

Dimensional Interchange with most Type E's

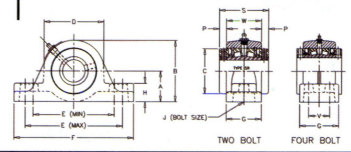

TWO BOLT   FOUR BOLT

SHAFT DIA. IN.	TYPE	A	B	C	D	E MIN	E MAX	F	G	H	J	P	S	V	W	UNIT WT.
TWO BOLT BASE																
1 3/16 / 1 1/4	SR	1 1/2	3	2 1/4	3	4 5/16	5	6 3/8	1 7/8	7/8	1/2	29/64	2 25/32	-	1 7/8	4.0
1 3/8 / 1 7/16	SR	1 7/8	3 3/4	2 3/4	3 5/8	5	6	7 3/8	2 1/8	1 1/8	1/2	7/16	3	-	2 1/8	6.9
1 1/2 / 1 5/8 / 1 11/16	SR	2 1/8	4 1/4	3 3/16	4 1/4	5 11/16	6 1/2	7 7/8	2 1/2	1 1/4	1/2	15/32	3 3/8	-	2 1/2	9.5
1 3/4 / 1 15/16 / 2	SR	2 1/4	4 1/2	3 7/16	4 1/2	6 3/16	7 1/4	8 7/8	2 1/2	1 1/4	5/8	1/2	3 1/2	-	2 1/2	11.0
2 3/16	SR	2 1/2	5	3 3/4	5	6 11/16	8	9 5/8	2 5/8	1 7/16	5/8	9/16	3 3/4	-	2 5/8	14.0
2 1/4 / 2 7/16 / 2 1/2	SR	2 3/4	5 1/2	4 1/16	5 3/8	7 1/8	8 3/4	10 1/2	2 7/8	1 5/8	5/8	9/16	4	-	2 7/8	19.0
2 11/16 / 2 3/4 / 2 15/16 / 3	SR	3 1/8	6 1/4	4 23/32	6 1/4	8 7/16	9 3/4	12	3	1 3/4	3/4	3/4	4 1/2	-	3	26.0
3 3/16 / 3 7/16 / 3 1/2	LN	3 3/4	7 1/2	5 7/16	7 5/8	9 3/4	11 1/2	14	3 5/8	2 1/8	7/8	11/16	5	-	3 5/8	44.0
FOUR BOLT BASE																
2 1/4 / 2 7/16 / 2 1/2	SR	2 3/4	5 1/2	4 1/16	5 3/8	7 1/8	8 3/4	10 1/2	3 1/2	1 5/8	5/8	9/16	4	1 7/8	2 7/8	19.0
2 11/16 / 2 3/4 / 2 15/16 / 3	SR	3 1/8	6 1/4	4 23/32	6 1/4	8	9 7/8	12	4	1 3/4	5/8	3/4	4 1/2	2 1/8	3	26.0
3 3/16 / 3 7/16 / 3 1/2	LN	3 3/4	7 1/2	5 7/16	7 1/2	9 11/16	11 7/16	14	4 1/2	2 1/8	3/4	11/16	5	2 3/8	3 5/8	44.0
3 15/16 / 4	LN	4 1/4	8 1/2	5 15/16	8 1/2	10 7/16	12 7/8	15 1/4	4 1/2	2 7/8	3/4	7/8	6 1/4	2 1/4	4 1/2	65.0
4 7/16 / 4 1/2	LN	4 3/4	9 3/8	6 13/32	9 1/8	11 1/4	13 7/8	16 5/8	4 5/8	2 3/4	3/4	1 1/16	6 3/4	2 1/2	4 5/8	81.0
4 15/16 / 5	LN	5 1/2	10 7/8	7 13/32	10 3/4	13	15 7/8	18 1/2	5 1/8	3 1/8	7/8	1 1/16	7 1/4	2 3/4	5 1/8	132.0

Part Numbers are specified by "PBE920" and bore size: Example, **PBE920x 1 3/16".**
Add "F" for Four Bolt Base: Example, PBE920Fx 2 7/16".

REV 2

Browning®

Tapered Roller Bearings

FBE920

FLANGE	**F**	
BLOCK	**B**	
TYPE 'E'	**E**	
	920	**9**
SET SCREW LOCK		**2**
TAPERED ROLLER BEARING		**0**

Dimensional Interchange with most Type E's

Lock:	Set Screw
Seal:	Contact
Housing:	Cast Iron
Temperatures:	-20° to 200°F
Fitting:	1/8" NPT
Bearing Insert:	Tapered Roller
Load Rating Tables:	Page 111
Interchange Tables:	Page 162

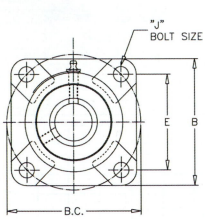

TYPE SR

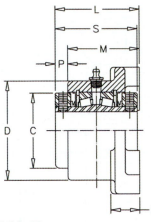

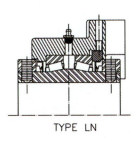

TYPE LN

SHAFT DIA. IN.	TYPE	Dimensions In Inches											UNIT WT.
		B	B.C	C	D	E	H	J	L	M	P	S	
1 3/16 1 1/4	SR	3 3/4	4 1/16	2 1/4	3	2 7/8	1 1/32	3/8	2 27/32	2 3/8	15/32	2 25/32	4.5
1 3/8 1 7/16	SR	4 5/8	4 61/64	2 3/4	3 5/8	3 1/2	1 1/16	1/2	3 5/64	2 5/8	29/64	3	6.7
1 1/2 1 5/8 1 11/16	SR	5 3/8	5 53/64	3 3/16	4 1/4	4 1/8	1 3/16	1/2	3 29/64	3	29/64	3 3/8	10.0
1 3/4 1 15/16 2	SR	5 5/8	6 3/16	3 7/16	4 1/2	4 3/8	1 3/16	1/2	3 5/8	3 1/8	1/2	3 1/2	12.0
2 3/16	SR	6 1/4	6 57/64	3 3/4	4 7/8	4 7/8	1 1/4	5/8	3 7/8	3 5/16	9/16	3 3/4	16.0
2 1/4 2 7/16 2 1/2	SR	6 7/8	7 39/64	4 1/16	5 1/4	5 3/8	1 1/2	5/8	4 3/16	3 5/8	9/16	4	21.0
2 11/16 2 3/4 2 15/16 3	SR	7 3/4	8 31/64	4 23/32	6 1/8	6	1 5/8	3/4	4 11/16	3 15/16	3/4	4 1/2	28.0
3 3/16 3 7/16 3 1/2	LN	9 1/4	9 29/32	5 7/16	7 1/2	7	1 7/8	3/4	5 1/4	4 9/16	11/16	5	46.0
3 15/16 4	LN	10 1/4	10 31/32	5 15/16	8 1/4	7 3/4	2 3/16	7/8	6 9/16	5 11/16	7/8	6 1/4	64.0

Part Numbers are specified by "FBE920" and bore size: Example, **FBE920x 1 3/16"**.

REV 1

Project 1: Nozzle Assembly Drawing

NOTES:

1. SPECIFICATIONS:
   JET DIA: 3.25
   BEAK DIA: 4.0625
   NEEDLE DIA: 4.875

   HYDRAULIC CYLINDER:
   Ø4.00 BORE
   8.0 STROKE

2. UNLESS OTHERWISE NOTED, BREAK ALL SHARP EDGES AND CORNERS.

TURBINE HOUSING INTERFACE

Ø20 O.D.

15X Ø.813 THRU
EQ. SP. ON A
Ø17.500 B.C.D. (REF.)

(Ø14 O.D.)

Ø14 300 LB
WELD-NECK FLANGE
(REF.)

HYD. PORTS
.5 NPT TYP.

13.66

73.38

44.61

42.37

19.71

14.78

(Ø4.063)

DETAIL A
SCALE 1:2

Parts List

ITEM QTY	DESCRIPTION	DWG/PART NO.	
22	15	WASHER, FLAT ANSI B18.221 NARROW	.75"	STANDARD
21	15	HEX-HEAD CAP SCREW ANSI B18.3	.75–10 X 2.50	STANDARD
20		O-RING	2-351	STANDARD
19		O-RING	6-222	STANDARD
18		BACKUP, O-RING	2-222	STANDARD
17	1	HYDRAULIC CYLINDER 4" BORE X 8" STROKE		CUNNINGHAM
16	1		.375–24 X .75	STANDARD
15	8	WASHER, FLAT ANSI B18.221 NARROW	.625"	STANDARD
14	8	HEX-HEAD CAP SCREW ANSI B18.3	.625–11 X 3.0	STANDARD
13		ALIGNMENT COUPLER	.50"	TRD
12	10	BUSHING, BRONZE	.50–13 X 1.75	STANDARD
11		O-RING	2-364	STANDARD
10	2	REAR BUSHING	N00008	CANYON IND.
9	1	SPACER, CYLINDER	N00007	CANYON IND.
8	1	INDICATOR ROD	N00006	CANYON IND.
7	1	SHAFT, NEEDLE	N00005	CANYON IND.
6	1	NEEDLE	N00004	CANYON IND.
5	1	BEAK	N00003	CANYON IND.
3	1	BELL HOUSING	N00002	CANYON IND.
2	1	LOWER NOZZLE HOUSING	N00001	VENDOR

PARTS LIST

LOWER NOZZLE ASSEMBLY
HYDROELECTRIC TURBINE

PROJECT 1—NOZZLE

FILE NAME
CONTRACT NO
DRAWN [YOUR NAME]
CHECK
APPR.
ISSUED

SIZE FSCM NO. DWG. NO. REV
SCALE 1:4 SHEET 1 OF 2

Project 1: Bell Housing Detail

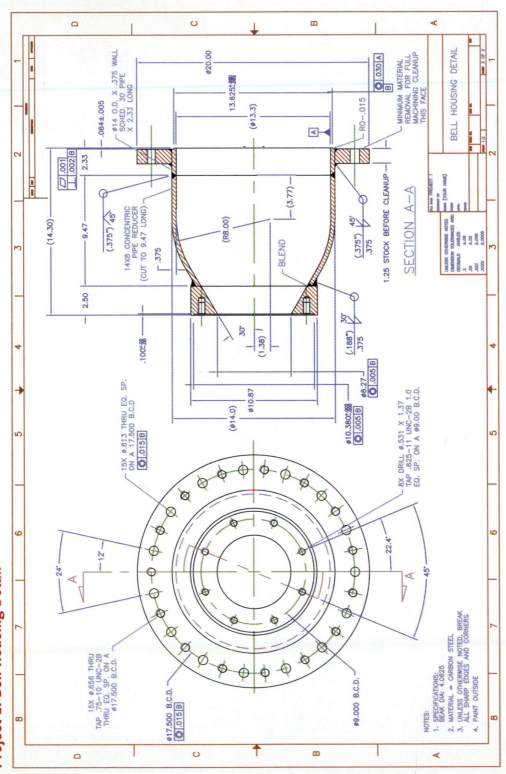

NOTES:
1. SPECIFICATIONS:
   BREAK DIA: 4.0625
2. MATERIAL = CARBON STEEL
3. UNLESS OTHERWISE NOTED, BREAK
   ALL SHARP EDGES AND CORNERS
4. PAINT OUTSIDE

SECTION A—A

BELL HOUSING DETAIL

1.25 STOCK BEFORE CLEANUP

MINIMUM MATERIAL
REMOVAL FOR FULL
MACHINING CLEANUP
THIS FACE

14X8 CONCENTRIC
PIPE REDUCER
(CUT TO 9.47 LONG)

Ø14 O.D. X .375 WALL
SCHED. 30 PIPE
X 2.33 LONG

BLEND

(R8.00)

15X Ø.813 THRU EQ. SP.
ON A 17.500 B.C.D

8X DRILL Ø.531 X 1.37
TAP .625—11 UNC—2B 1.0
EQ. SP. ON A Ø9.00 B.C.D.

15X Ø.656 THRU
TAP .75—10 UNC—2B
THRU EQ. SP. ON A
Ø17.500 B.C.D.

Ø17.500 B.C.D.

Ø9.000 B.C.D.

Project 2: Drawing 0020   Reference Assembly Drawing

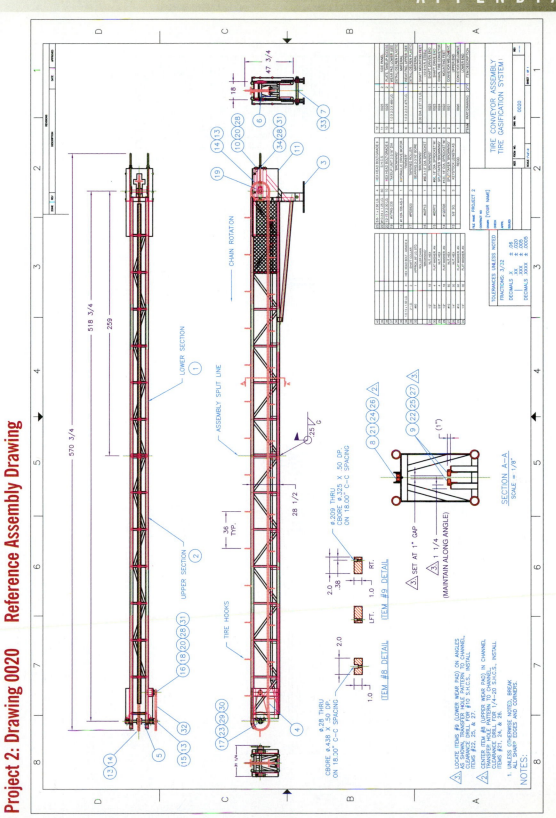

Project 2: Drawing 0021   Lower Weldment Drawing, Sheet 1

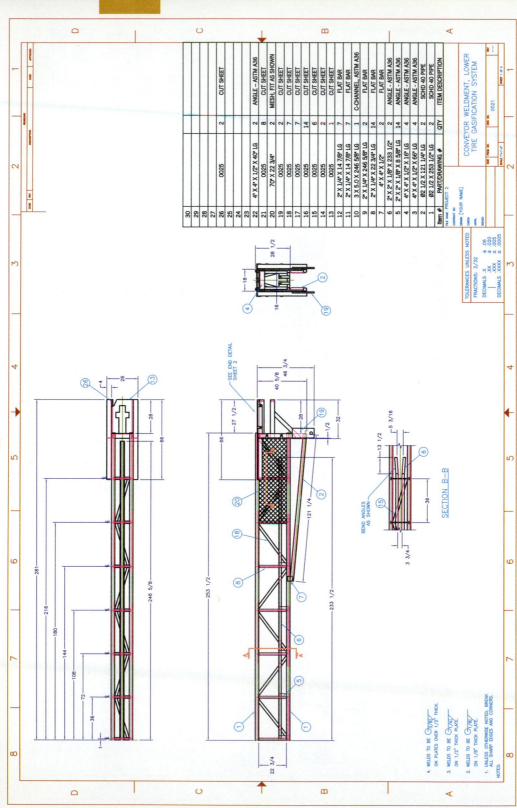

Item #	PART/DRAWING #		QTY	ITEM DESCRIPTION
30				
29				
28				
27				CUT SHEET
26	0025		2	CUT SHEET
25				
24				
23				CUT SHEET
22	4" X 4" X 1/2" X 40" LG		2	ANGLE – ASTM A36
21	0025		8	CUT SHEET
20	70" X 22 3/4"		2	MESH, FIT AS SHOWN
19	0025		2	CUT SHEET
18	0025		7	CUT SHEET
17	0025		7	CUT SHEET
16	0025		14	CUT SHEET
15	0025		6	CUT SHEET
14	0025		2	CUT SHEET
13	0025		1	FLAT BAR
12	2 X 1/4" X 14 7/8" LG		7	FLAT BAR
11	2 X 1/4" X 14 7/8" LG		7	FLAT BAR
10	3 X 5.0 X 246 5/8" LG		1	C-CHANNEL, ASTM A36
9	2 X 1/4" X 246 5/8" LG		2	FLAT BAR
8	2 X 1/4" X 22 3/4" LG		14	FLAT BAR
7	4" X 4" X 1/2"		2	ANGLE – ASTM A36
6	2" X 2" X 1/8" X 233 1/2"		2	ANGLE – ASTM A36
5	2" X 2" X 1/8" X 8 5/8" LG		14	ANGLE – ASTM A36
4	4" X 4" X 1/2" X 16" LG		4	ANGLE – ASTM A36
3	4" X 4" X 1/2" X 66" LG		4	ANGLE – ASTM A36
2	Ø2 1/2 X 121 1/4" LG		2	SCHD 40 PIPE
1	Ø2 1/2 X 253 1/2" LG		2	SCHD 40 PIPE

**CONVEYOR WELDMENT, LOWER
TIRE GASIFICATION SYSTEM**

DWG NO. 0021

SECTION B–B

BEND ANGLES
AS SHOWN

SEE END DETAIL
SHEET 2

TOLERANCES UNLESS NOTED
FRACTIONS: 3/32
DECIMALS X   ± .06
DECIMALS XXX   ± .020
DECIMALS XXXX   ± .0005

FILE NAME: PROJECT 2
CONTRACT NO.
DRAWN: [YOUR NAME]
CHECK:
APPR.:
ISSUED:

NOTES:
1. UNLESS OTHERWISE NOTED, BREAK
   ALL SHARP EDGES AND CORNERS.
2. WELDS TO BE 3/16"
   ON 1/8" THICK PLATE.
3. WELDS TO BE 1/4" THICK PLATE.
4. WELDS TO BE 3/8"
   ON PLATES OVER 1/2" THICK.

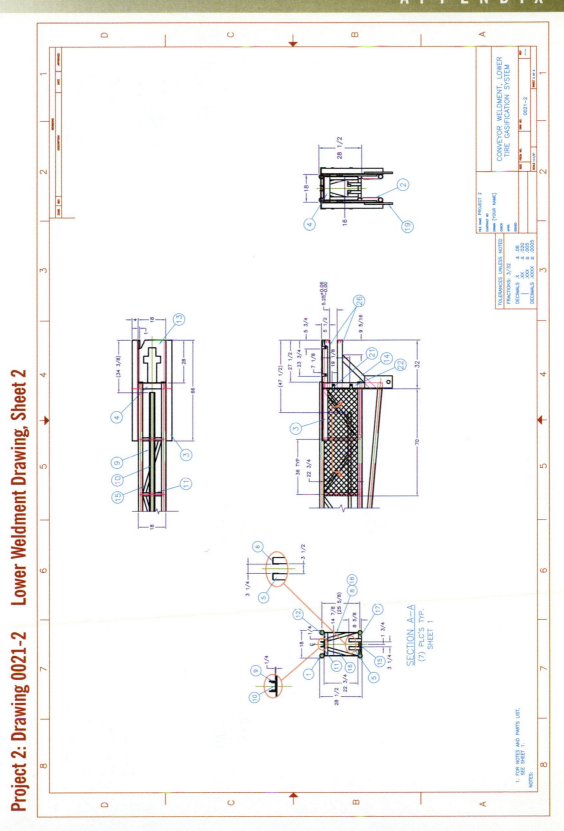

Project 2: Drawing 0021-2   Lower Weldment Drawing, Sheet 2

Project 2: Drawing 0022   Upper Weldment Drawing, Sheet 1

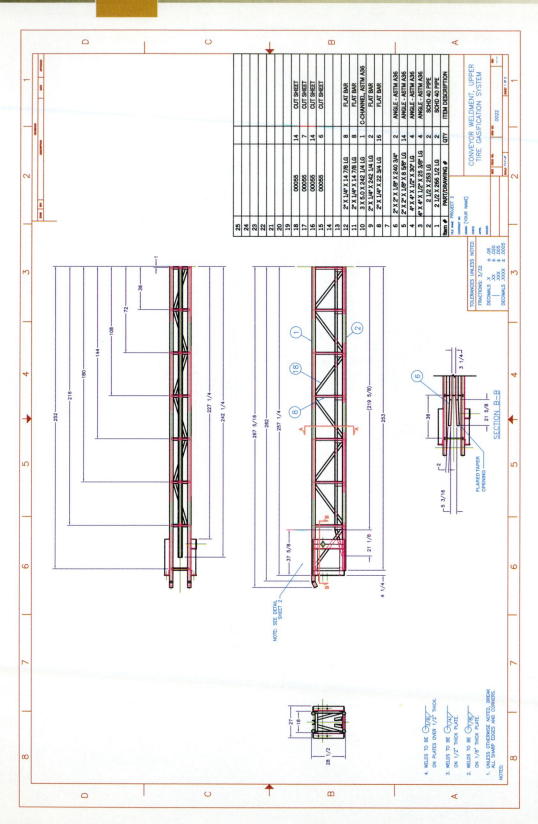

Project 2: Drawing 0022-2   Upper Weldment Drawing, Sheet 2

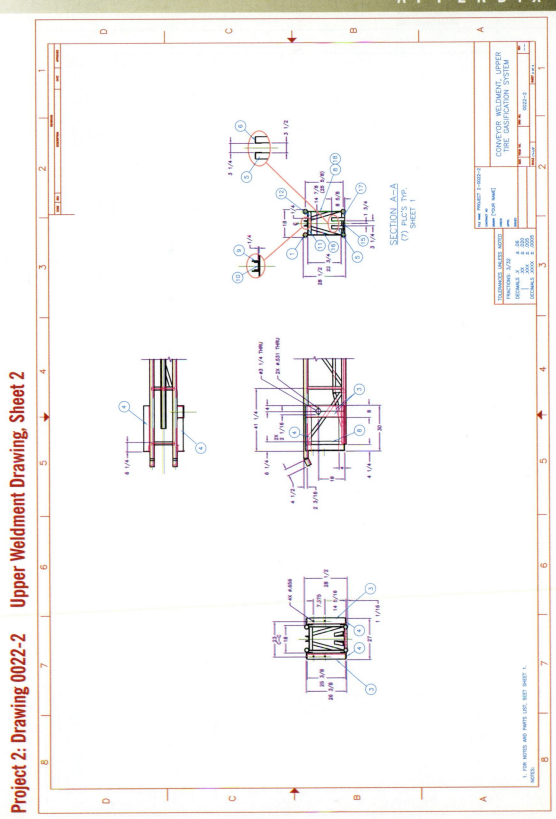

Project 2: Drawing 0023 Shaft Detail

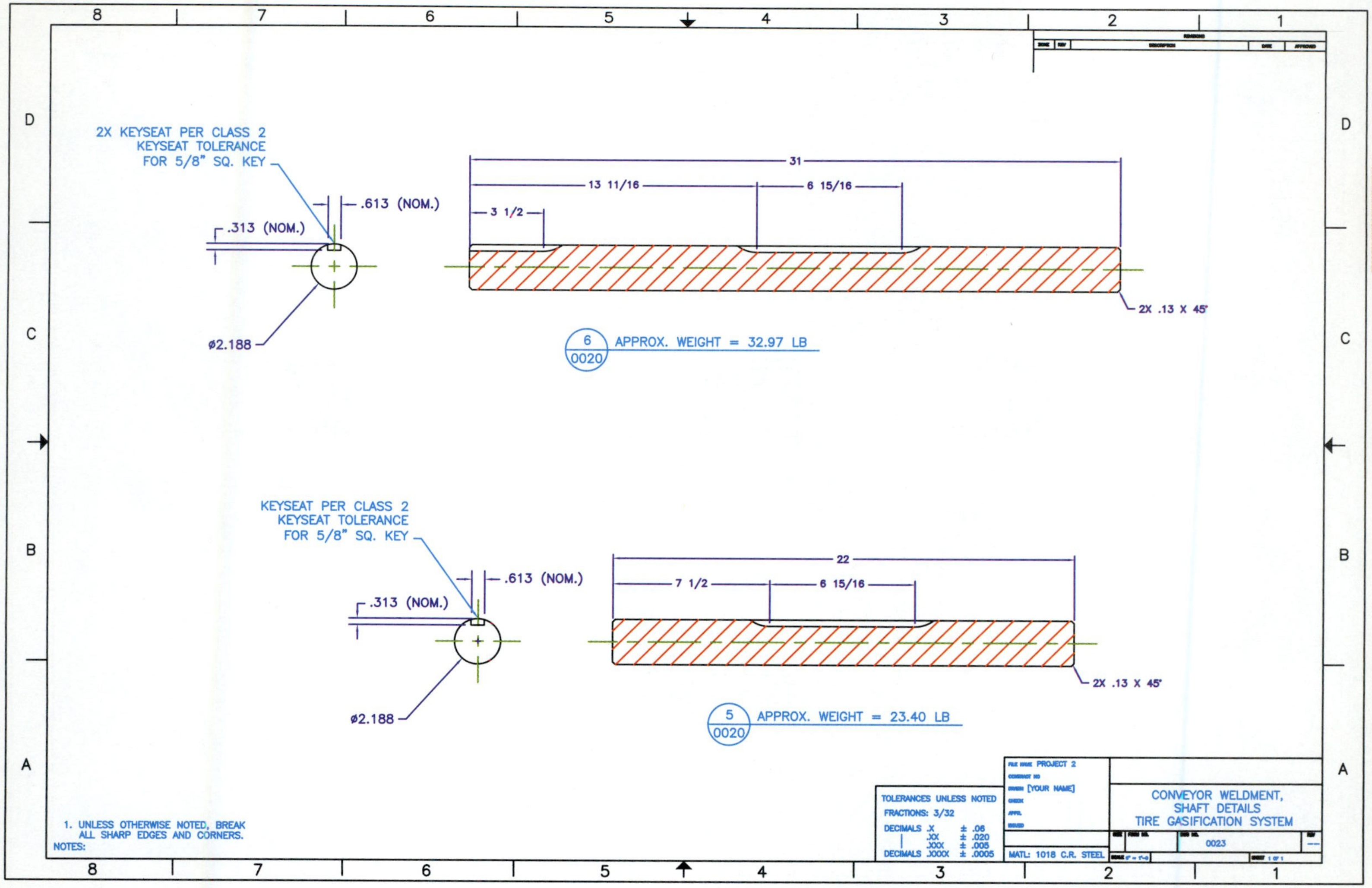

Project 2: Drawing 0024   Drive Chain Detail

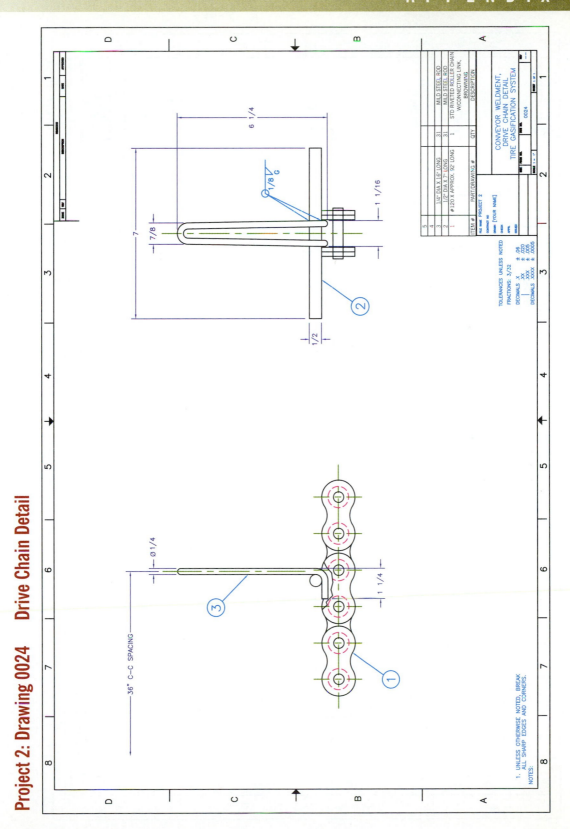

Project 2: Drawing 0025 Cut Sheet

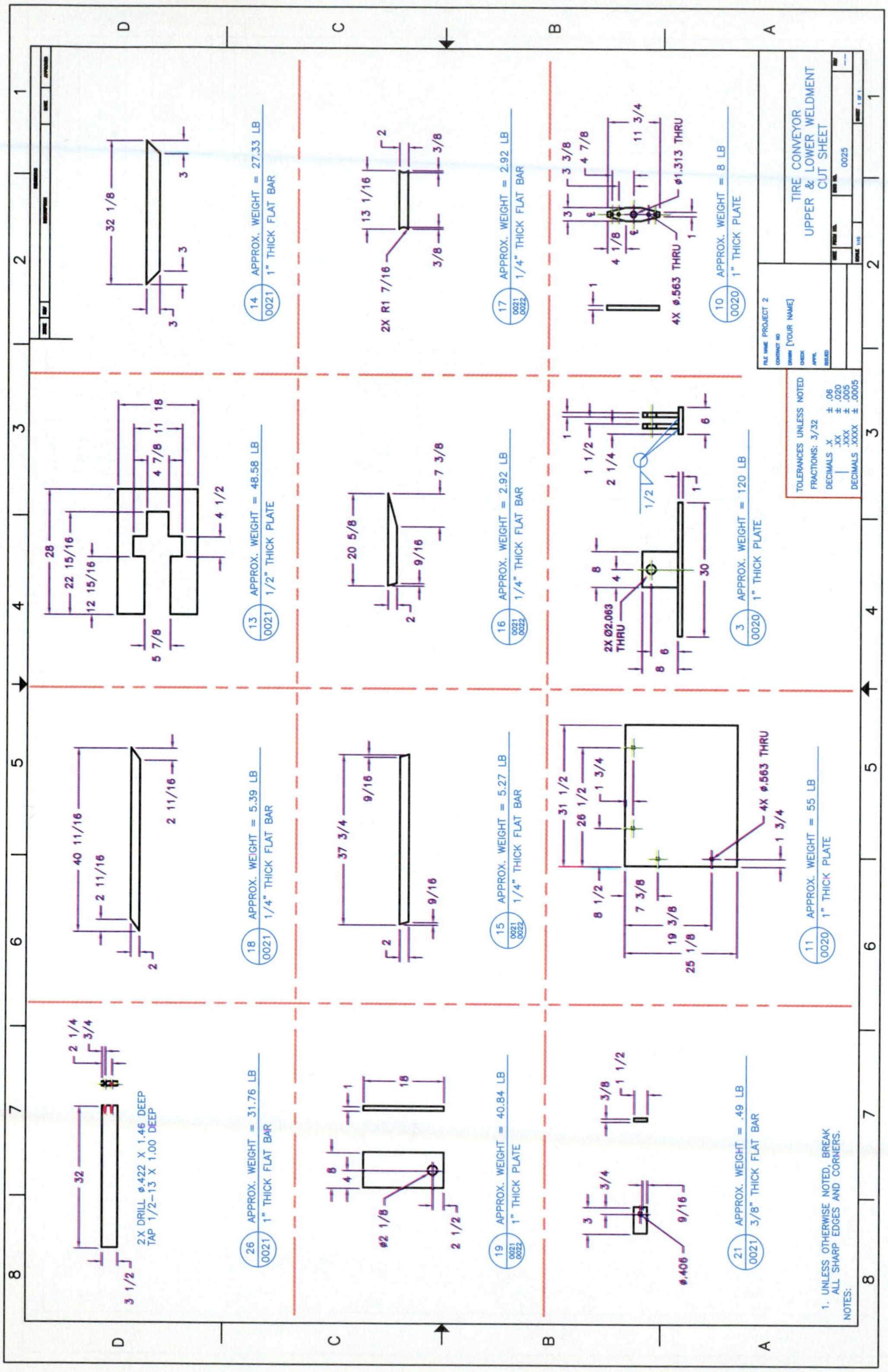

Project 3: Swivel Bracket Model

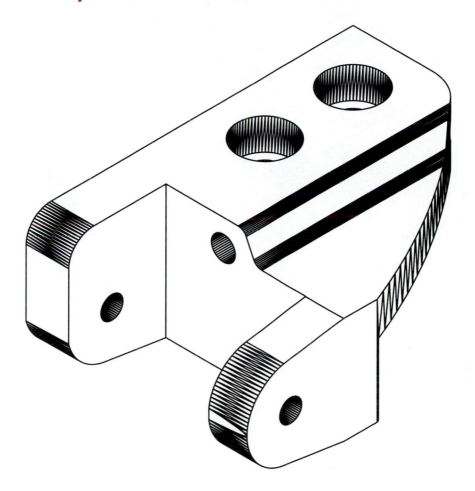

Project 3: Base Mold

Project 3: Top Mold

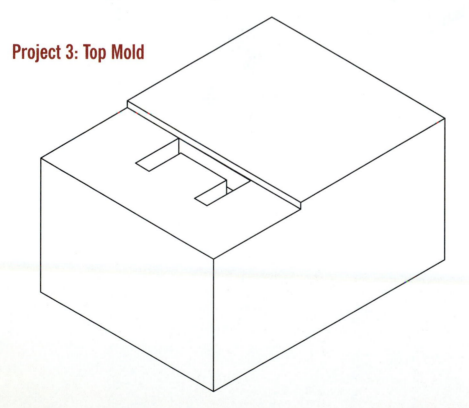

Project 4: Concept Model Created with AutoCAD Mechanical

Item	Qty	Description	FSCM NO	Standard	Material
11	1	Tapered Roller Bearing		ISO 355 – 4CB – 40 x 75 x 19	
10	1	Tapered Roller Bearing		ISO 355 – 2CC 25 x 50 x 17	
9	8	Cylinder Head Cap Screw		DIN 912 – M4 x 25	
8	8	Washer		ISO 7089 – 4 – 140 HV	
7	8	Washer		ISO 7089 – 4 – 140 HV	
6	8	Cylinder Head Cap Screw		DIN 912 – M4 x 35	
5	8	Cylinder Head Cap Screw		DIN 912 – M6 x 35	
4	8	Washer		ISO 7089 – 6 – 140 HV	
3	1	Shaft Seal Ring		DIN 3760 – AS – 50 x 68 x 8 – NBR	NBR
2	1	Shaft Seal Ring		DIN 3760 – AS – 20 x 35 x 7 – NBR	NBR
1	1	Parallel Key		B 5 x 5 x 12 – DIN 6885	

MECHANICAL SHAFT ASSEMBLY
CONCEPT DRAWING

DRIVEN END (INTERFACE WITH MIXER BLADE)

PULLEY

Project 4: Concept Model Created with Basic AutoCAD

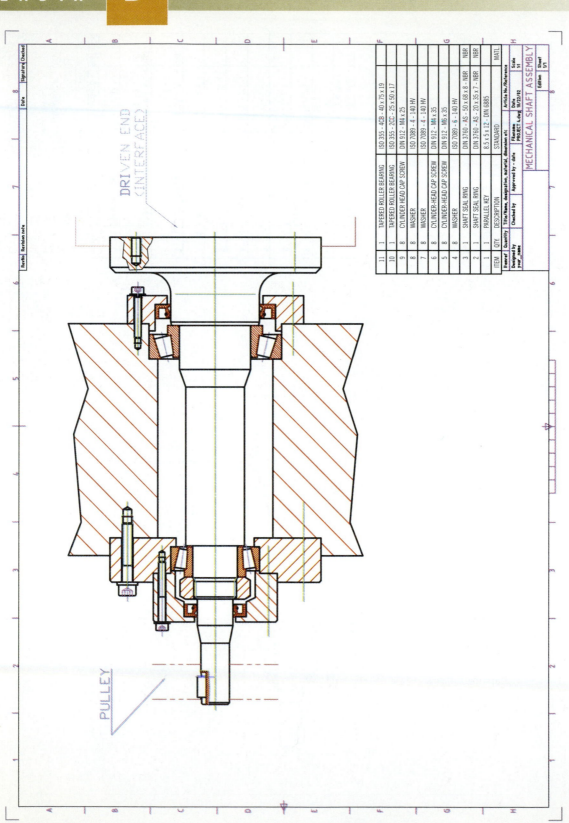

DRIVEN END
(INTERFACE)

PULLEY

ITEM	QTY.	DESCRIPTION	Article No./Reference	MATL.
11	1	TAPERED ROLLER BEARING	ISO 355 - 4CB - 40 x 75 x 19	
10	1	TAPERED ROLLER BEARING	ISO 355 - 2CC - 25 x 50 x 17	
9	8	CYLINDER HEAD CAP SCREW	DIN 912 - M4 x 25	
8	8	WASHER	ISO 7089 - 4 - 140 HV	
7	8	WASHER	ISO 7089 - 4 - 140 HV	
6	8	CYLINDER-HEAD CAP SCREW	DIN 912 - M4 x 35	
5	8	CYLINDER-HEAD CAP SCREW	DIN 912 - M6 x 35	
4	8	WASHER	ISO 7089 - 6 - 140 HV	
3	1	SHAFT SEAL RING	DIN 3760 - AS - 50 x 68 x 8 - NBR	NBR
2	1	SHAFT SEAL RING	DIN 3760 - AS - 20 x 35 x 7 - NBR	NBR
1	1	PARALLEL KEY	8.5 x 5 x 12 - DIN 6885	STANDARD
Item/ref	Quantity	Title/Name, designation, material, dimension etc.		

Designed by
your_name

Checked by

Approved by – date

Filename
PROJECT 4.dwg

Date
10/22/12

MECHANICAL SHAFT ASSEMBLY

Scale
1:1

Edition

Sheet
1/1

Revise Revision note   Date   Signature Checked

Index

Index

Index

scale
for standard parts, 46
in viewports, 61
of hatch patterns, 49
to plot a drawing, 70
schedule, of pipes, 21
scheduled dimensions, 21
section
hatching, 48-52
offset, 74
section lines, 48
sequence of notes on a drawing, 165
server application, in OLE, 161
SHADE command, 282
shaded image, of solid model, 273
Shaft Generator, in AutoCAD
Mechanical, 330
SKETCH command, 188
SKPOLY variable, 188
SmartShift high-performance bicycle,
267, 270, 271
SOLDRAW command, 315
SOLID command, 53
solid modeling, 267-318
solid models
advantages of, 269
design guidelines for, 290
manipulating, 280
practice creating, 275-279
types of, 275
uses for, 303
SOLPROF command, 315
sphere (solid model), 276
SPHERE command, 276
spotface, 296
spreadsheet, parts list, 161, 163-164
Standard dimension style, 81
standard parts, drawing practices for,
118
STRETCH command, 181, 216
structural field assembly project, 105-
266
STYLE command, 19
SUBTRACT command, 278
supplementary symbols, welding, 91
swivel bracket project, 267-318
symmetrical tolerances, 81
system level analysis, 310

T

tail, of welding symbol, 92
technical documentation, 108, 272
technical specifications, checking, 116
TEFC motor, 323
template, ISO, 327, 353, 354
text height, 18, 55
text styles, 18, 354
3D CAD, fundamentals of, 274-275
3D geometry, projecting from a 3D
model, 314-316
3D modeling, 267-318
3D Orbit, 293
3DORBIT command, 293, 298
3D views, working with, 292-293
title block, 66-67
tolerance block, 166
TOLERANCE command, 87, 89
tolerance style, setting up, 83
tolerances, 80
deviation, 81
general, 166
nonsymmetrical, 81
symmetrical, 81
tolerancing, 80-86
for manufacturability, 290
parametric, 288-289
plus-or-minus, 80, 81
tolerancing guidelines
for component manufacture, 288-
289, 290
ANSI Y14.5M, 80
torus (solid model), 277
TRIM command, 22
turbine efficiency, in hydroelectric
turbines, 13
turbine, hydroelectric, 13-14
2D views, extracting from a 3D solid
model, 267, 268, 292

U

U.S. Customary units, 17, 288
UCS command, 282
UCS icon, 276
unilateral tolerance, 289
UNITS command, 19

V

variables, dimension, 182
View toolbar, 293
viewports, 59-61
viewport scales, 61
views, predefined in AutoCAD, 293
visualization, 130

W

water power, determining, 13
wedge (solid model), 277
weld representations, 53-54
weld symbology, 91-92
weld symbols
and DesignCenter, 92
in AutoCAD, 92
libraries, 92
vs. welding symbol, 91
welding standards, 30
welding symbol
parts of, 91
placement of elements in, 92
vs. weld symbol, 91
welds
modeling, 290
showing on 2D drawing for reference, 53
working drawings, 15, 80, 105-266

X

XLINE command, 21, 78
XREF command, 272
Xrefs, 272-273

Z

zero line width, 68